AF847899

Progress in Molecular and Subcellular Biology

Series Editors: W.E.G. Müller (Managing Editor), Ph. Jeanteur,
I. Kostovic, Y. Kuchino, A. Macieira-Coelho, R.E. Rhoads

Springer

Berlin
Heidelberg
New York
Hong Kong
London
Milan
Paris
Tokyo

Progress in Molecular and Subcellular Biology

Volumes Published in the Series

Volume 14 **Biological Response Modifiers: Interferons, Double-Stranded RNA and 2′,5′-Oligoadenylates**
W.E.G. Müller and H.C. Schröder (Eds.)

Volume 15 **Invertebrate Immunology**
B. Rinkevich and W.E.G. Müller (Eds.)

Volume 16 **Apoptosis**
Y. Kuchino and W.E.G. Müller (Eds.)

Volume 17 **Signaling Mechanisms in Protozoa and Invertebrates**
G. Csaba and W.E.G. Müller (Eds.)

Volume 18 **Cytoplasmic Fate of Messenger RNA**
Ph. Jeanteur (Ed.)

Volume 19 **Molecular Evolution: Evidence for Monophyly of Metazoa**
W.E.G. Müller (Ed.)

Volume 20 **Inhibitors of Cell Growth**
A. Macieira-Coelho (Ed.)

Volume 21 **Molecular Evolution: Towards the Origin of Metazoa**
W.E.G. Müller (Ed.)

Volume 22 **Cytoskeleton and Small G Proteins**
Ph. Jeanteur (Ed.)

Volume 23 **Inorganic Polyphosphates: Biochemistry, Biology, Biotechnology**
H.C. Schröder and W.E.G. Müller (Eds.)

Volume 24 **Cell Immortalization**
A. Macieira-Coelho (Ed.)

Volume 25 **Signaling Through the Cell Matrix**
A. Macieira-Coelho (Ed.)

Volume 26 **Signaling Pathways for Translation: Insulin and Nutrients**
R.E. Rhoads (Ed.)

Volume 27 **Signaling Pathways for Translation: Stress, Calcium, and Rapamycin**
R.E. Rhoads (Ed.)

Volume 28 **Small Stress Proteins**
A.-P. Arrigo and W.E.G. Müller (Eds.)

Volume 29 **Protein Degradation in Health and Disease**
M. Reboud-Ravaux (Ed.)

Volume 30 **Biology of Aging**
A. Macieira-Coelho

Volume 31 **Regulation of Alternative Splicing**
Philippe Jeanteur (Ed.)

Volume 32 **Guidance Cues in the Developing Brain**
I. Kostovic (Ed.)

Volume 33 **Silicon Biomineralization**
W.E.G. Müller (Ed.)

Invertebrate Cytokines and the Phylogeny of Immunity
A. Beschin and W.E.G. Müller (Eds.)

RNA Trafficking and Nuclear Structure Dynamics
Ph. Jeanteur (Ed.)

Viruses and Apoptosis
C. Alonso (Ed.)

Covadonga Alonso (Ed.)

Viruses and Apoptosis

With 26 Figures and 5 Tables

Springer

[redacted]

Professor Dr. Covadonga Alonso
Dpt. Biotechnologia, INIA
Ctra. de la Coruña Km 7
28040 Madrid
Spain

11. JUNI 2015

Ausgesondert

Fachhochschule
Bonn-Rhein-Sieg
Hochschul- und
Kreisbibliothek
Bonn-Rhein-Sieg

ISSN 0079-6484
ISBN 3-540-20228-5 Springer-Verlag Berlin Heidelberg New York

Bibliographic information published by Die Deutsche Bibliothek
Die Deutsche Bibliothek lists this publication in the Deutsche Nationalbibliografie; detailed bibliographic data is available in the Internet at http://dnb.ddb.de

This work is subject to copyright. All rights reserved, whether the whole or part of the material is concerned, specifically the rights of translation, reprinting, reuse of illustrations, recitation, broadcasting, reproduction on microfilm or in any other way, and storage in data banks. Duplication of this publication or parts thereof is permitted only under the provisions of the German Copyright Law of September 9, 1965, in its current version, and permission for use must always be obtained from Springer-Verlag. Violations are liable for prosecution under the German Copyright Law.

Springer-Verlag is a part of Springer Science+Business Media
springeronline.com

© Springer-Verlag Berlin Heidelberg 2004
Printed in Germany

The use of general descriptive names, registered names, trademarks, etc. in this publication does not imply, even in the absence of a specific statement, that such names are exempt from the relevant protective laws and regulations and therefore free for general use.

Product liability: The publishers cannot guarantee the accuracy of any information about dosage and application contained in this book. In every individual case the user must check such information by consulting the relevant literature.

Production: Pro Edit GmbH, Heidelberg, Germany
Cover design: design and production GmbH, Heidelberg, Germany
Typesetting: SNP Best-set Typesetter Ltd., Hong Kong
Printed on acid-free paper 39/3150ML 5 4 3 2 1 0

Contents

HIV and Apoptosis: a Complex Interaction Between Cell Death and Virus Survival
J. Gil, M. Bermejo and J. Alcamí

Exploitation of Cell Cycle and Cell Death Controls by Adenoviruses: The Road to a Productive Infection
I. Alasdair Russell, J.A. Royds and A.W. Braithwaite

Induction of Transformed Cell-Specific Apoptosis by the Adenovirus E4orf4 Protein
T. Kleinberger

Preface

The pathogenesis of viral infections involves dynamic interactions between viruses and hosts, which can result in different outcomes including cell death, elimination of the virus or latent infection. Viruses deliver genomes and proteins with signaling potential into target cells, resulting in growth, proliferation and apoptosis. Viral infection modifies key cell regulatory elements involved in apoptotic pathways to successfully accomplish viral replication despite the toxicity of viral products and the immune response elicited against the virus. Also, the interplay between virus-induced apoptosis and cell survival is mediated by an accurate balance between pro-death and anti-apoptotic signals triggered by cellular and viral proteins. Hence, viral survival products causing a delay in the completion of apoptotic process can be critical in the replication and propagation of viruses.

Viral control of apoptosis can be a double-edged sword. In some viral infections, immune cell apoptosis will contribute to pathogenesis and determine disease evolution either by direct infection of immune cells or as a result of bystander cell apoptosis caused by viral proteins or mediators secreted from infected cells. The apoptotic mechanisms contributing to HIV pathogenesis, the functional consequences on the immune system and the effect of antiviral therapies on this process will be reviewed in depth in this Volume.

Upon viral infection, the host immune system can utilize apoptosis to limit the ability of the virus to replicate through cytotoxic T lymphocyte (CTL) attack against the infected cell. Virus-infected cells can be targeted for CTL lysis by death receptor engagement on the surface, initiating the extrinsic apoptosis pathway through Fas and TNF receptor death domains. Depending on the different virus models, viral proteins could elicit intracellular stress signals, such as DNA damage or alteration of ER homeostasis, which can trigger apoptosis through the intrinsic apoptosis pathway, regulated by the mitochondria. Changes in the permeability of the mitochondrial membrane result in the release of apoptosis-inducing factors from the mitochondrial matrix to the cytosol activating the caspase cascade. Caspase activation would, in turn, initiate the proteolytic processes leading to cytoplasmic alterations and DNA degradation. To circumvent this, viruses have evolved a variety of mechanisms to subvert host apoptotic pathways, ensuring continuous viral production and persistent viral infection. Moreover, some viruses are known to encode homologues of cellular anti-apoptotic Bcl-2 proteins which are central regulators of apoptosis at the mitochondrial level. Viral Bcl-2 homologues (vBcl-2) prevent premature death of the host cell, which would impair virus production and help to avoid the immune surveillance of the host. Some examples are vBcl-2s encoded by adenovirus, Epstein-Barr virus (EBV), human herpesvirus 8, herpesvirus saimiri, mouse herpesvirus, 68 homologues and African swine fever virus, among others. Lymphotropic γ-herpesvirus vBcl-2s are suspected to play a role in the oncogenic potential of these viruses.

On the other hand, viral infections may induce signaling, altering cell proliferation, which would ultimately lead to the induction of apoptosis. Often, viruses infect quiescent cells that do not provide an optimal environment for viral DNA synthesis due to rate-limiting levels of deoxynucleotides and proteins involved in DNA synthesis. Therefore, various viruses, including adenoviruses (AD), have evolved different means to overcome this obstacle by deregulating the cell cycle machinery. This interference with normal cell cycle control may lead to the induction of apoptosis, which is counteracted by viral apoptosis inhibitor genes. Viral modulation of apoptosis will contribute to the knowledge of the complex regulation of cell survival and death mechanisms, also providing useful information related to potential therapeutic targets. The finding that oncogenic transformation sensitizes cells to the apoptotic effect of some virus genes (AD E4orf4) heightens the exciting potential of using these genes as therapeutic tools.

Also, SV40 gene products disrupt the control of the cell cycle, resulting in oncogenic transformation that targets major tumor suppressor pathways of p53 and pRb. Potentially, viral proteins can elicit pro- and anti-apoptotic effects. Tissue homeostasis is maintained by a net balance of cell death and replication, and the rupture of this delicate balance may contribute to neoplastic growth. Genes that control cell proliferation and cell death are candidates to suffer alterations that ultimately lead to oncogenesis, which is another important aspect discussed in this Volume. Determining the exact role of signal transduction pathways involved in cell cycle, cell migration and cell survival and their contribution to the different steps that occur during tumor progression has led to the identification of targets for anticancer therapies. Tumor suppressors, such as p53, are general sensors of cellular damage that control the cell cycle and apoptosis, which are mutated in a high percentage of tumors. p53 restoration and antisense Bcl-2 are examples of the antitumor strategies recently developed against increased tumor cell survival.

Madrid, January 2004 Covadonga Alonso

Cell Death Suppressors Encoded by Cytomegalovirus

V.S. Goldmacher[1]

1 Introduction

Apoptosis of virally infected cells is a host defense mechanism designed to arrest viral replication by eliminating infected cells (O'Brien 1998). Apoptosis induced by virus infection can be triggered intrinsically by initiating an apoptotic pathway in response to the metabolic changes caused by the infection. The cytotoxic effector cells of the immune system can induce apoptosis in virally infected target cells via multiple mechanisms such as ligation of cell surface death receptors (Wallach et al. 1999), or the release of proteases (granzymes) on contact with target cells (Smyth et al. 2001). A number of viruses evolved with elaborate, often several, strategies to prevent or delay apoptotic host responses (O'Brien 1998; Tschopp et al. 1998). In this article I will review the current status of our knowledge on the cell-death suppressing strategies employed by cytomegaloviruses, a subset of β-herpesviruses.

Until recently, virtually nothing was known about the mechanisms of cell-death suppression by β-herpesviruses. There were reports that productive infection with human cytomegalovirus (CMV) rendered cells resistant to apoptosis induced by diverse pro-apoptotic stimuli [reviewed in Goldmacher et al. (1999)], but the nature of these phenomena remained obscure. Homology searches of the genome of human CMV failed to reveal any genes homologous to any known cell-death suppressor.

A functional screen of a CMV genomic library (Goldmacher et al. 1999; Skaletskaya et al. 2001) enabled us to identify two CMV-encoded cell-death suppressors, vICA (viral inhibitor of caspase-8-induced apoptosis) and vMIA (viral mitochondria-localized inhibitor of apoptosis). vICA and vMIA have functional similarities with c-FLIP or Bcl-2, respectively, but share no sequence homology with these proteins. These cell-death suppressors are rather unique: to date, no cellular or viral homologues of vICA or vMIA have been identified except those encoded by other β-herpesviruses (McCormick et al. 2003).

In addition to vICA and vMIA, several other CMV-encoded proteins, IE1 and IE2 of human CMV, and pM45 of mouse CMV, were reported to be capable

[1]ImmunoGen Inc., Cambridge, Massachusetts, 02139 USA,
e-mail: victor.goldmacher@immunogen.com

Progress in Molecular and Subcellular Biology
C. Alonso (Ed.): Viruses and Apoptosis
© Springer-Verlag Berlin Heidelberg 2004

of protecting cells from apoptosis. However, the evidence that these proteins are bona fide cell-death suppressors, and are involved in protection of CMV-infected cells from apoptosis, is inconclusive (see below).

2 vICA, a Cell-Death Suppressor That Blocks Caspase-8 Activation

2.1 Cell-Death Suppressing Activity of vICA

vICA, also known as pUL36, is encoded by the *UL36* gene of human CMV. The anti-apoptotic activity of vICA was demonstrated in stably transfected BJAB and HeLa cells, which are type I and type II cells, respectively (Skaletskaya et al. 2001). During death-receptor ligation-mediated apoptosis in type I cells, activation of caspase-8 leads directly to activation of caspase-3 bypassing the mitochondrial apoptotic signaling pathway, while in type II cells activation of caspase-3 can only be achieved through the mitochondrial apoptotic pathway (Scaffidi et al. 1998, 1999b; Schmitz et al. 1999). Ectopic expression of vICA protected both cell lines against Fas-mediated apoptosis. In this respect, vICA is similar to cFLIP, a cell-death suppressor that blocks caspase-8 activation. cFLIP is active both in type I (Scaffidi et al. 1999a) and type II cells (Scaffidi et al. 1999b; Engels et al. 2000). vICA protects cells against apoptosis triggered by ligands of other death receptors (tumor necrosis factor receptor I and Apo-2), but appears to only marginally protect cells against apoptosis induced by either the cytotoxic drugs doxorubicin or maytansine, or by infection with an E1B19k-deficient adenovirus.

2.2 vICA Prevents Caspase-8 Activation and Forms a Complex with Pro-Caspase-8

Ligation of cell-surface Fas triggers the recruitment of pro-caspase-8 to the death-inducing signaling complex (DISC) through the death domain-containing adapter protein FADD, and sets off activation of pro-caspase-8 by means of proteolytic processing (Krueger et al. 2001). Biochemical data indicate that vICA interrupts the Fas-ligation-mediated apoptotic pathway at, or upstream of, caspase-8 activation: vICA inhibits Fas-ligation-mediated activation of caspase-8 and downstream events of the apoptotic signaling pathway: proteolytic processing of BID, a substrate of caspase-8, and proteolytic processing of PARP, a substrate of downstream caspases (Skaletskaya et al. 2001). vICA has affinity to pro-caspase-8, specifically to its pro-domain region that contains two non-identical death-effector domains ($C8DED_2$; Skaletskaya et al. 2001). Taken together, these findings provide evidence that vICA directly inhibits proteolytic processing of caspase-8 (Fig. 1), the same step of the death-receptor apoptotic pathway that is blocked by cellular and viral FLIPs (Krueger et al. 2001). FLIPs contain a segment consisting of two non-identical death-effector domains ($FDED_2$) that is highly homologous to that of pro-caspase-8. FLIPs associate with pro-caspase-8 through $C8DED_2$–$FDED_2$ interaction (Krueger et al. 2001). vICA does not share any sequence similarity with FLIPs, and, specifically, it does not contain DED sequence motifs. This dissimilarity between vICA and FLIPs suggests (but does not prove) that vICA and FLIPs block caspase-8 activation by distinct mechanisms. Interestingly, vICA does not associate with FADD (Skaletskaya et al. 2001), an adaptor protein that also has a DED_2 segment in its amino acid sequence. Unlike vICA, c-FLIP is capable of forming a complex with either pro-caspase-8 or FADD (Irmler et al. 1997). While vICA associates with pro-caspase-8 constitutively in the absence of death-receptor ligation, it appears that under physiologically relevant conditions cFLIP forms a complex with FADD and pro-caspase-8 only at the DISC following ligation of a death receptor (Scaffidi et al. 1999a).

2.3
Homologues of vICA in Other β-Herpesviruses

Searches of DNA sequence databases have thus far failed to reveal any cellular or viral homologues of *UL36*-encoded proteins. The only exceptions are genomes of other β-herpesviruses. These viruses encode highly conserved vICA homologues (McCormick et al. 2003). Two of these vICA-like proteins, the *M36* protein product (M-vICA) of mouse CMV, and the *Rh36* protein product (Rh-vICA) of rhesus macaque CMV, were examined and found to be functionally active as cell-death suppressors (McCormick et al. 2003). These data indicate that the ability of vICA to suppress apoptosis is a general function of vICA homologues encoded by β-herpesviruses.

A segment of the guinea pig CMV genome that is, in general, co-linear with the human CMV *UL32-UL37* region does not contain a vICA homologue ORF (Liu and Biegalke 2001), an unexpected result, since all other sequenced genomes of β-herpesviruses were found to encode a vICA homologue in this region along with other genes found in human CMV (McCormick et al. 2003). It will be of interest to determine if guinea pig CMV encodes a vICA homologue in another part of its genome, or misses it altogether.

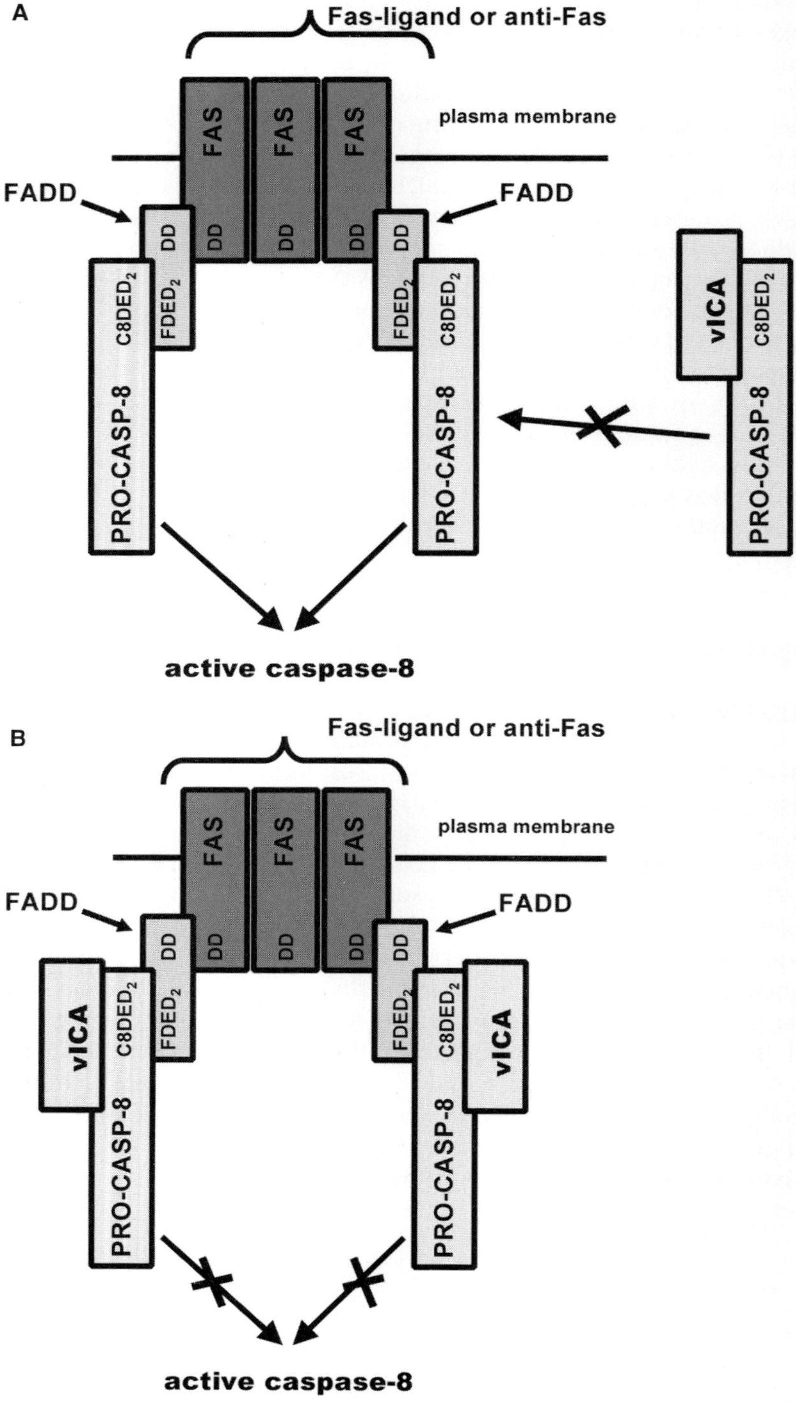
A
Fas-ligand or anti-Fas
plasma membrane
FAS
FAS
FAS
FADD
FADD
DD
DD
DD
DD
DD
FDED2
FDED2
C8DED2
C8DED2
PRO-CASP-8
PRO-CASP-8
vICA
C8DED2
PRO-CASP-8
active caspase-8
B
Fas-ligand or anti-Fas
plasma membrane
FAS
FAS
FAS
FADD
FADD
DD
DD
DD
DD
DD
vICA
C8DED2
FDED2
FDED2
C8DED2
vICA
PRO-CASP-8
PRO-CASP-8
active caspase-8

3
vMIA, a Cell-Death Suppressor Targeting Mitochondria

3.1
Cell-Death Suppressing Activity of vMIA

The *UL37* gene of human CMV encodes vMIA, also known as pUL37×1, and its two splice variants, gpUL37 and gpUL37$_M$. vMIA protects HeLa cells against apoptosis triggered by a wide variety of apoptotic stimuli: ligands of death receptors, various cytotoxic compounds, and infection with an E1B19K-deficient adenovirus mutant (Goldmacher 2002). The longer splice variants of vMIA, gpUL37 and gpUL37$_M$ are, like vMIA, functional cell-death suppressors. The ability of vMIA to protect cells from apoptosis appears to be limited to type II cells (Skaletskaya et al. 2001): vMIA protects HeLa cells and human fibroblasts (both behave as type II cells) from apoptosis, but is inactive as a cell-death suppressor in the type I human lymphoid cell line BJAB. In this respect, vMIA is similar to Bcl-2 that protects type II cells but not type I cells from apoptosis (Foghsgaard and Jaattela 1997; Scaffidi et al. 1998), consistent with the mode of action of Bcl-2, which targets the mitochondrial apoptotic signaling pathway.

3.2
Intracellular Localization of *UL37* Protein Products

Intracellular localization of the three protein products of *UL37*, vMIA, gpUL37, and gpUL37$_M$, was examined in infected and in transfected cells by immunofluorescence and by electron microscopy (Goldmacher et al. 1999; Colberg-Poley et al. 2000). vMIA is predominantly localized to mitochondria of CMV-infected cells or stably transfected cells, where it decorates the outer mitochondrial membrane. A fraction of vMIA expressed in transiently transfected human cells was observed in the endoplasmic reticulum and the Golgi apparatus (Colberg-Poley et al. 2000). The significance of this phenomenon is at present unclear.

Fig. 1A,B. Two possible models for the cell-death suppressing activity of vICA in Fas-ligation-mediated apoptosis. Ligation of Fas results in recruitment of the adaptor protein FADD [through death domain (DD) homotopic interactions] and pro-caspase-8 [through death effector domain (DED) homotopic interactions] to the death-inducing signaling complex (DISC). Pro-caspase-8 is then processed into active caspase-8. Caspase-8 proteolytically cleaves its substrates such as BID and (in type I cells) pro-caspase-3. vICA binds to the death effector domains of pro-caspase-8 and thus prevents its activation. The precise mechanism of this prevention has not been elucidated. The formation of vICA-pro-caspase-8 complex can either prevent the recruitment of pro-caspase-8 to the DISC (**A**) or prevent the proteolytic processing of pro-caspase-8 (**B**)

The amino acid sequence of gpUL37 predicts that it is an integral membrane protein (Kouzarides et al. 1988). gpUL37 co-localizes with mitochondria (Colberg-Poley et al. 2000) and with the endoplasmic reticulum and the Golgi apparatus of CMV-infected cells, where it is *N*-glycosylated (Al-Barazi and Colberg-Poley 1996). gpUL37 is also present on the plasma membrane of transiently transfected human cells, presumably its final destination, following its trafficking through the endoplasmic reticulum and the Golgi apparatus (Colberg-Poley et al. 2000).

3.3
Impact of vMIA on Apoptotic Signaling Pathways

Recent studies have shown (Kelekar and Thompson 1998; Cheng et al. 2001; Han et al. 2001; Chittenden 2002; Letai et al. 2002) that exposure of cells to diverse apoptotic stimuli, such as ligands of death receptors, withdrawal of growth factors, DNA damage, or various cytotoxic compounds, induces activation of BH3-only pro-apoptotic Bcl-2 family members (e.g. caspase-8-mediated processing of BID, or serum withdrawal-induced upregulation of BBC3/PUMA) which then triggers oligomerization of Bax and Bak, followed by permeabilization of mitochondria and the efflux of cytochrome c and other mitochondrial apoptogenic proteins into the cytoplasm. The release of cytochrome c sets off a chain of downstream events such as activation of caspase-9 and caspase-3, and cleavage of PARP. vMIA does not inhibit Fas-ligation-mediated caspase-8 activation, or BID processing (Goldmacher et al. 1999), and it does not seem to associate with VDAC (Vieira et al. 2001). On the other hand, vMIA blocks permeabilization of mitochondria in cells that have been exposed to anti-Fas, blocks downstream events in type II cells, and blocks permeabilization of cell-free mitochondria exposed to truncated BID (tBID; Goldmacher et al. 1999; Goldmacher 2002). Taken together, these results indicate that vMIA interrupts death-receptor-mediated apoptosis downstream of BID processing, but upstream, or at the step of permeabilization of mitochondria induced by tBID.

3.4
Interaction of vMIA with Adenine Nucleotide Translocator

vMIA has affinity towards adenine nucleotide translocator (ANT), an integral protein of the inner mitochondrial membrane (Goldmacher et al. 1999; Vieira et al. 2001). This finding is of interest since there is some circumstantial evidence that ANT may be involved in the control of apoptosis at the mitochondrial level (Belzacq et al. 2002). Because of the high degree of sequence similarity among the three isotypes of ANT it seems likely that all three have affinity towards vMIA, although only peptides unique for ANT-2 and ANT-3 were unambiguously detected in the ANT samples bound to vMIA, while the data for ANT 1 were inconclusive (Goldmacher 2002). At present, we can only speculate on the nature and on the functional significance of the vMIA-ANT interaction and whether vMIA associates with ANT directly or through an adaptor protein.

3.5
Sequence-Function Analysis of vMIA

The three *UL37*-encoded proteins share the first 162 N-terminal amino acid sequence of vMIA, essentially its entire length (Goldmacher et al. 1999), indicating that this common segment is responsible for the anti-apoptotic activities of these proteins, and, by extension, that these proteins suppress cell death by the same mechanism. Deletion mutagenesis of vMIA identified two segments within its ORF that are required for its anti-apoptotic activity, Tyr^{5}-Leu^{30} and Asp^{118}-Arg^{147} (Goldmacher 2002). Remarkably, these two domains are, together, sufficient for the anti-apoptotic function of vMIA, and a mini-protein, Δamino acids 35–112/Δamino acids148–163, in which most of the amino acids outside of these two segments are missing, is active as a cell death suppressor. The first domain is necessary and sufficient for the mitochondrial targeting of vMIA (Hayajneh et al. 2001). This segment contains a previously unreported mitochondria-targeting motif.

3.6
A Hypothetical Mechanism of Cell-Death Suppression by vMIA

Taken together, observations that vMIA does not prevent caspase-8 activation or BID processing during Fas-ligation-mediated apoptosis, but does block Fas-ligation-mediated or tBID-mediated cytochrome c efflux from mitochondria as well as downstream events, that vMIA localizes predominantly to mitochondria, and that vMIA forms a complex with the mitochondrial protein ANT, provide strong evidence that vMIA interrupts Fas-mediated apoptosis either at the step of mitochondrial permeabilization or just upstream (Fig. 2). The exact mechanism of cell-death suppression by vMIA remains elusive, and, at present, we do not know whether its mitochondrial targeting domain facilitates its association with ANT, or, alternatively, whether it targets vMIA to another mitochondrial molecule, or assists association of vMIA with a chaperone protein. The molecular function of the second domain required for the anti-apoptotic function of vMIA is also unclear.

There are similarities between the functional properties of vMIA and those of Bcl-2 family cell-death suppressors (reviewed in Goldmacher 2002): (1) vMIA, Bcl-2 and Bcl-x_L protect type II cells, but not type I cells against Fas-mediated apoptosis; (2) all three proteins are predominantly localized at mito-

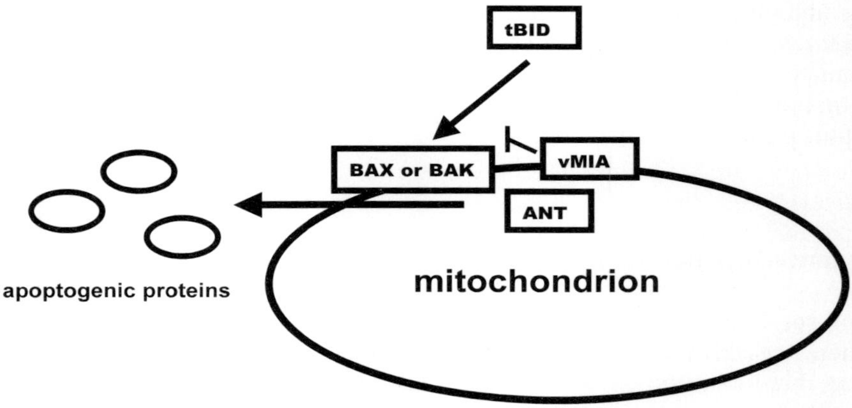

Fig. 2. A model for the cell-death suppressing activity of vMIA in Fas-ligation-mediated apoptosis. Activation of BID into tBID results in Bax- or Bak-dependent permeabilization of mitochondria and release of cytochrome c and other mitochondrial apoptogenic proteins into the cytoplasm. vMIA prevents this permeabilization. The exact mechanism of this blocking effect of vMIA is not clear. vMIA may act by either preventing the interaction of tBID with Bax or with Bak, or by preventing apoptosis-induced oligomerization of Bax and/or Bax (e.g. by binding and sequestering Bax and/or Bak), or by acting downstream of this aggregation. It has not been established yet if vMIA-ANT interaction plays any role in this blocking activity of vMIA

chondria; and (3) all three proteins suppress permeabilization of mitochondria. However, unlike Bcl-2 or Bcl-x_L, vMIA: (1) does not possess homology domains BH1, BH2, BH3, or BH4; and (2) blocks tBID-induced permeabilization of mitochondria even at high concentrations of tBID, while Bcl-2 and Bcl-x_L can only block permeabilization of mitochondria at low concentrations of tBID (Cheng et al. 2001; Goldmacher 2002). The ability of tBID to overcome, at a high enough concentration, the protection by Bcl-2 and Bcl-x_L is consistent with the recent proposal that Bcl-2 and Bcl-x_L block apoptosis primarily by sequestering BH3-only proteins and thus preventing their interaction with Bax or Bak and subsequent permeabilization of mitochondria (Cheng et al. 2001; Letai et al. 2002). The inability of tBID to overcome the protective effect of vMIA suggests that vMIA blocks mitochondrial permeabilization in a different manner. vMIA does not seem to associate with tBID (Goldmacher, unpublished), and we are currently investigating if vMIA can associate with Bax or Bak.

3.7 Homologues of *UL37* Gene Products Encoded by β-Herpesviruses

Amino acid sequences of human CMV *UL37*-encoded proteins do not seem to have any cellular or viral homologues, except for those encoded by other β-herpesviruses. Three non-human primate CMVs, chimpanzee CMV, rhesus macaque CMV, and African green monkey CMV-related virus, are homologous to human CMV in all three exons of *UL37*, and encode highly conserved homologues of vMIA (McCormick et al. 2003). The remarkable conservation of the vMIA functional domains in these proteins suggests that these vMIA homologues are functional cell-death suppressors, a prediction experimentally confirmed with the rhesus macaque CMV homologue of vMIA (Rh-vMIA; McCormick et al. 2003).

Positional homologues of *UL37* within the genomes of mouse CMV, rat CMV, guinea pig CMV, tupaia herpesvirus, HHV-6 and HHV-7 contain segments homologous to *UL37 exon 3*, but lack any homology to *UL37 exon 1*. Therefore, proteins produced from these genes cannot be predicted to function as cell-death suppressors based on sequence alone. Indeed, pM37 of mouse CMV lacked any anti-apoptotic activity in our experiments (McCormick et al. 2003). This finding is consistent with the previous data that an *M37* deletion mutant CMV virus replicated in cultured cells as efficiently as the wild-type virus (Lee et al. 2000), a feature similar to the dispensability of the human CMV *UL37 exon 3* region (Borst et al. 1999; Goldmacher et al. 1999). In contrast, the *UL37 exon 1* region encoding vMIA is indispensable for replication of human CMV in cell culture (Hahn et al. 2001).

4 Evidence That There Are Other Cell-Death Suppressing Mechanisms Employed by Cytomegaloviruses

4.1 Immediate Early Genes *IE1* and *IE2* of Human CMV

It has been reported (Zhu et al. 1995) that HeLa cells transfected with either *IE1* or *IE2* acquired resistance to apoptosis induced by either TNF-receptor ligation, or by infection with an E1B19k-deficient adenovirus. In our experiments, neither *IE1* nor *IE2* protected transiently transfected HeLa cells against TNF-receptor I-mediated apoptosis or Fas-mediated apoptosis (Goldmacher et al. 1999). The reason for this discrepancy is unclear, and more work will be needed to reconcile the data and to understand these phenomena better. IE2 (but not IE1) ectopically expressed in coronary artery smooth muscle cells protected these cells against apoptosis induced by ectopically expressed p53 (Tanaka et al. 1999). It is not known, however, if elevation of p53 induces apoptosis in human fibroblasts or other cell types that can be productively infected by human CMV.

4.2
A Ribonucleotide Reductase Homologue of Mouse CMV

Brune et al. (2001) reported that infection of cultured mouse cells with *M45*-null mouse CMV induces apoptosis in mouse endothelial cells, but not in fibroblasts, bone marrow stromal cells, or hepatocytes. The nature of the *M45* pro-survival function has not been elucidated, and it is not clear if it stems from a direct suppression of an apoptotic pathway, or if this gene protects cells by providing a vital physiological function unrelated to direct blocking of an apoptotic pathway, such as regulation of deoxyribonucleotide pools in infected cells (Lembo et al. 2000). *M45* failed to protect HeLa cells against Fas-mediated apoptosis in transient transfection assays (Lembo, Skaletskaya, and Goldmacher, unpublished). This result is consistent with pM45 either not being a cell-death suppressor, or being a cell-death suppressor functional only in some types of mouse cells. The corresponding gene in human CMV, *UL45*, is dispensable for replication of human CMV in both endothelial and fibroblast cells, and its inactivation does not induce any detectable apoptosis in either cell type (Hahn et al. 2002).

4.3
Induction of NF-κB, a Transcription Factor Capable of Suppressing Apoptosis During CMV Infection

Activation of NF-κB suppresses apoptosis induced by various stimuli (Beg and Baltimore 1996; Liu et al. 1996; Van Antwerp et al. 1996; Wang et al. 1996). NF-κB is elevated in cells infected with human CMV or mouse CMV (Yurochko et al. 1995; Gribaudo et al. 1996), which raises the question whether activation of NF-κB in CMV-infected cells contributes to the suppression of virus-induced apoptosis, and it will be of interest to examine these phenomena further.

4.4
Possible Sequestration of p53 During CMV Infection

A number of viruses inactivate p53 by means of its sequestration, blocking its function, or degradation [reviewed in Fortunato and Spector (1998)]. These phenomena, while far from being understood, suggest that p53 is involved in the cellular defense against these viruses. Specifically, p53 protein concentration is highly elevated in cells infected with human CMV (Muganda et al. 1994; Speir et al. 1994). There are only fragmentary and somewhat contradictory data available concerning the mechanisms of inactivation of p53 in cells infected with CMV. It was reported that p53 was sequestered outside the nucleus in cells infected with human CMV (Kovacs et al. 1996), apparently through blocking its nuclear localization signal (Wang et al. 2001), and that the immediate early protein IE2 formed an inactivating complex with p53 (Speir et al. 1994; Tsai et al. 1996; Tanaka et al. 1999). However, transfection of cells with *IE2* did not result in any change of p53 localization; p53 was still found in the nucleus (Wang et al. 2000). In another study (Fortunato and Spector 1998), p53 was observed sequestered not in the cytoplasm but rather within the nuclei of human fibroblasts infected with human CMV.

5
Cell-Death Suppressing Activities of vICA and vMIA During CMV Replication

5.1
Expression of vICA and vMIA in Virally Infected Cells

UL36 and *UL37* are immediate early genes, and their transcripts start being expressed in infected cells within several hours post-infection, and then persist throughout infection (Colberg-Poley 1996; Goldmacher et al. 1999; Mocarski and Courcelle 2001). The protein product of *UL36*, vICA, is, in addition, a component of the virion, providing infected cells with the protein immediately after viral adsorption, although at a very low level (Patterson and Shenk 1999). The onset of mitochondrial localization of vMIA is between 24 and 48 h post-infection (Goldmacher et al. 1999).

5.2
The Anti-apoptotic Function of vICA Is Dispensable

vICA seems to readily undergo inactivating mutations during propagation of human CMV in cultured fibroblasts, and the resulting vICA-deficient CMV strains replicate as efficiently as those encoding a functional vICA (Patterson and Shenk 1999; Skaletskaya et al. 2001), and do not seem to induce more cell death in these cells than vICA^{+}-viral strains[1]. This mutability and dispensability of vICA seems to be a general feature of cytomegaloviruses: Rh-vICA is dispensable for in vitro and in vivo replication of rhesus macaque CMV (Chang et al. 2002; McCormick et al. 2003), and M-vICA is dispensable for in vitro replication of mouse CMV (U. Koszinowski, pers. comm.). A likely explanation for the dispensability of vICA for replication of human and rhesus macaque CMV is that its anti-apoptotic function is redundant in the presence of vMIA. This hypothesis would not explain the dispensability of M-vICA for replication

[1]It is difficult to accurately access the extent of cell death in human fibroblasts infected with human CMV, but during the first 3–4 days of infection the fraction of viable cells, as observed under a phase microscope, it seems to always exceed 60–80% (V.S. Goldmacher, unpubl. observ.).

of mouse CMV that lacks vMIA, unless one invokes that mouse CMV encodes another yet unidentified cell-death suppressor, or, a less likely alternative, that mouse CMV infection does not trigger apoptosis.

Despite the evidence that vICA is dispensable for replication of at least some cytomegaloviruses, its amino acid sequence is highly conserved throughout the β-herpesvirus subfamily. These data suggest that the cell-death suppressing function of vICA, while not required for viral replication, provides an advantage for replication of these viruses in their hosts in vivo.

5.3 The Anti-apoptotic Function of vMIA Appears to Be Indispensable

There are several lines of evidence indicating that the anti-apoptotic function of vMIA is required for replication of human and closely related primate CMVs: (1) functional domains of vMIA are fully conserved (invariant) in all examined clinical isolates and laboratory strains of human CMV (Hayajneh et al. 2001); (2) both domains are highly conserved in chimpanzee CMV, rhesus macaque CMV, and African green monkey CMV-related virus (McCormick et al. 2003); and (3) vMIA-deficient CMV mutants fail to replicate in normal fibroblasts, and induce their massive apoptosis, but replicate well in fibroblasts that ectopically express either vMIA, or an anti-apoptotic Bcl-2 homologue (Hahn et al. 2001).

Unlike vMIA, its two longer splice variants, gpUL37 and gpUL37$_M$, are dispensable for replication of CMV in cultured fibroblasts, as long as vMIA is expressed (Borst et al. 1999; Goldmacher et al. 1999). Why then is the *UL37 exon 3* region retained in the CMV genome? pM37 of mouse CMV, a protein homologous to a *UL37 exon 3*-encoded segment of gpUL37 and gpUL37$_M$, is also dispensable for in vitro replication of mouse CMV, but is essential for replication of this virus in vivo (Lee et al. 2000). These data suggest that pM37, and gpUL37 and gpUL37$_M$, its counterparts in human CMV, may have a yet unknown function required for CMV replication in vivo, a function that is unrelated to cell-death suppression.

5.4 Involvement of vMIA and vICA in Suppression of Apoptosis During CMV Infection

Fibroblasts infected with human CMV remain sensitive towards Fas-mediated apoptosis until about 24 h post-infection and then acquire resistance (Skaletskaya et al. 2001). This acquired resistance roughly coincides with the onset of vICA expression by the majority of infected cells. Cells infected with vICA-deficient strains of CMV remain sensitive towards Fas-mediated apoptosis for an additional 24-h period, and acquire resistance at about 48 h post-infection, which coincides with the onset of vMIA expression in mitochondria (Goldmacher et al. 1999; Skaletskaya et al. 2001). These data support the notion that the resistance of CMV-infected cells to Fas-mediated apoptosis is caused by the expression of vICA and vMIA. However, it is not clear if the chain of events triggered during Fas-mediated apoptosis is similar to those during CMV infection-induced apoptosis, and it was only recently shown that infection of human fibroblasts with double vMIA/vICA-deficient human CMV caused their massive apoptosis (Hahn et al. 2001).

Taken together, the findings that the anti-apoptotic function of vMIA is indispensable for viral replication, and that the amino acid sequences of the two essential domains of vMIA are invariant among all sequenced strains of human CMV, provide strong evidence that the anti-apoptotic function of vMIA is largely responsible for the suppression of apoptosis in cells infected with human CMV. The importance of vICA in suppression of CMV infection-induced apoptosis remains unclear. One possible role for vICA during in vivo CMV infection may be to protect infected type I cells of the host, since vMIA may not be functional in such cells. It is not known, however, if there are any type I cells permissive for CMV replication, and if vMIA would or would not protect type I cells against CMV-induced apoptosis. Another possible role of vICA might be to protect CMV-infected cells against apoptosis induced by the immune response of the host. Regardless of what the function of vICA turns out to be, the recent finding that the anti-apoptotic activity of Rh-vICA appears to be dispensable for in vivo pathogenesis of a closely related rhesus macaque CMV (Chang et al. 2002; McCormick et al. 2003) argues that vICA is dispensable for replication of human CMV in vivo as well.

A schematic representation of what is known about the involvement of vICA and vMIA in the suppression of apoptosis is illustrated in Fig. 3.

6
A Speculation on Possible Strategies Employed by β-Herpesviruses to Suppress Apoptosis

The question whether there are any other CMV-encoded cell-death suppressors encoded by human or animal cytomegaloviruses remains open. Human CMV and the three sequenced non-human primate CMVs encode both vMIA and vICA. From the known data on the properties of human and rhesus macaque CMVs it seems likely that all four closely related viruses use similar strategies, and rely mostly on vMIA to suppress apoptosis of infected cells, with vICA playing only a secondary role.

Other β-herpesviruses, such as rodent CMVs, tupaia herpesvirus, HHV-6 and HHV-7, encode homologues of vICA, but do not encode homologues of vMIA and, therefore, their strategies to suppress apoptosis must differ from those of primate CMVs. As mentioned above, the anti-apoptotic function of M-vICA, the only known cell-death suppressor encoded by mouse CMV, is

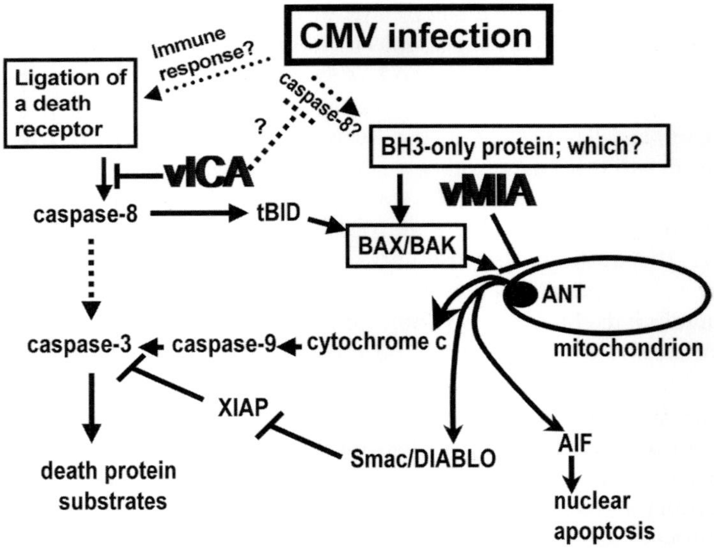

Fig. 3. Schematic representation of cell-death suppressing activities of vICA and vMIA. vICA acts by blocking caspase-8 activation caused by ligation of a death receptor. It is active in both type I cells in which caspase-8 can directly activate caspase-3, and type II cells in which activation of the mitochondrial apoptotic pathway is required for activation of caspase-3. However, it is not known if death-receptor-mediated apoptosis is initiated during human CMV infection, or if vICA is capable of blocking human CMV-induced apoptosis. Cytotoxic effector cells can induce apoptosis of infected cells both through ligation of Fas on the surface of target cells, and through injection of granzymes into target cells with subsequent activation of caspases. It will be important to examine if vICA is involved in blocking apoptosis of infected cells induced by these stimuli. vMIA blocks apoptosis by preventing mitochondrial permeabilization, and, accordingly, is only active in type II cells during death-receptor-mediated apoptosis. It is not known if vMIA can or cannot block human CMV-induced apoptosis in type I cells, and if there are any type I cells that are permissive for human CMV infection. Since apoptosis induced in cells infected with vMIA-deficient human CMV can be prevented by Bcl-2, and Bcl-2 appears to block apoptosis primarily by sequestering BH3-only pro-apoptotic proteins, one can argue that apoptosis triggered by human CMV infection proceeds through a step of activation of a BH3-only pro-apoptotic protein (or proteins), and it will be important to identify this protein(s)

dispensable for its replication, which raises a question if this virus, and perhaps other β-herpesviruses that do not encode vMIA, possess another yet unknown cell-death suppressor.

Similarities between the strategies of suppression of apoptosis by β- and γ-herpesviruses may shed light on which part of the apoptotic pathway is a likely target for this hypothetical cell-death suppressor. It is becoming increasingly evident that both γ-herpesviruses and primate cytomegaloviruses encode cell-death suppressors that block caspase-8 activation, and those that block mitochondrial permeabilization, and that blocking the latter event may be more important than the former for replication of these viruses. Herpesvirus saimiri encodes a functional Bcl-2 homologue (Nava et al. 1997; Derfuss et al. 1998), and a functional cFLIP homologue, vFLIP, and the latter is dispensable for replication and for pathogenicity of this virus (Glykofrydes et al. 2000). Other γ-herpesviruses also encode homologues of Bcl-2 (reviewed in Table 4 in Goldmacher 2002), and some of them encode homologues of cFLIP (Thome et al. 1997; Wang et al. 1997; Tschopp et al. 1998). From these data one could predict that mouse CMV is likely to encode a mitochondrial inhibitor of apoptosis.

7 vMIA as a Promising Target for Drug Development

Human cytomegalovirus is widely spread in human populations and is highly pathogenic for immunocompromised individuals such as organ transplant recipients, some cancer patients, and patients with AIDS. The finding that the anti-apoptotic function of vMIA is indispensable for viral replication indicates that vMIA is a promising target for development of an anti-CMV drug. An agent that blocks the anti-apoptotic function of vMIA is likely to induce apoptosis of CMV-infected cells and thus halt infection at an early stage. vMIA is not homologous to any known human protein, which suggests that drugs targeting vMIA will not target any cellular protein and will not be harmful to patients.

Acknowledgements. I am grateful to Drs. U. Koszinowski and G. Hahn for the permission to publish their unpublished data, and Drs. A. Colberg-Poley and S. Landolfo for helpful advice.

References

Al-Barazi HO, Colberg-Poley AM (1996) The human cytomegalovirus UL37 immediate-early regulatory protein is an integral membrane N-glycoprotein which traffics through the endoplasmic reticulum and Golgi apparatus. J Virol 70:7198–7208
Beg AA, Baltimore D (1996) An essential role for NF-kappaB in preventing TNF-alpha-induced cell death. Science 274:782–784
Belzacq AS, Vieira HL, Kroemer G, Brenner C (2002) The adenine nucleotide translocator in apoptosis. Biochimie 84:167–176
Borst EM, Hahn G, Koszinowski UH, Messerle M (1999) Cloning of the human cytomegalovirus (HCMV) genome as an infectious bacterial artificial chromosome in *Escherichia coli*: a new approach for construction of HCMV mutants. J Virol 73:8320–8329
Brune W, Menard C, Heesemann J, Koszinowski UH (2001) A ribonucleotide reductase homolog of cytomegalovirus and endothelial cell tropism. Science 291:303–305
Chang WL, Tarantal AF, Zhou SS, Borowsky AD, Barry PA (2002) A recombinant rhesus cytomegalovirus expressing enhanced green fluorescent protein retains the wild-type phenotype and pathogenicity in fetal macaques. J Virol 76:9493–9504
Cheng EH, Wei MC, Weiler S, Flavell RA, Mak TW, Lindsten T, Korsmeyer SJ (2001) BCL-2, BCL-X(L) sequester BH3 domain-only molecules preventing Bax- and Bak-mediated mitochondrial apoptosis. Mol Cell 8:705–711

Chittenden T (2002) BH3 domains: intracellular death-ligands critical for initiating apoptosis. Cancer Cell 2:165–166
Colberg-Poley AM (1996) Functional roles of immediate early proteins encoded by the human cytomegalovirus UL36-38, UL115-119, TRS1/IRS1 and US3 loci. Intervirology 39:350–360
Colberg-Poley AM, Patel MB, Erezo DP, Slater JE (2000) Human cytomegalovirus UL37 immediate-early regulatory proteins traffic through the secretory apparatus and to mitochondria. J Gen Virol 81:1779–1789
Derfuss T, Fickenscher H, Kraft MS, Henning G, Lengenfelder D, Fleckenstein B, Meinl E (1998) Antiapoptotic activity of the herpesvirus saimiri-encoded Bcl-2 homolog: stabilization of mitochondria and inhibition of caspase-3-like activity. J Virol 72:5897–5904
Engels IH, Stepczynska A, Stroh C, Lauber K, Berg C, Schwenzer R, Wajant H, Janicke RU, Porter AG, Belka C, Gregor M, Schulze-Osthoff K, Wesselborg S (2000) Caspase-8/FLICE functions as an executioner caspase in anticancer drug-induced apoptosis. Oncogene 19:4563–4573
Foghsgaard L, Jaattela M (1997) The ability of BHRF1 to inhibit apoptosis is dependent on stimulus and cell type. J Virol 71:7509–7517
Fortunato EA, Spector DH (1998) p53 and RPA are sequestered in viral replication centers in the nuclei of cells infected with human cytomegalovirus. J Virol 72:2033–2039
Glykofrydes D, Niphuis H, Kuhn EM, Rosenwirth B, Heeney JL, Bruder J, Niedobitek G, Muller-Fleckenstein I, Fleckenstein B, Ensser A (2000) Herpesvirus saimiri vFLIP provides an antiapoptotic function but is not essential for viral replication, transformation, or pathogenicity. J Virol 74:11919–11927
Goldmacher VS (2002) vMIA, a viral inhibitor of apoptosis targeting mitochondria. Biochimie 84:177–185
Goldmacher VS, Bartle LM, Skaletskaya A, Dionne CA, Kedersha NL, Vater CA, Han JW, Lutz RJ, Watanabe S, Cahir McFarland ED, Kieff ED, Mocarski ES, Chittenden T (1999) A cytomegalovirus-encoded mitochondria-localized inhibitor of apoptosis structurally unrelated to Bcl-2. Proc Natl Acad Sci USA 96:12536–12541
Gribaudo G, Ravaglia S, Guandalini L, Cavallo R, Gariglio M, Landolfo S (1996) The murine cytomegalovirus immediate-early 1 protein stimulates NF-kappa B activity by transactivating the NF-kappa B p105/p50 promoter. Virus Res 45:15–27
Hahn G, Eichhorst ST, Korn B, Krammer PN, Greaves R (2001) An anti-apoptotic protein of human cytomegalovirus enables virus replication. 26th International Herpesvirus Workshop. Regensburg, Germany, 2001
Hahn G, Khan H, Baldanti F, Koszinowski UH, Revello MG, Gerna G (2002) The human cytomegalovirus ribonucleotide reductase homolog UL45 is dispensable for growth in endothelial cells, as determined by a BAC-cloned clinical isolate of human cytomegalovirus with preserved wild-type characteristics. J Virol 76:9551–9555
Han J, Flemington C, Houghton AB, Gu Z, Zambetti GP, Lutz RJ, Zhu L, Chittenden T (2001) Expression of bbc3, a pro-apoptotic BH3-only gene, is regulated by diverse cell death and survival signals. Proc Natl Acad Sci USA 98:11318–11323
Hayajneh WA, Colberg-Poley AM, Skaletskaya A, Bartle LM, Lesperance MM, Contopoulos-Ioannidis DG, Kedersha NL, Goldmacher VS (2001) The sequence and antiapoptotic functional domains of the human cytomegalovirus UL37 exon 1 immediate early protein are conserved in multiple primary strains. Virology 279:233–240
Irmler M, Thome M, Hahne M, Schneider P, Hofmann K, Steiner V, Bodmer JL, Schroter M, Burns K, Mattmann C, Rimoldi D, French LE, Tschopp J (1997) Inhibition of death receptor signals by cellular FLIP. Nature 388:190–195
Kelekar A, Thompson CB (1998) Bcl-2-family proteins: the role of the BH3 domain in apoptosis. Trends Cell Biol 8:324–330
Kouzarides T, Bankier AT, Satchwell SC, Preddy E, Barrell BG (1988) An immediate early gene of human cytomegalovirus encodes a potential membrane glycoprotein. Virology 165:151–164
Kovacs A, Weber ML, Burns LJ, Jacob HS, Vercellotti GM (1996) Cytoplasmic sequestration of p53 in cytomegalovirus-infected human endothelial cells. Am J Pathol 149:1531–1539

Krueger A, Baumann S, Krammer PH, Kirchhoff S (2001) FLICE-inhibitory proteins: regulators of death receptor-mediated apoptosis. Mol Cell Biol 21:8247–8254

Lee M, Xiao J, Haghjoo E, Zhan X, Abenes G, Tuong T, Dunn W, Liu F (2000) Murine cytomegalovirus containing a mutation at open reading frame M37 is severely attenuated in growth and virulence in vivo. J Virol 74:11099–11107

Lembo D, Gribaudo G, Hofer A, Riera L, Cornaglia M, Mondo A, Angeretti A, Gariglio M, Thelander L, Landolfo S (2000) Expression of an altered ribonucleotide reductase activity associated with the replication of murine cytomegalovirus in quiescent fibroblasts. J Virol 74:11557–11565

Letai A, Bassik M, Walensky L, Sorcinelli M, Weiler S, Korsmeyer S (2002) Distinct BH3 domains either sensitize or activate mitochondrial apoptosis, serving as prototype cancer therapeutics. Cancer Cell 2:183–192

Liu Y, Biegalke BJ (2001) Characterization of a cluster of late genes of guinea pig cytomegalovirus. Virus Genes 23:247–256

Liu ZG, Hsu H, Goeddel DV, Karin M (1996) Dissection of TNF receptor 1 effector functions: JNK activation is not linked to apoptosis while NF-kappaB activation prevents cell death. Cell 87:565–576

McCormick AL, Skaletskaya A, Barry PA, Mocarski ES, Goldmacher VS (2003) Differential function and expression of the viral inhibitor of caspase 8-induced apoptosis (vICA) and the viral mitochondria-localized inhibitor of apoptosis (vMIA) cell death suppressors conserved in primate and rodent cytomegaloviruses. Virology 316:221–233

Mocarski ES, Courcelle CT (2001) In: Knipe DM, Howley PM (eds) Fields virology. Lippincott-Raven, New York, pp 2629–2673

Muganda P, Mendoza O, Hernandez J, Qian Q (1994) Human cytomegalovirus elevates levels of the cellular protein p53 in infected fibroblasts. J Virol 68:8028–8034

Nava VE, Cheng EH, Veliuona M, Zou S, Clem RJ, Mayer ML, Hardwick JM (1997) Herpesvirus saimiri encodes a functional homolog of the human bcl-2 oncogene. J Virol 71:4118–4122

O'Brien V (1998) Viruses and apoptosis. J Gen Virol 79:1833–1845

Patterson CE, Shenk T (1999) Human cytomegalovirus UL36 protein is dispensable for viral replication in cultured cells. J Virol 73:7126–7131

Scaffidi C, Fulda S, Srinivasan A, Friesen C, Li F, Tomaselli KJ, Debatin KM, Krammer PH, Peter ME (1998) Two CD95 (APO-1/Fas) signaling pathways. EMBO J 17:1675–1687

Scaffidi C, Schmitz I, Krammer PH, Peter ME (1999a) The role of c-FLIP in modulation of CD95-induced apoptosis. J Biol Chem 274:1541–1548

Scaffidi C, Schmitz I, Zha J, Korsmeyer SJ, Krammer PH, Peter ME (1999b) Differential modulation of apoptosis sensitivity in CD95 type I and type II cells. J Biol Chem 274:22532–22538

Schmitz I, Walczak H, Krammer PH, Peter ME (1999) Differences between CD95 type I and II cells detected with the CD95 ligand. Cell Death Differ 6:821–822

Skaletskaya A, Bartle LM, Chittenden T, McCormick AL, Mocarski ES, Goldmacher VS (2001) A cytomegalovirus-encoded inhibitor of apoptosis that suppresses caspase-8 activation. Proc Natl Acad Sci USA 98:7829–7834

Smyth MJ, Kelly JM, Sutton VR, Davis JE, Browne KA, Sayers TJ, Trapani JA (2001) Unlocking the secrets of cytotoxic granule proteins. J Leukoc Biol 70:18–29

Speir E, Modali R, Huang ES, Leon MB, Shawl F, Finkel T, Epstein SE (1994) Potential role of human cytomegalovirus and p53 interaction in coronary restenosis. Science 265:391–394

Tanaka K, Zou JP, Takeda K, Ferrans VJ, Sandford GR, Johnson TM, Finkel T, Epstein SE (1999) Effects of human cytomegalovirus immediate-early proteins on p53-mediated apoptosis in coronary artery smooth muscle cells. Circulation 99:1656–1659

Thome M, Schneider P, Hofmann K, Fickenscher H, Meinl E, Neipel F, Mattmann C, Burns K, Bodmer JL, Schroter M, Scaffidi C, Krammer PH, Peter ME, Tschopp J (1997) Viral FLICE-inhibitory proteins (FLIPs) prevent apoptosis induced by death receptors. Nature 386:517–521

Tsai HL, Kou GH, Chen SC, Wu CW, Lin YS (1996) Human cytomegalovirus immediate-early protein IE2 tethers a transcriptional repression domain to p53. J Biol Chem 271:3534–3540

Tschopp J, Thome M, Hofmann K, Meinl E (1998) The fight of viruses against apoptosis. Curr Opin Genet Dev 8:82–87

Van Antwerp DJ, Martin SJ, Kafri T, Green DR, Verma IM (1996) Suppression of TNF-alpha-induced apoptosis by NF-kappaB. Science 274:787–789

Vieira HL, Belzacq AS, Haouzi D, Bernassola F, Cohen I, Jacotot E, Ferri KF, El Hamel C, Bartle LM, Melino G, Brenner C, Goldmacher V, Kroemer G (2001) The adenine nucleotide translocator: a target of nitric oxide, peroxynitrite, and 4-hydroxynonenal. Oncogene 20:4305–4316

Wallach D, Varfolomeev EE, Malinin NL, Goltsev YV, Kovalenko AV, Boldin MP (1999) Tumor necrosis factor receptor and Fas signaling mechanisms. Annu Rev Immunol 17:331–367

Wang CY, Mayo MW, Baldwin AS Jr (1996) TNF- and cancer therapy-induced apoptosis: potentiation by inhibition of NF-kappaB. Science 274:784–787

Wang GH, Bertin J, Wang Y, Martin DA, Wang J, Tomaselli KJ, Armstrong RC, Cohen JI (1997) Bovine herpesvirus 4 BORFE2 protein inhibits Fas- and tumor necrosis factor receptor 1-induced apoptosis and contains death effector domains shared with other gamma-2 herpesviruses. J Virol 71:8928–8932

Wang J, Marker PH, Belcher JD, Wilcken DE, Burns LJ, Vercellotti GM, Wang XL (2000) Human cytomegalovirus immediate early proteins upregulate endothelial p53 function. FEBS Lett 474:213–216

Wang J, Belcher JD, Marker PH, Wilcken DE, Vercellotti GM, Wang XL (2001) Cytomegalovirus inhibits p53 nuclear localization signal function. J Mol Med 78:642–647

Yurochko AD, Kowalik TF, Huong SM, Huang ES (1995) Human cytomegalovirus upregulates NF-kappa B activity by transactivating the NF-kappa B p105/p50 and p65 promoters. J Virol 69:5391–5400

Zhu H, Shen Y, Shenk T (1995) Human cytomegalovirus IE1 and IE2 proteins block apoptosis. J Virol 69:7960–7970

Apoptosis Regulator Genes Encoded by Poxviruses

M. Barry[1], S.T. Wasilenko[1], T.L. Stewart[1] and J.M. Taylor[1]

1 Introduction

The Poxviridae family is a large family of double-stranded DNA viruses that infects both vertebrates (Chordopoxvirinae) and invertebrates (Entomopoxvirinae). The Chordopoxvirinae, which infect vertebrates, are subdivided into eight genera: *Orthopoxvirus*, *Parapoxvirus*, *Avipoxvirus*, *Capripoxvirus*, *Leporipoxvirus*, *Suipoxvirus*, *Molluscipoxvirus*, and *Yatapoxvirus*. The most famous member of the family is variola virus, a member of the *Orthopoxvirus* genus and the causative agent of smallpox disease (Smith and McFadden 2002). Smallpox was eradicated from the human population in 1977 through an aggressive vaccination program headed by the World Health Organization. Vaccinia virus, the prototypic member of the poxvirus family belonging to the *Orthopoxvirus* genus, was utilized in the vaccination program and is still widely studied today. Aside from variola virus, the only other member of the family that naturally elicits disease in humans is *Molluscum contagiosum* virus (MCV), which causes benign lesions on the skin of infected individuals (Epstein 1992). Other members of the family cause disease in a wide range of animals from rodents to birds to primates. Notable members include: ectromelia virus, which causes a lethal disease in mice; myxoma virus, the causative agent of myxomatosis in rabbits; swinepox virus; monkeypox virus; and fowlpox virus.

Poxviruses are unique in that they replicate autonomously in the cytoplasm of the infected cell. The genomes are composed of a large double-stranded DNA molecule ranging in size from 145 kbp to greater than 288 kbp and encode upwards of 200 proteins. At present, the genomes of 25 members of the Poxviridae family have been completely sequenced and are available online. The central region of the genome contains genes necessary for virus replication, assembly and propagation whereas genes encoding virulence factors and immune evasion proteins are located at the terminal ends of the genome. It is these terminal regions of the genome that display the greatest divergence among family members.

[1]Department of Medical Microbiology and Immunology, University of Alberta, 671 Heritage Medical Research Center, Edmonton, Alberta, T6G 2S2 Canada,
e-mail: michele.barry@ualberta.ca

Progress in Molecular and Subcellular Biology
C. Alonso (Ed.): Viruses and Apoptosis
© Springer-Verlag Berlin Heidelberg 2004

To survive and replicate within a host, it is well known that viruses possess specific strategies to circumvent the multifaceted immune response. Numerous studies have shown that the genomes of poxviruses encode an impressive array of proteins that function to evade the host immune response (Smith et al. 1997; Nash et al. 1999). For example, proteins encoded by poxviruses target multiple antiviral cytokines including chemokines, interleukins, interferon and tumor necrosis factor. Additionally, this family of viruses modulates the complement cascade, antigen presentation, and apoptosis.

1.1 Poxviruses and Apoptosis

Apoptosis, or programmed cell death, is a controlled process of cellular suicide that is important for development, tissue homeostasis and the elimination of damaged or pathogen-infected cells. Apoptotic cells display a number of characteristic features such as DNA fragmentation, chromatin condensation, mitochondrial disruption and plasma membrane alterations (Hengartner 2000). Although our understanding of the process is not complete, it is clear that the structured dismantling of a cell by the induction of apoptosis is an inherent ability of all cells and can be triggered by diverse stimuli. These include the activation by immune cells such as cytotoxic T lymphocytes (CTLs) and natural killer (NK) cells as well as virus infection itself (Everett and McFadden 1999; Barry and Bleackley 2002; Russell and Ley 2002). As such, it is not surprising that viruses have in turn evolved multiple strategies to modulate the apoptotic response. In some instances, viruses have evolved strategies to maximize apoptosis thereby facilitating virus spread, while in other instances viruses have evolved specific mechanisms to inhibit apoptosis (Barry and McFadden 1998; Tschopp et al. 1998; Roulston et al. 1999; Tortorella et al. 2000).

The ability to inhibit apoptosis is a recurring theme among many virus families and the poxvirus family is no exception. Poxvirus proteins that inhibit apoptosis were initially identified by functional assays in combination with gene disruptions. More recently, the combination of our increasing knowledge of apoptotic cascades with complete poxvirus genome sequencing has led to the identification of poxvirus proteins with homology to known cellular and pathogen-encoded anti-apoptotic molecules. At present, poxviruses are known to interfere with apoptosis by a variety of mechanisms, as summarized in Table 1. In this chapter, we review the current status of poxvirus-encoded proteins that regulate the apoptotic response.

Table 1. Poxvirus-encoded anti-apoptotic proteins. *AMV* Amsacta moorei virus; *CMPV* camelpox virus; *CPV* cowpox virus; *EV* ectromelia virus; *FPV* fowlpox virus; *MC* molluscum contagiosum virus; *Myx* myxoma virus; *SFV* Shope fibroma virus; *Vac* vaccinia virus

Protein	Gene	Genus	Function
TNF receptor decoy	Myx-MT-2, CrmB, CrmC, CrmD, CrmE	Leporipoxvirus, Orthopoxvirus	Sequester TNF
vFLIPS	MC159	Molluscipoxvirus	Inhibits death receptor signaling
MV-LAP	Myx-M153R	Leporipoxvirus, Capripoxvirus, Suipoxvirus	Modulation of cell surface MHC class I
CrmA/SPI-2	Vac(WR)-B13R, CPV-B12R	Orthopoxvirus	Inhibits caspase 1 and caspase 8
Serp2	Myx-M151R	Leporipoxvirus, Capripoxvirus, Yatapoxvirus	Caspase inhibitor
IAP	AMV-021	Entomopoxvirus	Predicted caspase inhibitor
P28/NIR	EV-012, SFV-N1R	Orthopoxvirus, Leporipoxvirus, Capripoxvirus, Suipoxvirus, Yatapoxvirus	Unknown
vBcl-2	FPV-039	Avipoxvirus	Predicted to inhibit cytochrome c release
M11L	Myx-M11L	Leporipoxvirus, Capripoxvirus, Suipoxvirus	Inhibits cytochrome c release
F1L	Vac(Cop)-F1L	Orthopoxvirus	Inhibits cytochrome c release
Bax-inhibitor 1	CMPV-6L, CPV-S1R	Orthopoxvirus	Unknown
Glutathione peroxidase	MC-066L	Molluscipoxvirus	Inhibits UV-induced apoptosis

2 Interference with Death Receptor Signaling

2.1 TNF Receptor Decoys

Apoptosis can be induced by stimulation of a family of death receptors found on the surface of cells. One such death receptor mechanism involves the binding of tumor necrosis factor (TNF) to tumor necrosis factor receptors (TNFR; Ashkenazi and Dixit 1998; Wallach et al. 1999). TNF is secreted from macrophages and T cells as part of the cell-mediated antiviral immune response (Tracey and Cerami 1994). The binding of TNF to its receptor induces receptor trimerization resulting in the recruitment of two adapter molecules to the cytoplasmic domain of the TNFR: TNF receptor-associated death domain (TRADD) and Fas-associated death domain (FADD). In turn, the recruitment of TRADD and FADD results in the activation of caspase 8 which is able to activate other members of the caspase family leading to apoptotic death of the virus-infected cell (Ashkenazi and Dixit 1998; Wallach et al. 1999).

To avoid the TNF-induced antiviral response, poxviruses produce secreted versions of TNFR thereby allowing them to sequester TNF ligand prior to TNF engagement with its cellular receptor. The expression of secreted cytokine receptors is a recurring theme among many members of the poxvirus family and secreted receptors that interfere with interferon (IFN)-γ, IFN-α β, interleukin-1, and chemokines are encoded by many members of the family (Smith et al. 1997, 1999; Nash et al. 1999).

Myxoma virus, the causative agent of myxomatosis in European rabbits, encodes a secreted TNFR that is a potent inhibitor of TNF-induced death (Upton et al. 1991; Schreiber and McFadden 1994). This protein, M-T2, is expressed early during myxoma virus infection and deletion of the M-T2 gene from myxoma virus results in decreased virus virulence in European rabbits (Upton et al. 1991). M-T2 displays a significant level of homology to the ligand-binding domain of human TNFR and is secreted both as a monomer and a dimer, with the dimer being a more effective TNF inhibitor (Schreiber et al. 1996). M-T2 inhibits TNF-induced apoptosis in a species-specific manner clearly demonstrating specificity for the natural host, the rabbit (Schreiber and McFadden 1994). Interestingly, M-T2 has been shown to have a second anti-apoptotic function (Schreiber et al. 1997). Infection of rabbit T cells with the M-T2 deletion virus results in an apoptotic response (Macen et al. 1996). Inhibition of this apoptotic response does not require TNF binding or secretion of M-T2, indicating a unique intracellular anti-apoptotic role for M-T2 (Schreiber et al. 1997).

Examples of TNFR homologues have been discovered in other members of the poxvirus family. Cowpox virus encodes four well-characterized TNFR homologues: CrmB, CrmC, CrmD and CrmE (Hu et al. 1994; Smith et al. 1996; Loparev et al. 1998; Saraiva and Alcami 2001; Reading et al. 2002). CrmB, CrmC, and CrmD all share sequence homology within their ligand-binding domains but differ in both their ligand specificity and their expression profile. CrmB binds TNF and lymphotoxin α and is produced early during virus infection (Hu et al. 1994). CrmD also binds both TNF and lymphotoxin α, but is expressed late during infection, while CrmC binds only TNF and is expressed late during infection (Smith et al. 1996; Loparev et al. 1998). CrmB-like proteins are found in the genomes of variola virus and camelpox virus, while ectromelia virus is predicted to encode a CrmD protein. Strains of vaccinia virus encode CrmC and CrmB homologues. The fourth member of the poxvirus TNFR family is CrmE (Saraiva and Alcami 2001; Reading et al. 2002). CrmE binds to human, mouse and rat TNF but is only capable of protecting cells from human TNF (Saraiva and Alcami 2001; Reading et al. 2002). CrmE homologues have been found in other orthopoxviruses, but, due to the presence of frame shifts or the introduction of stop codons, these are predicted to be inactive (Saraiva and Alcami 2001; Reading et al. 2002).

Recently, a fifth member of the TNFR family has been discovered which is a soluble secreted CD30 homologue (Panus et al. 2002; Saraiva et al. 2002). This viral CD30 homologue has 51–59% identity to mice and human CD30 and binds to the CD30 ligand (CD153), suggesting a possible role in inhibition of the host immune response (Panus et al. 2002; Saraiva et al. 2002).

2.2 Poxvirus vFLIPS

Aside from the now extinct variola virus, MCV is the only other member of the family that naturally infects humans. Interestingly, MCV does not encode a secreted TNFR homologue or the common orthopoxvirus caspase inhibitor CrmA (Senkevich et al. 1996; Senkevich et al. 1997). The absence of common death receptor inhibitors in the genome of MCV clearly suggested the possibility that MCV may encode a unique array of apoptotic inhibitors.

In order to interfere with death-receptor-induced apoptosis, MCV has evolved an alternative strategy of interference. MCV encodes two proteins, MC159 and MC160, which contain regions of homology to death effector domains (DEDs) found both in the prodomain of initiator caspases and in the adapter molecules TRADD and FADD (Bertin et al. 1997; Hu et al. 1997; Thome et al. 1997). DEDs contain a characteristic tertiary structure consisting of six alpha-helixes that are essential for the recruitment of procaspase 8 to the cytoplasmic domain of death receptors (Boldin et al. 1996; Muzio et al. 1996; Eberstadt et al. 1998). Various members of the herpesvirus family and rhesus rhadinovirus also contain DED proteins (Tschopp et al. 1998; Searles et al. 1999). This family of virus-encoded proteins are collectively referred to as viral FLICE (Fas-associated death-domain-like interleukin-1β converting enzyme) inhibitory proteins, vFLIPS. Cellular counterparts to vFLIPs exist, referred to as cellular FLIPs (cFLIPs), which contain two DED domains at their amino termini that interact with FADD and procaspase 8 thereby regulating apoptosis (Thome and Tschopp 2001). Interference with the recruitment of DED-containing proteins results in the inhibition of death-receptor-induced apoptosis illustrating the important role of these cellular DED-containing proteins in relaying extrinsic pro-apoptotic signals.

MC159 and MC160 contain two DED-like regions and co-immunoprecipitation studies have shown that MC159 and MC160 can interact with FADD and procaspase 8 (Bertin et al. 1997; Hu et al. 1997; Thome et al. 1997; Tsukumo and Yonehara 1999; Shisler and Moss 2001; Garvey et al. 2002). Both DED domains of MC159 are required for interaction with FADD and caspase 8 allowing MC159 to protect cells from death-receptor-induced apoptosis (Bertin et al. 1997; Hu et al. 1997; Thome et al. 1997; Tsukumo and Yonehara 1999; Shisler and Moss 2001; Garvey et al. 2002). MC160 expression on its own does not provide protection from apoptosis although MC160 retains the ability to interact with FADD and caspase 8 (Shisler and Moss 2001). Following the addition of anti-Fas to MC160-expressing cells, MC160 is degraded. This degradation can be inhibited by co-expression of MC159 or by pre-treating with the peptide-based caspase inhibitors zDEVD.fmk or zIETD.fmk, perhaps suggesting a cooperative role for MC159 and MC160 during virus infection (Shisler and Moss 2001). Although MC159 and MC160 both bind to FADD and procaspase 8, binding alone is clearly not sufficient for MC159 anti-apoptotic function (Shisler and Moss 2001; Garvey et al. 2002). Through the generation of a series of MC159 mutants, Garvey and colleagues (2002) recently identified several MC159 mutants that are unable to protect cells from apoptosis while retaining the ability to interact with FADD and caspase 8.

2.3 Myxoma Virus Leukemia-Associated Protein (MV-LAP)

The recognition of viral peptides displayed by MHC I molecules is the trigger for CTL-induced apoptotic death (Barry and Bleackley 2002; Russell and Ley 2002). CTLs play an important role in the detection and destruction of pathogen-infected cells. To evade this extremely efficient anti-viral component of the immune response, many viruses retaliate by specifically causing the downregulation of cell-surface MHC Class I. Ingeniously, viruses have targeted almost every step of the MHC Class I assembly pathway thereby short-circuiting CTL-induced apoptosis (Fruh et al. 1999; Tortorella et al. 2000).

Recently, a novel class of herpesvirus-encoded proteins involved in the downregulation of MHC class I has been identified. Kaposi's sarcoma-associated herpesvirus (KSHV or human herpesvirus 8) encodes two genes, K3 and K5, while murine gamma herpesvirus 68 (MVH68) encodes a single K3 gene, and their protein products can independently inhibit the surface expression of specific MHC Class I molecules (Coscoy and Ganem 2000; Ishido et al. 2000b). K3 and K5 expression induce the endocytosis of MHC Class I from the cell surface followed by lysosomal degradation (Coscoy and Ganem 2000; Ishido et al. 2000b). Although the precise mechanism of action of these proteins is still undefined it is believed to involve ubiquitination (Boname and Stevenson 2001; Coscoy et al. 2001; Lorenzo et al. 2002).

The loss of MHC Class I from poxvirus-infected cells is well documented but until recently no specific viral gene products had been implicated (Boshkov et al. 1992; Zuniga et al. 1999). Sequence analysis of the complete genome of myxoma virus revealed a K3/K5-like protein encoded by the M153 open reading frame (ORF; Cameron et al. 1999). M153, also referred to as MV-LAP (myxoma virus leukemia-associated protein), is required for the downregulation of MHC Class I from the surface of myxoma-virus-infected cells (Guerin et al. 2002). MV-LAP is expressed early during infection and confocal microscopy demonstrates that MV-LAP localizes to the endoplasmic reticulum

(Guerin et al. 2002). Two hydrophobic domains within the carboxy terminus of MV-LAP are necessary for ER localization. Like K3 and K5, MV-LAP contains a conserved plant homeodomain (PHD)/leukemia-associated protein (LAP) zinc finger domain at its amino terminus, which is necessary for K3 and K5 activity (Fruh et al. 2002). Deletion of MV-LAP from myxoma virus results in a severely attenuated disease in European rabbits, clearly indicating a role for MV-LAP in virus virulence (Guerin et al. 2002). In addition to stimulating the downregulation of MHC class I molecules, K5 also induces the internalization of B7.2 and ICAM-1 from the cell surface, but at present it is unknown if MV-LAP retains a similar ability (Ishido et al. 2000a; Coscoy and Ganem 2001). Fruh et al. (2002), however, have recently found that MV-LAP expression causes the downregulation of other as yet undefined cell-surface proteins. MV-LAP-related proteins are encoded by other members of the poxvirus family, including Shope fibroma virus, swinepox virus and Yaba-like disease virus. Interestingly, no MV-LAP-like proteins are encoded by members of the orthopoxvirus genus suggesting the existence of an alternative strategy for regulating MHC Class I on the surface of orthopoxvirus-infected cells.

3 Poxvirus-Encoded Caspase Inhibitors

3.1 Poxvirus SERPINS

One of the best-studied mechanisms of viral apoptotic control is the direct modulation of caspase activity. The caspases are a family of cysteine proteases related to the Ced3 gene product in *Caenorhabditis elegans* (Thornberry and Lazebnik 1998; Earnshaw et al. 1999). The caspase family consists of multiple members that can be subdivided into those necessary for inflammation and cytokine maturation and those necessary for the induction of apoptosis (Thornberry and Lazebnik 1998; Earnshaw et al. 1999). Since caspase activation is an essential element of apoptotic death, viruses have evolved mechanisms to directly inhibit members of the caspase family, and poxviruses are no exception (Roulston et al. 1999; Goyal 2001).

Without a doubt the cytokine response modifier A (CrmA) protein from cowpox virus is the most extensively investigated virus-encoded caspase inhibitor. CrmA is an intracellular protein with homology to members of the serine proteinase inhibitor (SERPIN) superfamily which inhibits serine protease function by forming an irreversible complex with the protease. Initial evidence regarding the role of CrmA in host immune evasion arose from the observation that CrmA inhibited the inflammatory process in chick chorioallentoic membranes and inhibited IL-1β-converting enzyme (ICE, also known as caspase 1) which converts the IL-1 precursor into its active form (Pickup et al. 1986; Ray et al. 1992). Subsequently, CrmA was also found to be an excellent inhibitor of caspase 8, the apical caspase activated during death-receptor ligation (Zhou et al. 1997). CrmA expression can prevent apoptosis induced by both Fas ligand and TNFα in a variety of cell lines, and poxviruses devoid of CrmA are more susceptible to Fas-mediated killing (Miura et al. 1995; Tewari and Dixit 1995; Tewari et al. 1995; Macen et al. 1996). In vitro, CrmA is also an effective inhibitor of CTL-released granzyme B which initiates apoptotic death by directly activating members of the caspase family (Darmon et al. 1995; Quan et al. 1995; Barry and Bleackley 2002). Expression of CrmA in whole cells, however, inhibits Fas-mediated death much more efficiently than granzyme-mediated death (Tewari et al. 1995).

While the orthopoxviruses all encode homologues of CrmA, the leporipoxviruses appear to have taken a slightly different evolutionary path. Myxoma virus encodes an intracellular 38 kDa SERPIN referred to as Serp2 (Cameron et al. 1999). Serp2 displays approximately 35% identity with CrmA (Petit et al. 1996). Expressed at all stages of the life cycle, Serp2 possesses an aspartic acid residue at the P1 position of its reactive site loop, similar to that of CrmA, suggesting that Serp2 may inhibit caspases. In fact, when assayed for activity against caspase 1 and granzyme B, Serp2 showed inhibitory activity, albeit at a dramatically reduced level than that of CrmA (Turner et al. 1999). Despite the similarities between CrmA and Serp2, Serp2 is unable to inhibit caspases 8, 9, and 2, and is also unable to inhibit virus-induced apoptosis initiated by infection with a CrmA-deleted cowpox virus (Turner et al. 1999). However, infection of European rabbits with a myxoma virus devoid of Serp2 results in disease attenuation and increased apoptosis of lymphocytes in the draining lymph node (Messud-Petit et al. 1998). Explanations for these results include the possibility that Serp2 may inhibit rabbit caspases more efficiently than human or murine caspases, or that Serp2 has a unique function.

3.2 Poxvirus-Encoded IAP Molecules

Inhibitors of apoptosis (IAPs) were initially identified in strains of baculovirus that were able to prevent virus-induced apoptosis (Crook et al. 1993; Birnbaum et al. 1994). IAPs are typically potent apoptotic inhibitors that are capable of providing protection from a variety of apoptotic stimuli through direct inhibition of caspases (Deveraux and Reed 1999; Salvesen and Duckett 2002). Structurally, IAPs are characterized by the presence of two or three conserved 60 to 70 amino acid baculovirus IAP repeat (BIR) motifs in the amino termini or central portion of the protein, and additionally some IAPs contain a carboxy-terminal RING finger domain (Deveraux and Reed 1999; Salvesen and Duckett 2002). Although the exact role of the two separate domains remains unexplained, both domains are essential for anti-apoptotic function of baculovirus IAPs (Clem and Miller 1993, 1994; Duckett et al. 1998). Following the identification of baculovirus IAPs, a number of cellular homologues were identified, suggesting that IAPs play a significant role in regulating apoptotic events in multicellular organisms (Deveraux and Reed 1999; Salvesen and Duckett 2002). The RING finger domains of several cellular IAPs have recently been shown to contain ubiquitin ligase activity resulting in caspase degradation (Huang et al. 2000; Yang et al. 2000; Suzuki et al. 2001; Li et al. 2002). Cellular IAPs have been implicated in several human diseases including spinal muscular atrophy and mucosa-associated lymphoid tissue lymphoma (Roy et al. 1995; Dierlamm et al. 1999).

Until recently, poxviruses were not known to encode IAPs. A search of the entomopoxvirus *Amsacta moorei* genome, however, has revealed the presence of a potential IAP-encoding gene (Bawden et al. 2000). *Amsacta moorei* does not contain a CrmA-like protein to inhibit caspase activity suggesting that the presence of an IAP protein may be essential for caspase inhibition. The *Amsacta moorei* IAP homologue contains two BIR domains and a RING finger-like domain (Bawden et al. 2000). To date, no other poxviruses possess known IAP homologues. However, African swine fever virus, a closely related virus which cycles between an arthropod vector and the animal host, encodes a functional IAP (Nogal et al. 2001).

4
p28/NIR

IAPs are not the only class of poxvirus-encoded RING finger-containing proteins reported to be involved in apoptotic modulation. The ectromelia virus protein p28 possesses a zinc-binding RING motif at the carboxy-terminal end of the protein (Senkevich et al. 1994). In infected cells, p28 localizes to sites of virus replication and is essential for virus virulence (Senkevich et al. 1994, 1995). With respect to anti-apoptotic function, p28 has also been implicated in protection against UV-induced cell death, but not against Fas- or TNF-induced cell death (Brick et al. 2000). Examination of caspase 3 levels suggests that prevention of UV-induced death occurs prior to caspase 3 activation (Brick et al. 2000). Shope fibroma virus also expresses a homologue to p28, designated N1R, which prevents UV-induced apoptosis, but not Fas- or TNF-mediated apoptosis (Brick et al. 1998, 2000). What remains to be discerned is the functional role of the RING finger in p28. Recently, RING finger proteins have been implicated in ubiquitination pathways and proteasomal degradation, suggesting the possibility of a novel protein degradation mechanism for p28/N1R (Hicke 2001; Weissman 2001).

5 Modulation of Mitochondria

5.1 Inhibitors of Cytochrome c Release

The importance of the mitochondrion in orchestrating cell death originated with the observation that cytochrome c released from mitochondria induces caspase activation (Liu et al. 1996). This led to an interest in the mitochondrion as an important component of the apoptotic death process. As such, mitochondria are now regarded as essential coordinating centers within apoptotic cells (Desagher and Martinou 2000; Kroemer and Reed 2000; Martinou and Green 2001; Wang 2001).

The induction of apoptosis results in both structural and physiological alterations to mitochondria. These alterations include loss of the mitochondrial membrane potential ($\Delta\Psi_m$), the production of reactive oxygen species, and the release of pro-apoptotic proteins such as cytochrome c, endonuclease G, apoptosis-inducing factor, and the IAP antagonist SMAC/DIABLO (Desagher and Martinou 2000; Kroemer and Reed 2000; Martinou and Green 2001; Wang 2001). Mitochondrial-released cytochrome c plays an essential role in apoptosis by promoting the activation of caspase 9 (Li et al. 1997). Members of the Bcl-2 family, both anti- and pro-apoptotic, regulate the mitochondrial checkpoint (Gross et al. 1999). The exact mechanism of cytochrome c release and its regulation by Bcl-2 family members is currently undefined and controversial, but two theories have been postulated. The first suggests that Bcl-2 family members form a pore through which cytochrome c is released, while the second theory suggests that cytochrome c release is controlled by a multiprotein complex known as the permeability transition pore complex (PTPC; Martinou and Green 2001).

Viruses recognize the importance of the mitochondrion in cell death by encoding specific apoptotic inhibitors that localize to the mitochondrion (Liu et al. 1996; Boya et al. 2001; Everett and McFadden 2001). For example, many viruses encode Bcl-2-like proteins that function to inhibit apoptosis by maintaining $\Delta\Psi_m$ and inhibiting the release of pro-apoptotic molecules from the mitochondrion (Tschopp et al. 1998; Cuconati and White 2002). Adenovirus, African swine fever virus and multiple members of the herpesvirus family all encode Bcl-2 homologous proteins that function to inhibit apoptosis (Tschopp et al. 1998; Cuconati and White 2002). At present the only poxvirus known to contain a Bcl-2 homologue is fowlpox virus (Afonso et al. 2000). The FPV039 ORF in fowlpox virus was initially identified as a Bcl-2 homologue based on the presence of Bcl-2 homology (BH) domains (Afonso et al. 2000; Cuconati and White 2002). The presence of conserved BH domains as well as a putative C-terminal transmembrane domain suggests FPV039 may localize to the mitochondrion where it could function to inhibit apoptosis.

Interestingly, no other members of the poxvirus family encode obvious Bcl-2 homologues; this suggests that perhaps some poxviruses may have evolved alternative mechanisms to control the mitochondrial component of the apoptotic pathway. In fact, M11L, a novel protein encoded by myxoma virus, has recently been shown to modulate apoptotic signals generated from the mitochondrion (Everett et al. 2000, 2002). Observations demonstrating that a recombinant myxoma virus devoid of M11L initiates an apoptotic response in rabbit T cells was the first indication that M11L may be essential for modulating apoptosis (Macen et al. 1996). Subsequently, it was found that, in the absence of other viral proteins, M11L can inhibit $\Delta\Psi_m$ and cytochrome c release from the mitochondrion (Everett et al. 2000, 2002). M11L contains a carboxy-terminal transmembrane sequence common in other mitochondrial-localized proteins, including Bcl-2 (Everett et al. 2000). Significantly, M11L interacts with the mitochondrial peripheral benzodiazepine receptor (PBR), which is a component of the mitochondrial PTPC, and M11L expression inhibits cytochrome c release stimulated by the addition of protoporphorin IX, a PBR ligand (Everett et al. 2002). M11L is the first example of a protein that interacts with the PBR and modulates apoptosis, suggesting that other pathogens may encode proteins with a similar function. Sequence analysis reveals that Shope fibroma virus and members of *Suipoxivirus*, *Capripoxvirus* and *Yatapoxivirus* all encode proteins homologous to M11L. Notably, however, no members of the orthopoxvirus genus encode M11L-like proteins.

When the genome of vaccinia virus was completely sequenced in 1990, no ORFs with homology to Bcl-2 were identified (Goebel et al. 1990). However, the importance of the mitochondrial checkpoint in apoptotic cells led us to speculate that vaccinia virus might employ a mechanism to modulate the mitochondrial arm of the apoptotic pathway. To this end, we found that Jurkat cells infected with vaccinia virus strain Copenhagen, which is naturally devoid of the caspase 8 inhibitor CrmA, were resistant to apoptosis induced by the addition of anti-Fas or staurosporine which acts directly at the mitochondria (Wasilenko et al. 2001). Furthermore, vaccinia virus strain Copenhagen infection inhibited $\Delta\Psi_m$ and release of cytochrome c clearly showing that vaccinia virus employed a mechanism to regulate the retention of cytochrome c (Wasilenko et al. 2001). By utilizing a panel of vaccinia virus deletion mutants we determined that the F1L ORF was essential for the inhibition of apoptosis (unpubl. data). F1L localizes to the mitochondrion and this localization is dependent on a carboxy-terminal transmembrane sequence similar to those present in Bcl-2 and M11L (unpublished data). The F1L ORF is present in all sequenced orthopoxviruses yet is noticeably absent in other poxvirus genera. Although the anti-apoptotic mechanism of F1L is currently unknown, vaccinia virus infection is able to inhibit cytochrome c release triggered by PTPC activators, suggesting that F1L may mediate its anti-apoptotic effect by controlling this multiprotein complex (Wasilenko et al. 2001). The existence of novel mitochondrial-localized proteins encoded by poxviruses, as well as the recently identified novel protein vMIA in human cytomegalaovirus, suggests that other pathogens may encode anti-apoptotic proteins that are unrelated to Bcl-2 (Goldmacher et al. 1999). Additionally, these novel proteins may contribute to better understanding of the molecular mechanisms involved in the loss of the mitochondrial integrity leading to cell death.

5.2 Bax-Inhibitor 1

In addition to encoding F1L homologues, two orthopoxviruses, camelpox and cowpox, possess an additional ORF that may function to inhibit apoptosis. Camelpox 6L and cowpox S1R display amino acid identity to human Bax inhibitor-1 protein (Shchelkunov et al. 1998; Gubser and Smith 2002). Bax inhibitor-1 was originally identified as a novel inhibitor of Bax-mediated death in yeast (Xu and Reed 1998). In mammalian cells, Bax inhibitor-1 also inhibits apoptosis induced by stimuli such as expression of human Bax, serum starvation, cytokine withdrawal, etoposide, and staurosporine (Xu and Reed 1998). Cellular Bax inhibitor-1 localizes mainly to the endoplasmic reticulum and minimally to the mitochondria and interacts with Bcl-2 and Bcl-XL. Whether these interactions are necessary for Bax inhibitor-1 function is unknown.

6 Anti-oxidant MC066L

The induction of cellular stress generates reactive oxygen species that are known to induce cell death (Green and Reed 1998; Kroemer and Reed 2000). Cellular glutathione peroxidases counteract the toxic affects of reactive oxygen species (Michiels et al. 1994). To counter the harmful effects of reactive oxygen species, MCV encodes MC066L, a selenoprotein with 75% sequence similarity to the human glutathione peroxidase (Senkevich et al. 1996, 1997). In a transient system, MC066L inhibits morphological features associated with apoptosis induced by UV light and peroxide but not Fas or TNF (Shisler et al. 1998). MCV-encoded MC066L may be involved in circumventing the harmful affects induced by UV light (Shisler et al. 1998).

7 Conclusions

To survive and replicate within the host, viruses possess specific strategies to circumvent and inhibit apoptosis. Our present understanding indicates that members of the poxvirus family, with the exception of the entomopoxviruses, encode multiple anti-apoptotic proteins. As indicated in this review these proteins are varied, as are their respective mechanisms of action (Fig. 1). Most notable is the extensive array of distinct mechanisms that has evolved within this family of viruses to inhibit apoptosis. This is undoubtedly reflective of the extensive array of extrinsic and intrinsic pro-apoptotic stimuli, the complexity of the apoptotic pathways and the wide range of cell types infected within the host. Several common themes, however, emerge. Firstly, poxviruses encode caspase inhibitors as a direct mechanism to interfere with caspase activity. A

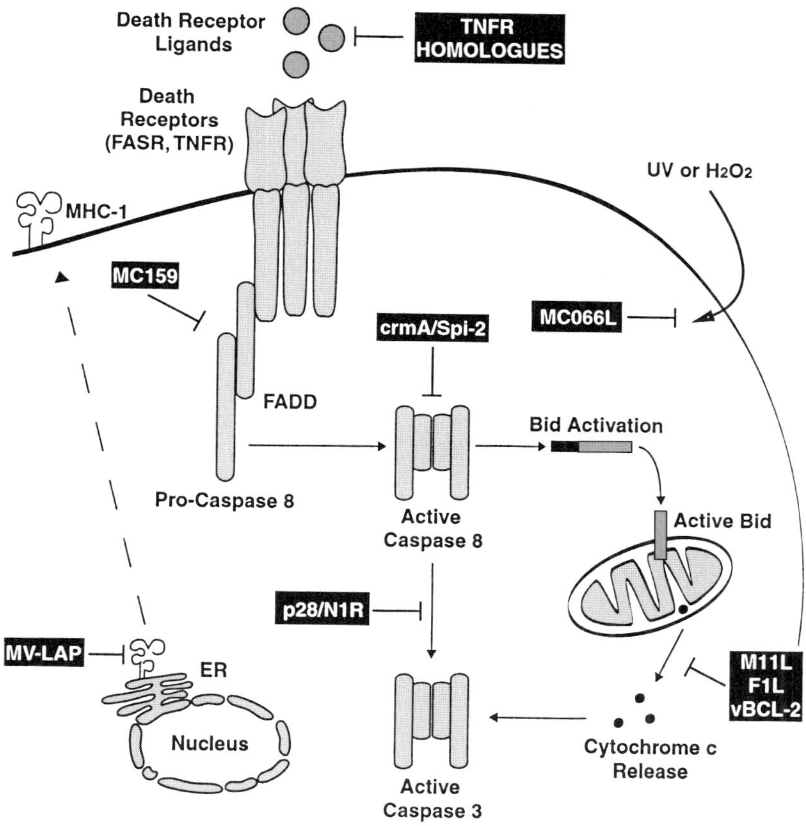

Fig. 1. Members of the poxvirus family have developed multiple strategies that target distinct components of the apoptotic cascade. In order to inhibit TNF-induced apoptosis poxviruses encode secreted TNFR homologous proteins that sequester extracellular TNF. *Molluscum contagiosum* virus inhibits death receptor signaling by expression of a DED-containing protein MC159. MC159 interacts with the cellular DED-containing proteins FADD and caspase 8. Members of the *Orthopoxvirus* genus inhibit caspase 8 activity via the production of CrmA/Spi-2. The MV-LAP protein encoded by myxoma virus inhibits CTL-induced apoptosis by preventing the expression of MHC class I on the surface of infected cells. Ectromelia virus p28, Shope fibroma virus N1R and *Molluscum contagiosum* virus MC066L all inhibit UV-induced apoptosis. M11L encoded by myxoma virus, F1L encoded by vaccinia virus and potentially vBcl-2 encoded by fowlpox virus all inhibit the release of cytochrome c from the mitochondria.

second emerging theme within the poxvirus family is the expression of proteins that interfere with death-receptor-mediated apoptosis. Finally, it is now obvious that poxviruses encode proteins that localize to the mitochondria to inhibit the mitochondrial arm of the apoptotic pathway.

In the last 10 years there has been an exponential increase in our understanding of the apoptotic death response and a significant amount of our understanding has arisen through the study of virus-encoded gene products. The recent identification of non-Bcl-2 proteins, such as M11L and F1L, which regulate the mitochondrial checkpoint, clearly suggests that poxviruses will continue to enlighten us.

Note Added in Proof: While this review was in press the data documenting that the Vaccinia encoded FIL protein inhibits apoptosis was publish (Wasiienko et al. Proc Natl Acad Sci USA 100:14345–14350).

References

Afonso CL, Tulman ER, Lu Z, Zsak L, Kutish GF, Rock DL (2000) The genome of fowlpox virus. J Virol 74:3815–3831

Ashkenazi A, Dixit VM (1998) Death receptors: signaling and modulation. Science 281:1305–1308

Barry M, Bleackley RC (2002) Cytotoxic T lymphocytes: all roads lead to death. Nat Rev Immunol 2:401–409

Barry M, McFadden G (1998) Apoptosis regulators from DNA viruses. Curr Opin Immunol 10:422–430

Bawden AL, Glassberg KJ, Diggans J, Shaw R, Farmerie W, Moyer RW (2000) Complete genomic sequence of the *Amsacta moorei* entomopoxvirus: analysis and comparison with other poxviruses. Virology 274:120–139

Bertin J, Armstrong RC, Ottilie S, Martin DA, Wang Y, Banks S, Wang GH, Senkevich TG, Alnemri ES, Moss B, Lenardo MJ, Tomaselli KJ, Cohen JI (1997) Death effector domain-containing herpesvirus and poxvirus proteins inhibit both Fas- and TNFR1-induced apoptosis. Proc Natl Acad Sci USA 94:1172–1176

Birnbaum MJ, Clem RJ, Miller LK (1994) An apoptosis-inhibiting gene from a nuclear polyhedrosis virus encoding a polypeptide with Cys/His sequence motifs. J Virol 68:2521–2528

Boldin MP, Goncharov TM, Goltsev YV, Wallach D (1996) Involvement of MACH, a novel MORT1/FADD-interacting protease, in Fas/APO-1- and TNF receptor-induced cell death. Cell 85:803–815

Boname JM, Stevenson PG (2001) MHC class I ubiquitination by a viral PHD/LAP finger protein. Immunity 15:627–636

Boshkov LK, Macen JL, McFadden G (1992) Virus-induced loss of class I MHC antigens from the surface of cells infected with myxoma virus and malignant rabbit fibroma virus. J Immunol 148:881–887

Boya P, Roques B, Kroemer G (2001) New EMBO members' review: viral and bacterial proteins regulating apoptosis at the mitochondrial level. EMBO J 20:4325–4331

Brick DJ, Burke RD, Schiff L, Upton C (1998) Shope fibroma virus RING finger protein N1R binds DNA and inhibits apoptosis. Virology 249:42–51

Brick DJ, Burke RD, Minkley AA, Upton C (2000) Ectromelia virus virulence factor p28 acts upstream of caspase-3 in response to UV light-induced apoptosis. J Gen Virol 81:1087–1097

Cameron C, Hota-Mitchell S, Chen L, Barrett J, Cao JX, Macaulay C, Willer D, Evans D, McFadden G (1999) The complete DNA sequence of myxoma virus. Virology 264:298–318

Clem RJ, Miller LK (1993) Apoptosis reduces both the in vitro replication and the in vivo infectivity of a baculovirus. J Virol 67:3730–3738

Clem RJ, Miller LK (1994) Control of programmed cell death by the baculovirus genes p35 and iap. Mol Cell Biol 14:5212–5222
Coscoy L, Ganem D (2000) Kaposi's sarcoma-associated herpesvirus encodes two proteins that block cell surface display of MHC class I chains by enhancing their endocytosis. Proc Natl Acad Sci USA 97:8051–8056
Coscoy L, Ganem D (2001) A viral protein that selectively downregulates ICAM-1 and B7-2 and modulates T cell costimulation. J Clin Invest 107:1599–1606
Coscoy L, Sanchez DJ, Ganem D (2001) A novel class of herpesvirus-encoded membrane-bound E3 ubiquitin ligases regulates endocytosis of proteins involved in immune recognition. J Cell Biol 155:1265–1273
Crook NE, Clem RJ, Miller LK (1993) An apoptosis-inhibiting baculovirus gene with a zinc finger-like motif. J Virol 67:2168–2174
Cuconati A, White E (2002) Viral homologs of BCL-2: role of apoptosis in the regulation of virus infection. Genes Dev 16:2465–2478
Darmon AJ, Nicholson DW, Bleackley RC (1995) Activation of the apoptotic protease CPP32 by cytotoxic T-cell-derived granzyme B. Nature 377:446–448
Desagher S, Martinou JC (2000) Mitochondria as the central control point of apoptosis. Trends Cell Biol 10:369–377
Deveraux QL, Reed JC (1999) IAP family proteins–suppressors of apoptosis. Genes Dev 13:239–252
Dierlamm J, Baens M, Wlodarska I, Stefanova-Ouzounova M, Hernandez JM, Hossfeld DK, De Wolf-Peeters C, Hagemeijer A, Van den Berghe H, Marynen P (1999) The apoptosis inhibitor gene API2 and a novel 18q gene, MLT, are recurrently rearranged in the t(11;18)(q21;q21) associated with mucosa-associated lymphoid tissue lymphomas. Blood 93:3601–3609
Duckett CS, Li F, Wang Y, Tomaselli KJ, Thompson CB, Armstrong RC (1998) Human IAP-like protein regulates programmed cell death downstream of Bcl-xL and cytochrome c. Mol Cell Biol 18:608–615
Earnshaw WC, Martins LM, Kaufmann SH (1999) Mammalian caspases: structure, activation, substrates, and functions during apoptosis. Annu Rev Biochem 68:383–424
Eberstadt M, Huang B, Chen Z, Meadows RP, Ng SC, Zheng L, Lenardo MJ, Fesik SW (1998) NMR structure and mutagenesis of the FADD (Mort1) death-effector domain. Nature 392:941–945
Epstein WL (1992) *Molluscum contagiosum*. Semin Dermatol 11:184–189
Everett H, McFadden G (1999) Apoptosis: an innate immune response to virus infection. Trends Microbiol 7:160–165
Everett H, McFadden G (2001) Viruses and apoptosis: meddling with mitochondria. Virology 288:1–7
Everett H, Barry M, Lee SF, Sun X, Graham K, Stone J, Bleackley RC, McFadden G (2000) M11L: a novel mitochondria-localized protein of myxoma virus that blocks apoptosis of infected leukocytes. J Exp Med 191:1487–1498
Everett H, Barry M, Sun X, Lee SF, Frantz C, Berthiaume LG, McFadden G, Bleackley RC (2002) The myxoma poxvirus protein, M11L, prevents apoptosis by direct interaction with the mitochondrial permeability transition pore. J Exp Med 196:1127–1140
Fruh K, Gruhler A, Krishna RM, Schoenhals GJ (1999) A comparison of viral immune escape strategies targeting the MHC class I assembly pathway. Immunol Rev 168:157–166
Fruh K, Bartee E, Gouveia K, Mansouri M (2002) Immune evasion by a novel family of viral PHD/LAP-finger proteins of gamma-2 herpesviruses and poxviruses. Virus Res 88:55
Garvey TL, Bertin J, Siegel RM, Wang GH, Lenardo MJ, Cohen JI (2002) Binding of FADD and caspase-8 to *Molluscum contagiosum* virus MC159 v-FLIP is not sufficient for its antiapoptotic function. J Virol 76:697–706
Goebel SJ, Johnson GP, Perkus ME, Davis SW, Winslow JP, Paoletti E (1990) The complete DNA sequence of vaccinia virus. Virology 179:247–266, 517–563
Goldmacher VS, Bartle LM, Skaletskaya A, Dionne CA, Kedersha NL, Vater CA, Han JW, Lutz RJ, Watanabe S, Cahir McFarland ED, Kieff ED, Mocarski ES, Chittenden T (1999) A cytomega-

lovirus-encoded mitochondria-localized inhibitor of apoptosis structurally unrelated to Bcl-2. Proc Natl Acad Sci USA 96:12536–12541
Goyal L (2001) Cell death inhibition: keeping caspases in check. Cell 104:805–808
Green DR, Reed JC (1998) Mitochondria and apoptosis. Science 281:1309–1312
Gross A, McDonnell JM, Korsmeyer SJ (1999) BCL-2 family members and the mitochondria in apoptosis. Genes Dev 13:1899–1911
Gubser C, Smith GL (2002) The sequence of camelpox virus shows it is most closely related to variola virus, the cause of smallpox. J Gen Virol 83:855–872
Guerin JL, Gelfi J, Boullier S, Delverdier M, Bellanger FA, Bertagnoli S, Drexler I, Sutter G, Messud-Petit F (2002) Myxoma virus leukemia-associated protein is responsible for major histocompatibility complex class I and Fas-CD95 down-regulation and defines scrapins, a new group of surface cellular receptor abductor proteins. J Virol 76:2912–2923
Hengartner MO (2000) The biochemistry of apoptosis. Nature 407:770–776
Hicke L (2001) Protein regulation by monoubiquitin. Nat Rev Mol Cell Biol 2:195–201
Hu FQ, Smith CA, Pickup DJ (1994) Cowpox virus contains two copies of an early gene encoding a soluble secreted form of the type II TNF receptor. Virology 204:343–356
Hu S, Vincenz C, Buller M, Dixit VM (1997) A novel family of viral death effector domain-containing molecules that inhibit both CD-95- and tumor necrosis factor receptor-1-induced apoptosis. J Biol Chem 272:9621–9624
Huang H, Joazeiro CA, Bonfoco E, Kamada S, Leverson JD, Hunter T (2000) The inhibitor of apoptosis, cIAP2, functions as a ubiquitin-protein ligase and promotes in vitro monoubiquitination of caspases 3 and 7. J Biol Chem 275:26661–26664
Ishido S, Choi JK, Lee BS, Wang C, DeMaria M, Johnson RP, Cohen GB, Jung JU (2000a) Inhibition of natural killer cell-mediated cytotoxicity by Kaposi's sarcoma-associated herpesvirus K5 protein. Immunity 13:365–374
Ishido S, Wang C, Lee BS, Cohen GB, Jung JU (2000b) Downregulation of major histocompatibility complex class I molecules by Kaposi's sarcoma-associated herpesvirus K3 and K5 proteins. J Virol 74:5300–5309
Kroemer G, Reed JC (2000) Mitochondrial control of cell death. Nat Med 6:513–519
Li P, Nijhawan D, Budihardjo I, Srinivasula SM, Ahmad M, Alnemri ES, Wang X (1997) Cytochrome c and dATP-dependent formation of Apaf-1/caspase-9 complex initiates an apoptotic protease cascade. Cell 91:479–489
Li X, Yang Y, Ashwell JD (2002) TNF-RII and c-IAP1 mediate ubiquitination and degradation of TRAF2. Nature 416:345–347
Liu X, Kim CN, Yang J, Jemmerson R, Wang X (1996) Induction of apoptotic program in cell-free extracts: requirement for dATP and cytochrome c. Cell 86:147–157
Loparev VN, Parsons JM, Knight JC, Panus JF, Ray CA, Buller RM, Pickup DJ, Esposito JJ (1998) A third distinct tumor necrosis factor receptor of orthopoxviruses. Proc Natl Acad Sci USA 95:3786–3791
Lorenzo ME, Jung JU, Ploegh HL (2002) Kaposi's sarcoma-associated herpesvirus K3 utilizes the ubiquitin-proteasome system in routing class major histocompatibility complexes to late endocytic compartments. J Virol 76:5522–5531
Macen JL, Graham KA, Lee SF, Schreiber M, Boshkov LK, McFadden G (1996) Expression of the myxoma virus tumor necrosis factor receptor homologue and M11L genes is required to prevent virus-induced apoptosis in infected rabbit T lymphocytes. Virology 218:232–237
Martinou JC, Green DR (2001) Breaking the mitochondrial barrier. Nat Rev Mol Cell Biol 2:63–67
Messud-Petit F, Gelfi J, Delverdier M, Amardeilh MF, Py R, Sutter G, Bertagnoli S (1998) Serp2, an inhibitor of the interleukin-1beta-converting enzyme, is critical in the pathobiology of myxoma virus. J Virol 72:7830–7839
Michiels C, Raes M, Toussaint O, Remacle J (1994) Importance of Se-glutathione peroxidase, catalase, and Cu/Zn-SOD for cell survival against oxidative stress. Free Radic Biol Med 17:235–248

Miura M, Friedlander RM, Yuan J (1995) Tumor necrosis factor-induced apoptosis is mediated by a CrmA-sensitive cell death pathway. Proc Natl Acad Sci USA 92:8318–8322
Muzio M, Chinnaiyan AM, Kischkel FC, O'Rourke K, Shevchenko A, Ni J, Scaffidi C, Bretz JD, Zhang M, Gentz R, Mann M, Krammer PH, Peter ME, Dixit VM (1996) FLICE, a novel FADD-homologous ICE/CED-3-like protease, is recruited to the CD95 (Fas/APO-1) death-inducing signaling complex. Cell 85:817–827
Nash P, Barrett J, Cao JX, Hota-Mitchell S, Lalani AS, Everett H, Xu XM, Robichaud J, Hnatiuk S, Ainslie C, Seet BT, McFadden G (1999) Immunomodulation by viruses: the myxoma virus story. Immunol Rev 168:103–120
Nogal ML, Gonzalez de Buitrago G, Rodriguez C, Cubelos B, Carrascosa AL, Salas ML, Revilla Y (2001) African swine fever virus IAP homologue inhibits caspase activation and promotes cell survival in mammalian cells. J Virol 75:2535–2543
Panus JF, Smith CA, Ray CA, Smith TD, Patel DD, Pickup DJ (2002) Cowpox virus encodes a fifth member of the tumor necrosis factor receptor family: a soluble, secreted CD30 homologue. Proc Natl Acad Sci USA 99:8348–8353
Petit F, Bertagnoli S, Gelfi J, Fassy F, Boucraut-Baralon C, Milon A (1996) Characterization of a myxoma virus-encoded serpin-like protein with activity against interleukin-1 beta-converting enzyme. J Virol 70:5860–5866
Pickup DJ, Ink BS, Hu W, Ray CA, Joklik WK (1986) Hemorrhage in lesions caused by cowpox virus is induced by a viral protein that is related to plasma protein inhibitors of serine proteases. Proc Natl Acad Sci USA 83:7698–7702
Quan LT, Caputo A, Bleackley RC, Pickup DJ, Salvesen GS (1995) Granzyme B is inhibited by the cowpox virus serpin cytokine response modifier A. J Biol Chem 270:10377–10379
Ray CA, Black RA, Kronheim SR, Greenstreet TA, Sleath PR, Salvesen GS, Pickup DJ (1992) Viral inhibition of inflammation: cowpox virus encodes an inhibitor of the interleukin-1 beta converting enzyme. Cell 69:597–604
Reading PC, Khanna A, Smith GL (2002) Vaccinia virus CrmE encodes a soluble and cell surface tumor necrosis factor receptor that contributes to virus virulence. Virology 292:285–298
Roulston A, Marcellus RC, Branton PE (1999) Viruses and apoptosis. Annu Rev Microbiol 53:577–628
Roy N, Mahadevan MS, McLean M, Shutler G, Yaraghi Z, Farahani R, Baird S, Besner-Johnston A, Lefebvre C, Kang X, et al. (1995) The gene for neuronal apoptosis inhibitory protein is partially deleted in individuals with spinal muscular atrophy. Cell 80:167–178
Russell JH, Ley TJ (2002) Lymphocyte-mediated cytotoxicity. Annu Rev Immunol 20:323–370
Salvesen GS, Duckett CS (2002) IAP proteins: blocking the road to death's door. Nat Rev Mol Cell Biol 3:401–410
Saraiva M, Alcami A (2001) CrmE, a novel soluble tumor necrosis factor receptor encoded by poxviruses. J Virol 75:226–233
Saraiva M, Smith P, Fallon PG, Alcami A (2002) Inhibition of type 1 cytokine-mediated inflammation by a soluble CD30 homologue encoded by ectromelia (mousepox) virus. J Exp Med 196:829–839
Schreiber M, McFadden G (1994) The myxoma virus TNF-receptor homologue (T2) inhibits tumor necrosis factor-alpha in a species-specific fashion. Virology 204:692–705
Schreiber M, Rajarathnam K, McFadden G (1996) Myxoma virus T2 protein, a tumor necrosis factor (TNF) receptor homolog, is secreted as a monomer and dimer that each bind rabbit TNFalpha, but the dimer is a more potent TNF inhibitor. J Biol Chem 271:13333–13341
Schreiber M, Sedger L, McFadden G (1997) Distinct domains of M-T2, the myxoma virus tumor necrosis factor (TNF) receptor homolog, mediate extracellular TNF binding and intracellular apoptosis inhibition. J Virol 71:2171–2181
Searles RP, Bergquam EP, Axthelm MK, Wong SW (1999) Sequence and genomic analysis of a Rhesus macaque rhadinovirus with similarity to Kaposi's sarcoma-associated herpesvirus/human herpesvirus 8. J Virol 73:3040–3053

Senkevich TG, Koonin EV, Buller RM (1994) A poxvirus protein with a RING zinc finger motif is of crucial importance for virulence. Virology 198:118–128

Senkevich TG, Wolffe EJ, Buller RM (1995) Ectromelia virus RING finger protein is localized in virus factories and is required for virus replication in macrophages. J Virol 69:4103–4111

Senkevich TG, Bugert JJ, Sisler JR, Koonin EV, Darai G, Moss B (1996) Genome sequence of a human tumorigenic poxvirus: prediction of specific host response-evasion genes. Science 273:813–816

Senkevich TG, Koonin EV, Bugert JJ, Darai G, Moss B (1997) The genome of *Molluscum contagiosum* virus: analysis and comparison with other poxviruses. Virology 233:19–42

Shchelkunov SN, Safronov PF, Totmenin AV, Petrov NA, Ryazankina OI, Gutorov VV, Kotwal GJ (1998) The genomic sequence analysis of the left and right species-specific terminal region of a cowpox virus strain reveals unique sequences and a cluster of intact ORFs for immunomodulatory and host range proteins. Virology 243:432–460

Shisler JL, Moss B (2001) *Molluscum contagiosum* virus inhibitors of apoptosis: the MC159 v-FLIP protein blocks Fas-induced activation of procaspases and degradation of the related MC160 protein. Virology 282:14–25

Shisler JL, Senkevich TG, Berry MJ, Moss B (1998) Ultraviolet-induced cell death blocked by a selenoprotein from a human dermatotropic poxvirus. Science 279:102–105

Smith CA, Hu FQ, Smith TD, Richards CL, Smolak P, Goodwin RG, Pickup DJ (1996) Cowpox virus genome encodes a second soluble homologue of cellular TNF receptors, distinct from CrmB, that binds TNF but not LT alpha. Virology 223:132–147

Smith GL, Symons JA, Alcami A (1999) Immune modulation by proteins secreted from cells infected by vaccinia virus. Arch Virol Suppl 15:111–129

Smith GL, McFadden G (2002) Smallpox: anything to declare? Nat Rev Immunol 2:521–527

Smith GL, Symons JA, Khanna A, Vanderplasschen A, Alcami A (1997) Vaccinia virus immune evasion. Immunol Rev 159:137–154

Suzuki Y, Nakabayashi Y, Takahashi R (2001) Ubiquitin-protein ligase activity of X-linked inhibitor of apoptosis protein promotes proteasomal degradation of caspase-3 and enhances its anti-apoptotic effect in Fas-induced cell death. Proc Natl Acad Sci USA 98:8662–8667

Tewari M, Dixit VM (1995) Fas- and tumor necrosis factor-induced apoptosis is inhibited by the poxvirus crmA gene product. J Biol Chem 270:3255–3260

Tewari M, Telford WG, Miller RA, Dixit VM (1995) CrmA, a poxvirus-encoded serpin, inhibits cytotoxic T-lymphocyte- mediated apoptosis. J Biol Chem 270:22705–22708

Thome M, Schneider P, Hofmann K, Fickenscher H, Meinl E, Neipel F, Mattmann C, Burns K, Bodmer JL, Schroter M, Scaffidi C, Krammer PH, Peter ME, Tschopp J (1997) Viral FLICE-inhibitory proteins (FLIPs) prevent apoptosis induced by death receptors. Nature 386:517–521

Thome M, Tschopp J (2001) Regulation of lymphocyte proliferation and death by FLIP. Nat Rev Immunol 1:50–58

Thornberry NA, Lazebnik Y (1998) Caspases: enemies within. Science 281:1312–1316

Tortorella D, Gewurz BE, Furman MH, Schust DJ, Ploegh HL (2000) Viral subversion of the immune system. Annu Rev Immunol 18:861–926

Tracey KJ, Cerami A (1994) Tumor necrosis factor: a pleiotropic cytokine and therapeutic target. Annu Rev Med 45:491–503

Tschopp J, Thome M, Hofmann K, Meinl E (1998) The fight of viruses against apoptosis. Curr Opin Genet Dev 8:82–87

Tsukumo SI, Yonehara S (1999) Requirement of cooperative functions of two repeated death effector domains in caspase-8 and in MC159 for induction and inhibition of apoptosis, respectively. Genes Cells 4:541–549

Turner PC, Sancho MC, Thoennes SR, Caputo A, Bleackley RC, Moyer RW (1999) Myxoma virus Serp2 is a weak inhibitor of granzyme B and interleukin-1beta-converting enzyme in vitro and unlike CrmA cannot block apoptosis in cowpox virus-infected cells. J Virol 73:6394–6404

Upton C, Macen JL, Schreiber M, McFadden G (1991) Myxoma virus expresses a secreted protein with homology to the tumor necrosis factor receptor gene family that contributes to viral virulence. Virology 184:370–382
Wallach D, Varfolomeev EE, Malinin NL, Goltsev YV, Kovalenko AV, Boldin MP (1999) Tumor necrosis factor receptor and Fas signaling mechanisms. Annu Rev Immunol 17:331–367
Wang X (2001) The expanding role of mitochondria in apoptosis. Genes Dev 15:2922–2933
Wasilenko ST, Meyers AF, Vander Helm K, Barry M (2001) Vaccinia virus infection disarms the mitochondrion-mediated pathway of the apoptotic cascade by modulating the permeability transition pore. J Virol 75:11437–11448
Weissman AM (2001) Themes and variations on ubiquitination. Nat Rev Mol Cell Biol 2:169–178
Xu Q, Reed JC (1998) Bax inhibitor-1, a mammalian apoptosis suppressor identified by functional screening in yeast. Mol Cell 1:337–346
Yang Y, Fang S, Jensen JP, Weissman AM, Ashwell JD (2000) Ubiquitin protein ligase activity of IAPs and their degradation in proteasomes in response to apoptotic stimuli. Science 288:874–877
Zhou Q, Snipas S, Orth K, Muzio M, Dixit VM, Salvesen GS (1997) Target protease specificity of the viral serpin CrmA. Analysis of five caspases. J Biol Chem 272:7797–7800
Zuniga MC, Wang H, Barry M, McFadden G (1999) Endosomal/lysosomal retention and degradation of major histocompatibility complex class I molecules is induced by myxoma virus. Virology 261:180–192

T-Cell-Mediated Control of Poxvirus Infection in Mice

A. Müllbacher[1] and R.V. Blanden[1]

1
Introduction

An orthopoxvirus, variola major, gave rise to immunology. Somewhere in Asia centuries ago someone must have made an extraordinary intellectual leap. Presumably, based upon the observation that people who had recovered from smallpox (caused by variola major) never suffered from the disease again, an ancient lateral thinker decided to test the proposition that exposure of a naive subject to scabs from the rash of smallpox sufferers might prevent or alleviate later naturally acquired disease. Even now, we do not fully understand why deliberate cutaneous or inhalation infection with virulent variola major from scabs should cause infection with much lower mortality than naturally acquired smallpox. However, this first experiment in immunology was so successful that the practice became widespread in India and China for centuries before it reached Europe.

Lady Mary Wortley Montagu, who had been left severely pockmarked after an attack of smallpox, spent time in Turkey as the wife of the British Ambassador to Constantinople in the early eighteenth century. She observed the local custom of using smallpox scabs to inoculate naive subjects to cause immunity against smallpox. She described the practice in a letter ironically dated 1st April 1717, but this was no April fool's joke. Lady Mary urged doctors in England to begin using the method and had her son and daughter immunised. Later, after a successful trial involving six convict "volunteers", even members of the Royal family were immunised (Fenner et al. 1988).

Edward Jenner was variolated (as the practice became known) and thus had first-hand motivation to avoid the potential mortality and uncomfortable morbidity caused by variolation. The knowledge that cowpox infection in milkmaids rendered them immune to smallpox encouraged him to replace variolation with cowpox inoculation (a much milder process) in 1796 (Sanderson 1988). Almost two centuries later, Jenner's method, with a few modern adaptations, led to arguably the greatest triumph in the history of preventative medicine, with the certification by a World Health Organisation committee,

[1]Division of Immunology and Genetics, John Curtin School of Medical Research, Australian National University, P.O. Box 334, Canberra, ACT 2601, Australia,
e-mail: arno.mullbacher@anu.edu.au

Progress in Molecular and Subcellular Biology
C. Alonso (Ed.): Viruses and Apoptosis
© Springer-Verlag Berlin Heidelberg 2004

chaired by Frank Fenner, that smallpox had been eradicated as a human infection in 1977 (Fenner et al. 1988).

2
Role of CD8$^+$ T Cells in Recovery from Mousepox

In the late 1940s, Frank Fenner undertook a comprehensive investigation of the pathogenesis of mousepox and made the case that the disease in mice was a model for generalised viral infection in humans (Fenner 1948, 1949a,b). The natural route of infection was through skin abrasions and, after local multiplication in the skin and draining lymph nodes, a primary viremia led to infection of visceral target organs, in particular the liver and spleen, by 3 days after infection. Mims had shown (1959) that ectromelia virus (EV), like any other blood-borne particle, is taken up by littoral macrophages in the liver and spleen, which then transmit infection to contiguous cells. Rapid multiplication of the virus in hepatocytes and splenocytes caused massive cellular destruction and death of the animal if unchecked (Fenner 1949a; Roberts 1964a). In survivors, a secondary viremia led to a skin rash by 7 or 8 days after infection (Fenner 1949a). In these mice, the rate of virus growth in the spleen diminished by 6 days after infection (Fenner 1949b), suggesting that anti-viral mechanisms were operating by this time, but, in the foot, the virus was not controlled for a further 4 days (Fenner 1949b).

All of this quantitative information, together with the precedent set in the early and mid-1960s by the classical studies of Mackaness (1962, 1964) on the importance of cell-mediated immunity in mice against infection with *Listeria* monocytogenes, a bacterial pathogen, led to the choice in the late 1960s of mousepox as a model of generalised viral infection caused by a natural pathogen to investigate the role of cell-mediated immunity in controlling primary infection (Blanden 1970, 1971a,b). The importance of investigating cell-mediated immunity comprehensively in an appropriate animal model was reinforced at this time by emerging evidence that cell-mediated mechanisms were critical for recovery of humans from viral infections (Fulginiti et al. 1968).

The pioneering studies of Fenner and Mackaness showed the importance of quantifying the course of infection together with measurements of potential anti-viral mechanisms that coincide with control of virus growth in target organs. In the first of a series of papers (Blanden 1970), it was shown that depletion of thymus-derived lymphocytes by anti-thymocyte serum led to both increased mortality in mice infected with EV and a failure to control virus growth in the liver and spleen. Neutralising antibody and interferon responses were not impaired in treated mice, but the cell-mediated response as measured by delayed hypersensitivity was significantly suppressed. Furthermore, in normal mice, the cell-mediated response first became evident at a time when control of viral growth in the target organs was initiated.

The importance of T-cell-mediated immunity was demonstrated in a second paper (Blanden 1971b) in which immune splenocytes were transferred intravenously into recipient animals that had been infected intravenously 24 h previously to seed infection in the critical visceral target organs (the liver and spleen). A further 24 h later, when viral titres were quantified in these target organs, it was clear that immune cell transfer caused a dramatic reduction in viral load. The use of anti-theta serum and complement unequivocally identified T cells as the critical effector cells that triggered viral clearance. The peak of donor T cell potency was reached 6 days after immunisation of donors and then declined. No neutralising antibody was present in the donors at that time. Furthermore, transfer of hyperimmune serum that caused readily detectable neutralising activity in the serum of recipients was far less efficient at reducing virus titres than immune cells that transferred no detectable neutralising antibody activity. Massive doses of transferred INF had no effect on recipient viral titres. Specificity was established with reciprocal cell transfers involving *Listeria* monocytogenes and EV.

In a third paper (Blanden 1971a) the histological correlation of clearance of infection was studied. The reduction in viral titres following T cell transfer coincided with an influx of mononuclear cells into foci of infection in the liver and with a loss of viral antigen as detected by direct immunofluorescence in those lesions. Since 800 rads of whole body irradiation immediately before intravenous infection of the recipients reduced the effectiveness of T cell transfer 24 h later, it was concluded that recruitment of blood monocytes, known to be reduced 24 h after irradiation, was an important component of the anti-viral mechanism.

The conclusions from these experiments were that effector T cells in the blood specifically recognised virus-infected cells in foci of infection in the liver and spleen via antigen receptors and migrated into the lesion. The potential anti-viral mechanisms they initiated included the lysis of infected cells prior to the completion of the virus replication cycle, the secretion of interferon (IFN), which may protect surrounding uninfected cells and the attraction of mononuclear phagocytes, which could ingest pre-existing virions resulting from completed viral replication in the lesion. Roberts had previously shown (1964b) that mouse macrophages, though infected by EV, were relatively unproductive. It was speculated that recruited monocytes, activated by T cell products within the lesion, might be relatively resistant to productive infection, thus resulting in the elimination of virus that they ingested.

The development of cytotoxicity assays in vitro for effector T cells against EV-infected target cells (Gardner et al. 1974) was critical to the further elucidation of T-cell-mediated anti-viral mechanisms. This enabled the development of methods to generate secondary effector T cells in vitro from memory cells taken from primed mice (Gardner and Blanden 1976), a technique essential for the later study of T-cell-mediated immunity in human infections (Beverley 1990) and also in the generation of primary $CD8^+$ cytotoxic T (Tc) cell responses in vitro (Blanden et al. 1977).

The observation that clearance of EV infections in vivo in the cell transfer model (Blanden 1971b) was MHC-restricted (Blanden 1974), later unequivocally mapped to MHC-class I loci (Kees and Blanden 1976), suggested that T cells responsible for viral clearance in vivo and cytotoxicity in vitro were one and the same (Blanden 1974; Gardner et al. 1975). We also established that the kinetics of generation of T cells that cleared virus after transfer into infected recipient mice were identical to the kinetics of generation of T cells that lysed EV-infected target cells in vitro (Blanden and Gardner 1976). Furthermore, the antigenic epitope recognised in vitro by Tc cells on EV-infected target cells was present by 3 h after infection (Ada et al. 1976), long before completion of the viral replication cycle. All of the above results obtained by the mid-1970s strongly suggested, but did not prove, that T-cell-mediated cytotoxicity was an important effector function in vivo.

In the last 25 years there has been a wealth of information published regarding cytotoxic pathways mediated by T cells and other lymphocytes (see below) and soluble factors produced and secreted by T cells and other lymphocytes. However, only a tiny fraction of this mountain of information is directly pertinent to updating our view of critical events in T-cell-mediated immunity against natural poxvirus pathogens since much of the work has been done with unnatural viral-host combinations, e.g. vaccinia infection of mice. Revealing experiments using the EV cell transfer model described above have been performed by Karupiah and collaborators (Ramshaw et al. 1997 and personal communication). In these experiments, effector T cells from IFN-γ knockout (KO) mice were just as effective as those from wild-type (wt) mice at reducing viral titres in the liver, lungs and spleen of wt-infected recipients whereas effector T cells from perforin (perf) KO mice were completely impotent. Furthermore, IFN-γ KO mice could be used as recipients of T cells from IFN-γ knockouts without impairment of anti-EV activity, denying any role for IFN-γ in viral clearance in this experimental format. These results emphasise the critical importance of a cytolytic pathway involving perf in the T-cell-mediated clearance of EV from critical visceral target organs, a conclusion reinforced by 100% mortality in perf KO mice infected with EV at doses that killed none of the wt control mice (Ramshaw et al. 1997; Müllbacher et al. 1999a). Surprisingly, knockout mice lacking IFN-γ or its receptor exhibited 100% mortality from doses of EV that caused no mortality in wt controls, a result that cannot be readily reconciled with the cell transfer data (Ramshaw et al. 1997). At face value, IFN-γ seems to play an essential role in protection mechanisms distinct from Tc-cell-mediated viral clearance.

Other work by Karupiah et al. (1996) has clearly shown that $CD8^+$ T cells are essential for survival from ectromelia infection in mice whereas $CD4^+$ T cells are not. However, in mice depleted of $CD4^+$ T cells, $CD8^+$ T-cell-mediated cytotoxicity responses were less efficient than normal and virus persisted longer than normal in both visceral target organs and particularly in the skin (Karupiah et al. 1996). The precise role of $CD4^+$ T cells in these phenomena remains to be resolved.

3 Function of CD8⁺ T Cells in Apoptosis

CD8$^+$ T (Tc) cells are pleiotropic in their effector function. Their two principal mechanisms of mediating immunity are release of cytokines and cellular cytotoxicity (Fig. 1). Both these effector functions are triggered as a consequence of TcR engagement of the Tc cell with antigen (cell surface MHC classI/peptide complexes) on the target cell. Release of cytokines can cause effects beyond the antigen-presenting/target cell whereas direct cytolytic function requires cell-to-cell contact.

3.1 Cytokines Released by Tc Cells

Tc cells, upon antigen recognition and depending on the milieu present, secrete a host of different cytokines (Sad et al. 1995). The two most prominent species are TNF-α and IFN-γ (Vassalli 1992; Boehm et al. 1997). Both of these cytokines are directly or indirectly associated with the induction of apoptosis of target cells. TNF-α induces apoptosis directly by binding to one of its two receptors (TNFR, p55 or p75; Locksley et al. 2001). Signal transduction occurs via death domain and caspase cascade leading to apoptosis.

IFN-γ, on the other hand, is a major regulator of gene transcription. Many products of IFN-γ-induced expression are involved in the MHC-class I antigen presentation pathway facilitating the expression of MHC-class I/peptide ligand for recognition by Tc cells on the target cell surface (Boehm et al. 1997). In addition, IFN-γ induces the expression of Fas (CD95/Apo1) on target cells, sensitising them to lysis and apoptosis via the Fas pathway (Watanabe-Fukunaga et al. 1992; Simon et al. 2000; Müllbacher et al. 2002). Induction of apoptosis by Fas and TNF-α, although occurring via engagement of different ligands with distinct cell surface molecular structures and intracellular death domains, however, activates a common effector caspase cascade (O'Reilly and Strasser 1999).

The role these two dominant cytokines play in control of poxvirus infections has been addressed in the EV/mouse model using specific receptor or ligand gene targeted knockout mice. Studies by Ruby et al. (1997) using mice deficient in either TNFR p55 and/or p75 showed significantly enhanced EV replication in spleen in the absence of either TNFR. Mice lacking p75 also showed enhanced replication in liver and increased mortality. No synergistic effect of p55 and p75 was observed, as double KO mice did not show increased sensitivity to EV infections, measured in either mortality or virus load in various organs, compared with p75 single KO mice. Although highly suggestive that this increased susceptibility to EV infection in the absence of TNFR is a direct consequence of an inability to execute TNF-α-mediated immune protection,

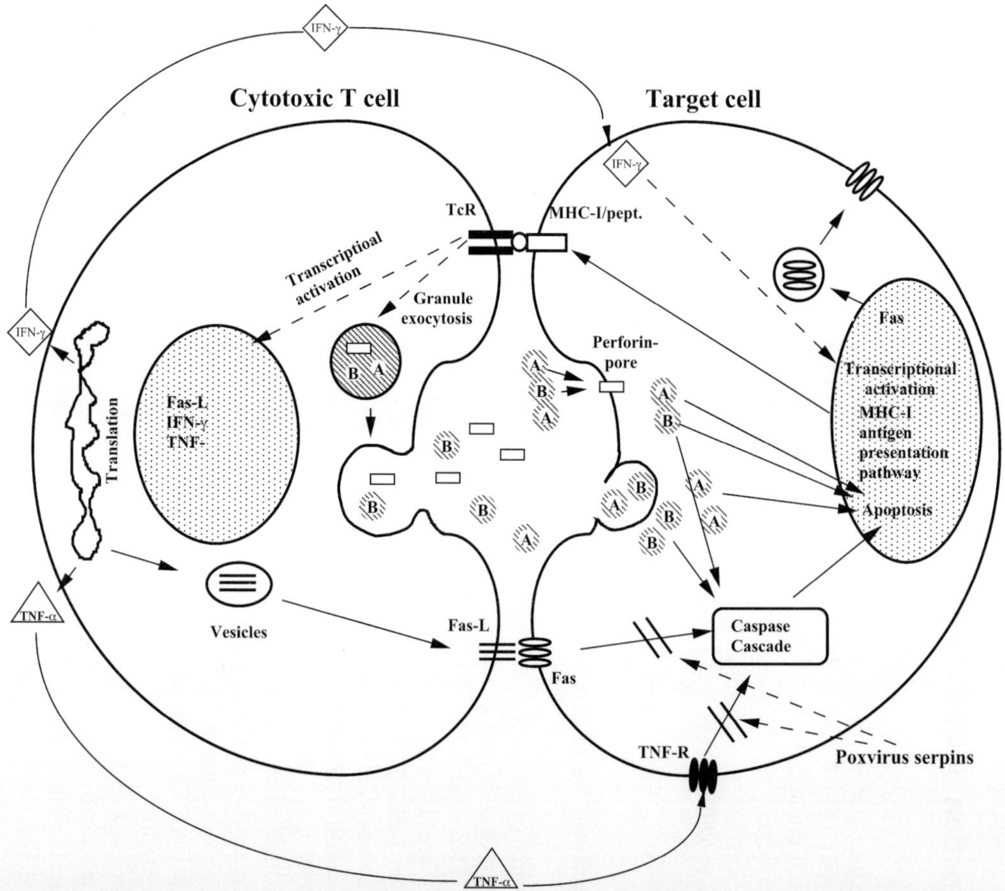

Fig. 1. Schematic representation of effector Tc cell engagement with antigen-presenting target cell. Consequences of TcR/MHC class I-peptide interaction triggering cytolytic and cytokine defence mechanisms

the possibility that a ligand other than TNF-α signalling via TNFR is responsible cannot be ruled out. Such a phenomenon has in fact been demonstrated in adenovirus infection in the mouse (Hayder et al. 1999). An additional factor which raises some doubts regarding a major role of TNF-α in resistance to EV is the evidence that EV, like most other orthopoxviruses (Buller and Palumbo 1991; Turner et al. 1995), encodes serpins (see below; Turner et al. 2000; Wallich et al. 2001) which interfere with Tc-cell-mediated apoptosis (Tewari et al. 1995b) by inhibiting caspases (Ray et al. 1992; Tewari et al. 1995a; Medema et al. 1997). For example, vaccinia virus (VV)-infected cells in vitro, when treated with TNF-α, die by a mechanism distinct from apoptosis (Li and Beg 2000).

As for a role of IFN-γ responses in recovery from primary EV infections, the lack of either IFN-γ or IFN-γ receptors renders mice increasingly more sensitive to EV (Ramshaw et al. 1997), but this is not due to the lack of Tc cells or impairment of their capacity to clear EV from liver, spleen and lung (see above). IFN-γ, although initially described as an anti-viral agent (Wheelock 1965), is now recognised as primarily an important immune regulator. Thus it is unlikely that IFN-γ exerts its protective role in poxvirus infections directly by inducing an anti-viral state in otherwise susceptible cells but rather by one or more secondary mechanisms which become activated as a result of IFN-γ-mediated signalling (Boehm et al. 1997).

3.2 Cytotoxicity

Tc cells exert their cytolytic function, leading to apoptosis and target cell lysis, by two quite distinct pathways. One, the Fas pathway, requires the ligation of Fas on the target cell with the Fas ligand (Fas, CD 95-L) on the effector cell. The second, the granule exocytosis pathway, is mediated by perforin (perf) and granzyme (gzms). Initially, it was thought that exocytosis-mediated cytotoxicity was primarily involved in the control of intracellular pathogens such as viruses (Kägi et al. 1995a) while the Fas pathway was thought to be in essence immunoregulatory (Rouvier et al. 1993; Nagata 1997). However, more recent evidence provides strong support for the notion that both cytolytic pathways are involved in recovery from infections and immunopathology, as well as regulation of the immune response (Doherty et al. 1997; Kägi et al. 1999; Parra et al. 2000; Balkow et al. 2001; Licon Luna et al. 2002).

3.2.1 The Fas Pathway of Cytotoxicity and the Effect of Poxvirus-Encoded Serpins

FasL, a type II membrane protein, is expressed on activated Tc cells and stored in specialised cytoplasmic vesicles. TcR engagement leads to transport to the cell surface and polymerisation (Bossi and Griffiths 1999). Ligation of Fas, a class I membrane protein and the receptor for FasL, on the target cell by FasL of the effector Tc cell leads to the assembly of the death domains of Fas and the adapter protein FADD/MORT1 which in turn leads to initiation of the caspase cascade, causing the target cell to undergo apoptosis (Nagata 1999; O'Reilly and Strasser 1999). That deficiency in the Fas pathway of cytotoxicity influences the outcome of infections with a variety of viruses only became apparent when a simultaneous deficiency in the granule exocytosis pathway was present (Parra et al. 2000; Balkow et al. 2001; Licon Luna et al. 2002).

The notable exceptions to involvement of the Fas pathway in recovery are primary infections with orthopoxviruses. Most or all viruses in this family encode proteins related to the family of serine protease inhibitors (serpins; Pickup et al. 1986; Buller and Palumbo 1991; Turner et al. 1995). The genes for at least three serpins, SPI-1, 2 and 3, of cowpox virus (CPV), rabbitpox (RPV), variola (small pox), Western Reserve strain of vaccinia virus (VV-WR) and EV exhibit extensive (>90%) sequence homology which indicates conserved function and survival advantage (Turner et al. 2000; Wallich et al. 2001). The gene of CPV, cytokine response modifier (crmA) or SPI-2, has been the most extensively studied. SPI-2 of CPV has been shown to inhibit a variety of inflammatory processes, including inhibition of caspase activities, in particular caspase 1 (ICE; Ray et al. 1992), caspase 3 (CPP32; Tewari et al. 1995a) and caspase 8 (FLICE; Medema et al. 1997). More importantly, it was shown that crmA inhibited apoptosis of target cells by alloreactive Tc cells, mediated predominantly via the Ca^{2+}-independent Fas pathway (Tewari et al. 1995b). In addition, Macen et al. (1996) showed that both cytolytic pathways, Fas and granule exocytosis, were inhibited in particular by SPI-2 of CPV and RPV as measured by ^{51}Cr release assay. The observation that infection of target cells with orthopoxvirus, in this instance EV, inhibits cytolysis of target cells by alloreactive Tc cells was originally made in 1975 by Gardner et al. However, importantly, targets infected with EV were highly susceptible to lysis by EV-immune Tc cells. This is consistent with the experimental evidence that Tc cells are absolutely required for recovery of mice from primary infections with EV. Conclusive evidence that EV, like CPV and RPV, inhibits cytolysis of target cells via the Fas pathway was obtained by using alloreactive Tc cells from perf-deficient mice (Müllbacher et al. 1999b). In this study it was also shown that cytolytic activity was strongly inhibited when alloreactive Tc cells were used and to a much lesser extent, if at all, when MHC class I restricted Tc cells were tested. The reason for this difference between Tc cell populations in their susceptibilities to inhibition of lysis by EV is not known. This observation makes the notion that poxviruses have evolved serpins to evade control by Tc cells less convincing.

The inhibition of the Fas pathway of cytotoxicity mediated by poxvirus-encoded serpins offers an explanation for two observations. Firstly, it explains the clear-cut susceptible phenotype seen in perf-deficient mice in the EV model (Müllbacher et al. 1999a; Table 1), where mice, in the absence of an exocytosis-mediated pathway of cytotoxicity, become unable to recover from

Table 1. Resistance of mouse strains to EV infections

Strain	LD_{50}[a]
C57BL/6[b]	$>2 \times 10^6$
A/J[c]	$<10^{-1}$
$Perf^{-/-}$[b]	$<10^{-1}$
gld (Fas-L^{neg})	$>10^6$

[a]PFU Ectromelia Moscow (fp).
[b]From Müllbacher et al. (1999a).
[c]From Blanden et al. (1989).

EV infection, similar to the genetically most sensitive mouse strain, A/J (Blanden et al. 1989). On the other hand, lack of the Fas pathway as a result of a mutation in FasL (gld; Nagata and Suda 1995) did not alter resistance to EV when compared with the wt strain B6. In all other virus models an effect of perf deficiency is either completely masked or changes to susceptibility are not absolute (see above). The second phenomenon that is potentially explained as a consequence of inhibition of the Fas pathway is the observed increase in virulence of recombinant poxviruses encoding IL-4 (Müllbacher and Lobigs 2001), first seen with VV (Andrew and Coupar 1992) and later also with EV (Jackson et al. 2001). Aung and Graham (2000) showed that the presence of IL-4 during the induction of a Tc cell response skewed the cytolytic effector mechanism towards the Fas pathway and diminished the efficiency of the perf pathway, thus allowing serpin-mediated inhibition of the Fas pathway to cause an apparent increase in viral virulence. The detailed mechanism of how this reprogramming of cytolytic pathways occurs under the influence of Il-4 has yet to be worked out.

3.2.2 The Exocytosis-Mediated Cytolytic Pathway

Engagement of the TcR with MHC class I/peptide complexes on the target cell leads to signal transduction culminating in exocytosis of preformed granular content from Tc cells (Fig. 1). The dominant constituents of Tc cell granules are perf, or pore-forming protein, and gzms, in particular gzmA and B. The precise details of how the gzms and perf interact and effect apoptosis and cell death is not known. Two recent reviews provide a concise description of the state of the art (Barry and Bleackley 2002; Trapani and Smyth 2002). It has now become clear that perf, beside its initial proposed role of facilitating entry of gzms into the target cell (Henkart 1994), is essential for gzm function by enabling gzms to be released from endosomes and possibly also for activation and nuclear translocation (Froelich et al. 1996; Trapani and Smyth 2002). Presence of perf in Tc effector cells, in the absence of gzms, still leads to target cell death, as measured by ^{51}Cr release and survival assays (Simon et al. 1997; Müllbacher et al. 1999c; Simon and Müllbacher 2000), but not classical apoptosis.

The two gzms, A and B, are members of a novel family of serine proteases with highly restricted tissue expression, predominantly in effector cytolytic lymphocytes, T and NK cells (Pasternack and Eisen 1985; Kramer et al. 1986; Masson et al. 1986; Young et al. 1986). The two gzms have different substrate specificities; gzmA is a tryptase while gzmB predominantly cleaves after aspartic acid (Simon and Kramer 1994; Tschopp 1994). Furthermore, their genes are located on separate chromosomes (Pham and Ley 1997; Shresta et al. 1997) and they differ in their structure; gzmA is a homodimer while gzmB is a single polypeptide chain (Simon and Kramer 1994; Tschopp 1994). All this strongly suggests that these two enzymes evolved independently. This conclusion is also supported by their differing mechanisms of inducing apoptosis (Sarin et al. 1997; Trapani et al. 1998; Beresford et al. 1999) and the fact that they can function synergistically (Nakajima et al. 1995).

To evaluate the role of the individual components of the exocytosis pathway in recovery from primary viral infections, the use of mice (generated by knockout technology) deficient in one or more of the effector molecules was most instructive. The perf KO mouse ($perf^{-/-}$) was the first to validate the concept that the exocytosis pathway of cytotoxicity was crucial for recovery from virus infections (Kägi et al. 1994). However, the same group experimenting with cytopathic viruses that were not mouse pathogens observed no increased susceptibility in perf-deficient mice leading them to conclude that perf was primarily important in defence against non-cytopathic viruses (Kägi et al. 1995b).

In contrast, using the natural mouse pathogen EV, a highly cytopathic virus, perf was shown to be absolutely necessary for recovery (Müllbacher et al. 1999a). Susceptibility to EV increased by more than 6 $\log_{10}$ in the $perf^{-/-}$ compared with the wt counterpart (see Table 1) providing the most clear-cut phenotype of any pathogen model and the role of perf. Somewhat unexpectedly, $perf^{-/-}$ mice also lost their resistance to the attenuated strain of EV, Hampstead Egg (Wallich et al. 2001). However, very surprising observations were made when the role of perf was investigated using very closely related viruses such as CPV and VV (Esposito and Fenner 2001). VV did not behave differently in $perf^{-/-}$and wt mice (Kägi et al. 1995b), which was not unexpected in light of an earlier finding that $CD8^{+}$T cells were not required for mice to survive VV infections (Spriggs et al. 1992). For CPV on the other hand, whose natural host is also a murine species, the rat, absence of perf actually increased resistance in the mouse (Müllbacher et al. 1999a). This is compelling evidence that, even when using closely related pathogens, reliable predictions about factors affecting pathogenesis cannot be made.

EV infections provided the first evidence of a biological role for gzmA. Mice deficient in gzmA ($gzmA^{-/-}$) are partially defective in recovery from EV infections (Müllbacher et al. 1996; Table 2). This manifested itself in more severe lesions in liver and spleen, increased viral titres in both organs and increased

Table 2. Increase in sensitivity to mousepox of granzyme-deficient mouse strains

Strain	LD_{50}[a]
C57BL/6[b]	$>2 \times 10^6$
$Perf^{-/-}$[b]	$<10^{-1}$
$gzmA^{-/-}$[c]	10^5
$gzmB^{-/-}$[d]	10^4
$gzmA \times B^{-/-}$[d]	$<10^{-1}$

[a]PFU Ectromelia Moscow (fp).
[b]From Müllbacher et al. (1999a).
[c]From Müllbacher et al. (1996).
[d]From Müllbacher et al. (1999c).

liver enzyme levels in blood. Tc and NK cells of these mice have normal cytolytic activity in vitro (^{51}Cr release assays) and are indistinguishable from their wt B6 counterparts in their ability to recover from LCMV infections (Ebnet et al. 1995). Use of $gzmA^{-/-}$ mice has also provided evidence that this enzyme plays a role in resistance to herpes infections (Pereira et al. 2000) and two fungal infections, *Candida albicans* (Ashman and Müllbacher, unpublished) and *Aspergillus fumigatus* (Simon and Müllbacher, unpublished), all pathogens with DNA genomes. No effect of gzmA deficiency has been observed using viruses with an RNA genome, such as LCMV (Ebnet et al. 1995), flavi-, alpha-, orthomyxo- and paramyxoviruses (our unpubl. results).

GzmB deficiency increased susceptibility to EV infection to an even larger extent than gzmA deficiency. The LD_{50} was more than 2 log_{10} less than in wt mice (Table 2; Müllbacher et al. 1999c), tissue destruction in liver and spleen was significantly increased and so were virus titres in both organs. There is evidence that gzmB has significant binding affinity for SPI-2 (Quan et al. 1995) suggesting that SPI-2 may have evolved in poxviruses to evade gzmB-mediated defence mechanisms. However, since $gzmB^{-/-}$ mice are significantly less resistant to EV infections than wt mice, it is clear that binding of SPI-2 to gzmB in vivo does not neutralize its activity against EV.

Mice deficient in both gzmA and B ($gzmAxB^{-/-}$) exhibited a phenotype not unlike $perf^{-/-}$ mice in response to EV infections (Müllbacher et al. 1999c). Mortality was 100% with doses as low as 1pfu given via the footpad (fp), a dose $>10^6$-fold lower than the lethal dose for wt B6 mice. Virus could be detected 2 days post-infection in spleen and liver and titres increased till death. Histological evidence from spleen and liver showed increased severity of pathology with $gzmAxB^{-/-} > gzmB^{-/-} > gzmA^{-/-} >$ wt (Müllbacher et al. 1999c). This synergy between gzmA and gzmB, which also implies a degree of redundancy, was unexpected given their structural and functional dissimilarities (see above).

One difference between $perf^{-/-}$ and $gzmAxB^{-/-}$ mice at low doses of EV infections was that in $gzmAxB^{-/-}$ mice death was delayed by up to 4 days

(Müllbacher et al. 1999c). On the other hand, deletions of one or both gzms in addition to perf had no affect on the mean time of death. This is compatible with evidence that perf is involved in gzm function and/or activation (Trapani and Smyth 2002), and gzms are the downstream effector molecules in recovery from primary EV infections (Müllbacher et al. 1999c). The precise mechanism(s) by which these gzms execute their effector function is not yet clear. Tc and NK cytolytic effector cells of gzmAxB$^{-/-}$ mice have unimpaired cytolytic function as measured in the ^{51}Cr release assay but do not induce apoptosis (^{125}I release assay; Simon et al. 1997). Furthermore, cell death as measured by ^{51}Cr release could be verified by assays showing lack of target cell colony formation and proliferation (Müllbacher et al. 1999c; Simon and Müllbacher 2000). In addition, the evidence that virus titres in spleen are elevated in gzmAxB$^{-/-}$ mice compared with wt B6 mice at a time (2 days post-infection) when mature Tc cells are undetectable but NK cells have normal cytolytic function suggests that gzms act against EV in ways as yet undiscovered, but distinct from their role as effector molecules of cytolytic lymphocytes in the granule exocytosis pathway.

4 Conclusions

The orthopoxviruses EV, CPV and VV have been used extensively as disease models in mice. Only EV is a natural mouse pathogen with a very restrictive host range (mus-species), whereas VV and CPV have a wide host range. CPV is a natural pathogen of rats. All three viruses induce a potent Tc cell response in spleen with effector cells lysing target cells infected with either the homologous or heterologous viruses equally well. Lysis of infected target cells is predominantly via the exocytosis-mediated pathway of cytotoxicity as the Fas pathway is inhibited by poxvirus-encoded serpins. This inhibition of the Fas pathway may explain the absolute necessity of perf in recovery from EV infections. The lack of perf does not render mice infected with VV or CPV more susceptible, in spite of the presence of serpins that inhibit the Fas pathway. This implies that, in VV and CPV infections of the mouse, cytolytic effector function is not required for resolving the infections. The reason for this is not known. This emphasises the caution needed in making generalisations of disease outcomes even when dealing with extremely closely related viral pathogens.

The findings of the importance of gzms in recovery from EV infections, especially the observation that EV-resistant B6 mice succumb to low dose of EV infection when neither gzmA nor B is present, despite possessing fully functionally cytolytic activity due to perf, suggest that gzms are involved in anti-viral defence by mechanisms unrelated to their role as effector molecules in the exocytosis pathway of Tc and NK cells.

References

Ada GI, Jackson DC, Blanden RV, Hla RT, Bowern NA (1976) Changes in the surface of virus-induced cells recognized by cytotoxic T cells. I. Minimal requirements for lysis of ectromelia-infected P-815 cells. Scand J Immunol 5:23–30

Andrew ME, Coupar BEH (1992) Biological effects of recombinant vaccinia virus-expressed interleukin 4. Cytokine 4:281–286

Aung S, Graham BS (2000) IL-4 diminishes perforin-mediated and increases fas ligand-mediated cytotoxicity in vivo. J Immunol 164:3487–3493

Balkow S, Kersten A, Tran TT, Stehle T, Grosse P, Museteanu C, Utermohlen O, Pircher H, von Weizsacker F, Wallich R, Müllbacher A, Simon MM (2001) Concerted action of the FasL/Fas and perforin/granzyme A and B pathways is mandatory for the development of early viral hepatitis but not for recovery from viral infection. J Virol 75:8781–8791

Barry M, Bleackley RC (2002) Cytotoxic T lymphocytes: all roads lead to death. Nat Rev Immunol 2:401–409

Beresford PJ, Xia Z, Greenberg AH, Lieberman J (1999) Granzyme A loading induces rapid cytolysis and a novel form of DNA damage independently of caspase activation. Immunity 10:585–594

Beverley PCL (1990) Human T-cell memory. Curr Top Microbiol Immunol 159:111–122

Blanden RV (1970) Mechanisms of recovery from a generalized viral infection: mousepox. I. The effects of anti-thymocyte serum. J Exp Med 132:1035–1054

Blanden RV (1971a) Mechanisms of recovery from a generalized viral infection: mousepox. 3. Regression infectious foci. J Exp Med 133:1090–1104

Blanden RV (1971b) Mechanisms of recovery from a generalized viral infection: mousepox. 2. Passive transfer of recovery mechanisms with immune lymphoid cells. J Exp Med 133:1074–1089

Blanden RV (1974) Mechanisms of cell-mediated immunity in viral infection. In: Brent L, Holborow J (eds) Progress in immunology II. Clinical aspects, vol 4, North Holland, Amsterdam, pp 117–125

Blanden RV, Gardner ID (1976) The cell-mediated immune response to ectromelia virus infection. I. Kinetics and characteristics of the primary effector T cell response in vivo. Cell Immunol 22:271–282

Blanden RV, Kees U, Dunlop MBC (1977) In vitro primary induction of cytotoxic T cells against virus-infected syngeneic cells. J Immunol Methods 16:73–89

Blanden RV, Deak BD, McDevitt HO (1989) Strategies in virus-host interactions. In: Blanden RV (ed) Immunology of virus diseases. Brolga Press, Curtin, ACT 2605, Australia, pp 125–138

Boehm U, Klamp T, Groot M, Howard JC (1997) Cellular responses to interferon-γ. Annu Rev Immunol 15:749–795

Bossi G, Griffiths GM (1999) Degranulation plays an essential part in regulating cell surface expression of Fas ligand in T cells and natural killer cells. Nat Med 5:90–96

Buller RML, Palumbo GJ (1991) Poxvirus pathogenesis. Microbiol Rev 55:81–122

Doherty PC, Topham DJ, Tripp RA, Cardin RD, Brooks JW, Stevenson PG (1997) Effector CD4+ and CD8+ T-cell mechanisms in the control of respiratory virus infections. Immunol Rev 159:105–117

Ebnet K, Hausmann M, Lehmann-Grube F, Müllbacher A, Kopf M, Lamers M, Simon MM (1995) Granzyme A-deficient mice retain potent cell-mediated cytotoxicity. EMBO J 14:4230–4239

Esposito JJ, Fenner F (2001) Poxviruses. In: Knipe DM, Howley PM (eds) Fields virology, vol 2. Lippincott Williams & Wilkins, Philadelphia, pp 2885–2921

Fenner F (1948) The pathogenesis of the acute exanthems. The Lancet 11(Dec):915–920

Fenner F (1949a) Mouse-pox (infectious ectromelia of mice): a review. J Immunol 63:341–373

Fenner F (1949b) Studies in mousepox (infectious ectromelia in mice). IV. Quantitative investigations on the spread of virus through the host in actively and passively immunized animals. Aust J Exp Biol Med Sci 28:1

Fenner F, Henderson DA, Arita I, Jezek Z, Ladnyi ID (1988) Smallpox and its eradication. World Health Organization, Geneva
Froelich CJ, Orth K, Turbov J, Seth P, Gottlieb R, Babior B, Shah GM, Bleackley RC, Dixit VM, Hanna W (1996) New paradigm for lymphocyte granule-mediated cytotoxicity. Target cells bind and internalize granzyme B, but an endosomolytic agent is necessary for cytosolic delivery and subsequent apoptosis. J Biol Chem 271:29073–29079
Fulginiti VA, Kempe CH, Hathaway WE, Pearlman DS, Sieber OF, Eller JJ, Joyner JJ, Robinson A (1968) Progressive vaccinia in immunologically deficient individuals. In: Bergsma D (ed) Immunologic deficiency diseases in man, vol IV. The National Foundation-March of Dimes, pp 129–144
Gardner ID, Blanden RV (1976) The cell-mediated immune response to ectromelia virus infection. II. Secondary response in vitro and kinetics of memory T cell production in vivo. Cell Immunol 22:283–296
Gardner I, Bowern NA, Blanden RV (1974) Cell-mediated cytotoxicity against ectromelia virus-infected target cells. II. Identification of effector cells and analysis of mechanisms. Eur J Immunol 4:68–72
Gardner I, Bowern NA, Blanden RV (1975) Cell-mediated cytotoxicity against ectromelia virus-infected target cells. III. Role of the H-2 gene complex. Eur J Immunol 5:122–127
Hayder H, Blanden RV, Korner H, Riminton DS, Sedgwick JD, Mullbacher A (1999) Adenovirus-induced liver pathology is mediated through TNF receptors I and II but is independent of TNF or lymphotoxin. J Immunol 163:1516–1520
Henkart PA (1994) Lymphocyte-mediated cytotoxicity: two pathways and multiple effector molecules. Immunity 1:343–346
Jackson RJ, Ramsay AJ, Christensen CD, Beaton S, Hall DF, Ramshaw IA (2001) Expression of mouse interleukin-4 by a recombinant ectromelia virus suppresses cytolytic lymphocyte responses and overcomes genetic resistance to mousepox. J Virol 75:205–1210
Kägi D, Ledermann B, Burki K, Seiler P, Odermatt B, Olsen KJ, Podack ER, Zinkernagel RM, Hengartner H (1994) Cytotoxicity mediated by T cells and natural killer cells is greatly impaired in perforin-deficient mice. Nature 369:31–37
Kägi D, Ledermann B, Bürki K, Zinkernagel RM, Hengartner H (1995a) Lymphocyte-mediated cytotoxicity in vitro and in vivo: mechanisms and significance. Immunol Rev 146:95–115
Kägi D, Odermatt B, Mak TW (1999) Homeostatic regulation of $CD8^+$ T cells by perforin. Eur J Immunol 29:3262–3272
Kägi D, Seiler P, Pavlovic P, Ledermann B, Bürki K, Zinkernagel RM, Hengartner H (1995b) The roles of perforin- and fas-dependent cytotoxicity in protection against cytopathic and non-cytopathic viruses. Eur J Immunol 25:3256–3262
Karupiah G, Buller RM, Van Rooijen N, Duarte CJ, Chen J (1996) Different roles for CD4+ and CD8+ T lymphocytes and macrophage subsets in the control of a generalized virus infection. J Virol 70: 8301–8309
Kees U, Blanden RV (1976) A single genetic element in H-2K affects mouse T-cell antiviral function in poxvirus infection. J Exp Med 143:450–455
Kramer MD, Binninger L, Schirrmacher V, Moll H, Prester M, Nerz G, Simon MM (1986) Characterization and isolation of a trypsin-like serine protease from a long-term culture cytolytic T cell line and its expression by functional distinct T cells. J Immunol 136:4644–4651
Li M, Beg AA (2000) Induction of necrotic-like cell death by tumor necrosis factor alpha and caspase inhibitors: novel mechanism for killing virus-infected cells. J Virol 74:7470–7477
Licon Luna RM, Lee E, Müllbacher A, Blanden RV, Langman R, Lobigs M (2002) Lack of both Fas ligand and perforin protects from flavivirus-mediated encephalitis in mice. J Virol 76:3202–3211
Locksley RM, Killeen N, Lenardo MJ (2001) The TNF and TNF receptor superfamilies: integrating mammalian biology. Cell 104:487–501
Macen JL, Garner RS, Musy PY, Brooks MA, Turner PC, Moyer RW, McFadden G, Bleackley RC (1996) Differential inhibition of Fas- and granule-mediated cytolysis pathways by the ortho-

poxvirus cytokine response modifier A/SPI-2 and SPI-1 protein. Proc Natl Acad Sci USA 93:9108–9113
Mackaness GB (1962) Cellular resistance to infection. J.Exp Med 116:81
Mackaness GB (1964) The immunological basis of acquired cellular resistance. J Exp Med 120:105
Masson D, Zamai M, Tschopp J (1986) Identification of granzyme A isolated from cytotoxic T-lymphocyte-granules as one of the proteases encoded by CTL-specific genes. FEBS Lett 208:84–88
Medema J, Scaffidi C, Kischkel F, Shevchenko A, Mann M, Krammer P, Peter M (1997) FLICE is activated by association with the CD95 death-inducing signaling complex (DISC). EMBO J 16:2794–2804
Mims CA (1959) The response of mice to large intravenous injections of ectromelia virus. I. The fate of injected virus. Brit J Exp Pathol 40:543
Müllbacher A, Lobigs M (2001) Creation of killer poxvirus could have been predicted. J Virol 75:8353–8355
Müllbacher A, Ebnet K, Blanden RV, Stehle T, Museteanu C, Simon MM (1996) Granzyme A is essential for recovery of mice from infection with the natural cytopathic viral pathogen, ectromelia. Proc Natl Acad Sci USA 93:5783–5787
Müllbacher A, Tha Hla R, Museteanu C, Simon MM (1999a) Perforin is essential for the control of ectromelia virus but not related poxviruses in mice. J Virol 73:1665–1667
Müllbacher A, Wallich R, Moyer RW, Simon MM (1999b) Poxvirus-encoded serpins do not prevent cytolytic T cell-mediated recovery from primary infections. J Immunol 162:7315–7321
Müllbacher A, Waring P, Tha Hla R, Tran T, Chin S, Stehle T, Museteanu C, Simon MM (1999c) Granzymes are the essential downstream effector molecules for the control of primary virus infections by cytolytic leukocytes. Proc Natl Acad Sci USA 96:13950–13955
Müllbacher A, Lobigs M, Hla RT, Tran T, Stehle T, Simon MM (2002) Antigen-dependent release of IFN-gamma by cytotoxic T cells up-regulates Fas on target cells and facilitates exocytosis-independent specific target cell lysis. J Immunol 169:145–150
Nagata S (1997) Apoptosis by death factor. Cell 88:355–365
Nagata S (1999) Fas ligand-induced apoptosis. Annu Rev Genet 33:29–55
Nagata S, Suda T (1995) Fas and Fas ligand: lpr and gld mutations. Immunol Today 16:39–43
Nakajima H, Park HL, Henkart PA (1995) Synergistic roles of granzymes A and B in mediating target cell death by rat basophilic leukemia mast cell tumors also expressing cytolysin/perforin. J Exp Med 181:1037–1046
O'Reilly LA, Strasser A (1999) Apoptosis and autoimmune disease. Inflamm Res 48:5–21
Parra B, Lin MT, Stohlman SA, Bergmann CC, Atkinson R, Hinton DR (2000) Contributions of fas-Fas ligand interactions to the pathogenesis of mouse hepatitis virus in the central nervous system [In Process Citation]. J Virol 74:2447–2450
Pasternack MS, Eisen HN (1985) A novel serine esterase expressed by cytotoxic T lymphocytes. Nature 314:743–745
Pereira RA, Simon MM, Simmons A (2000) Granzyme A, a noncytolytic component of CD8(+) cell granules, restricts the spread of herpes simplex virus in the peripheral nervous systems of experimentally infected mice. J Virol 74:1029–1032
Pham CT, Ley TJ (1997) The role of granzyme B cluster proteases in cell-mediated cytotoxicity. Sem Immunol 9:127–133
Pickup DJ, Ink BS, Hu W, Ray CA, Joklik WK (1986) Hemorrhage in lesions caused by cowpox virus is induced by a viral protein that is related to plasma protein inhibitors of serine proteases. Proc Natl Acad Sci USA 83:7698–7702
Quan LT, Caputo A, Bleackley RC, Pickup DJ, Salvesen GS (1995) Granzyme B is inhibited by the cowpox virus serpin cytokine response modifier A. J Biol Chem 270:10377–10379
Ramshaw IA, Ramsay AJ, Karupiah G, Rolph MS, Mahalingam S, Ruby JC (1997) Cytokines and immunity to viral infections. Immunol Rev 159:119–135

Ray CA, Black RA, Cronheim SR, Greenstreet TA, Sleath PR, Salwesen GS, Pickup DJ (1992) Viral inhibition of inflammation: cowpox virus encodes an inhibitor of the interleukin-1 beta converting enzyme. Cell 69:597–604
Roberts JA (1964a) Enhancement of the virulence of attenuated ectromelia virus in mice maintained in a cold environment. Aust J Exp Biol Med Sci 42:657
Roberts JA (1964b) Growth of virulent and attenuated ectromelia virus in cultured macrophages from normal and ectromelia-immune mice. J Immunol 92:837
Rouvier R, Luciani M-F, Golstein P (1993) Fas involvement in Ca^{2+}-independent T-cell-mediated cytotoxicity. J Exp Med 177:195–200
Ruby J, Bluethmann H, Peschon JJ (1997) Antiviral activity of tumor necrosis factor (TNF) is mediated via p55 and p75 TNF receptors. J Exp Med 186:1591–1596
Sad S, Marcotte R, Mosmann TR (1995) Cytokine-induced differentiation of precursor mouse CD8+ T cells into cytotoxic CD8+ T cells secreting Th1 or Th2 cytokines. Immunity 2:271–279
Sanderson A (1988) Smallpox is dead. Farrand Press, London
Sarin A, Williams MS, Alexander-Miller MA, Berzofsky JA, Zacharchuk CM, Henkart PA (1997) Target cell lysis by CTL granule exocytosis is independent of ICE/Ced-3 family proteases. Immunity 6:209–215
Shresta S, Goda P, Wesselschmidt R, Ley TJ (1997) Residual cytotoxicity and granzyme K expression in granzyme A-deficient cytotoxic lymphocytes. J Biol Chem 272:20236–20244
Simon MM, Kramer MD (1994) Granzyme A. Methods Enzymol 244:68–79
Simon MM, Müllbacher A (2000) Role of granzymes in target cell lysis and viral infections. In: Sitkovsky MV, Henkart PA (eds) Cytotoxic cells: basic mechanisms and medical applications, Lippincott Williams & Wilkins, Philadelphia, pp 197–211
Simon MM, Hausmann M, Tran T, Ebnet K, Tschopp J, ThaHla R, Müllbacher A (1997) In vitro and ex vivo-derived cytolytic leukocytes from granzyme AxB double knockout mice are defective in granule-mediated apoptosis but not lysis of target cells. J Exp Med 186:1781–1786
Simon MM, Waring P, Lobigs M, Nil A, Tran T, Hla RT, Chin S, Müllbacher A (2000) Cytotoxic T cells specifically induce Fas on target cells, thereby facilitating exocytosis-independent induction of apoptosis. J Immunol 165:3663–3672
Spriggs MK, Koller BH, Sato T, Morrissey PJ, Fanslow WC, Smithies O, Voice RF, Widmer MB, Maliszewski CR (1992) Beta 2-microglobulin-, CD8+ T-cell-deficient mice survive inoculation with high doses of vaccinia virus and exhibit altered IgG responses. Proc Natl Acad Sci USA 89:6070–6074
Tewari M, Quan LT, O'Rourke K, Desnoyers S, Zeng Z, Beidler DR, Poirier GG, Salvesen GS, Dixit VM (1995a) Yama/CPP32 beta, a mammalian homolog of CED-3, is a CrmA-inhibitable protease that cleaves the death substrate poly(ADP-ribose) polymerase. Cell 81:801–809
Tewari M, Telford WG, Miller RA, Dixit VM (1995b) CrmA, a poxvirus-encoded serpin, inhibits cytotoxic T-lymphocyte-mediated apoptosis. J Biol Chem 270:2705–2708
Trapani JA, Smyth MJ (2002) Functional significance of the perforin/granzyme cell death pathway. Nat Rev Immunol 2:735–747
Trapani JA, Jans DA, Jans PJ, Smyth MJ, Browne KA, Sutton VR (1998) Efficient nuclear targeting of granzyme B and the nuclear consequences of apoptosis induced by granzyme B and perforin are caspase-dependent, but cell death is caspase-independent. J Biol Chem 273:27934–27938
Tschopp J (1994) Granzyme B. Methods Enzymol 244:80–87
Turner PC, Musy PY, Moyer RW (1995) In: McFadden G (ed) Poxvirus serpins. Viroceptors, virokines and related immune modulators encoded by DNA viruses Springer, Berlin Heidelberg New York
Turner SJ, Silke J, Kenshole B, Ruby J (2000) Characterization of the ectromelia virus serpin, SPI-2. J Gen Virol 81:2425–2430
Vassalli P (1992) The pathophysiology of tumor necrosis factor. Annu Rev Immunol 10:411–452
Wallich R, Simon MM, Müllbacher A (2001) Virulence of mousepox virus is independent of serpin-mediated control of cellular cytotoxicity. Viral Immunol 14:71–81

Watanabe-Fukunaga R, Brannan CI, Itoh N, Yonehara S, Copeland NG, Jenkins NA, Nagata S (1992) The cDNA structure, expression, and chromosomal assignment of the mouse Fas antigen. J Immunol 148:1274–1279
Wheelock E (1965) Interferon-like virus inhibitor induced in human leukocytes by phytohemagglutinin. Science 149:310–311
Young JD-E, Leong LG, Liu C-C, Damiano A, Wall DA, Cohn ZA (1986) Isolation and characterization of a serine esterase from cytolytic T cell granules. Cell 47:183–194

Switching On and Off the Cell Death Cascade: African Swine Fever Virus Apoptosis Regulation

B. Hernáez[1], J.M. Escribano[1] and C. Alonso[1]

1 African Swine Fever Virus

African swine fever virus (ASFV) is a double-stranded DNA virus that infects domestic and wild swine, warthogs and bushpigs, causing a devastating disease in affected geographical regions (Dixon et al. 2000). Soft ticks of the genus *Ornithodoros* are also infected, with no disease signs, acting as a vector. African swine fever (ASF) is a very high mortality hemorrhagic syndrome characterized clinically by anorexia, fever, leukopenia and disseminated intravascular coagulation (Tulman and Rock 2001). Widespread cell death caused by apoptosis occurs in lymphoid tissues affecting macrophages, T and B lymphocytes, and endothelial cells (Ramiro-Ibanez et al. 1996, 1997; Oura et al. 1998a).

ASFV is the only member of the new family Asfarviridae (Dixon et al. 2000). Macrophages and cells of the mononuclear phagocytic system and reticuloendothelial lineage are the major host cells of ASFV. Binding and internalization occurs by receptor-mediated endocytosis, and the viral proteins p54, p30 (Gomez-Puertas et al. 1998) and p12 (Carrascosa et al. 1991) are the viral attachment proteins. Following internalization, uncoating takes place in the cytoplasm, starting the viral replication, protein synthesis and morphological processes in perinuclear, cytoplasmic areas known as *virus factories*. Finally, ASFV exit from infected cell proceeds either by cell destruction or budding through cell membrane. The ASFV genome consists of a single molecule of linear DNA of about 170–190 Kpb containing about 150 major open reading frames and sharing several structural features and replication strategies with poxviruses. The virus encodes not only for structural proteins, but also for many life cycle economy genes, such as proteins involved in DNA replication and repair, transcription or post-translational protein modifications (Dixon et al. 2000).

ASFV, like many other large DNA viruses, develops different strategies to evade host response to virus infection including apoptosis mechanisms. ASFV encodes for apoptosis inhibitor homologue genes such as a Bcl-2 homologue (Neilan et al. 1993; Brun et al. 1996, 1998), an IAP homologue (Nogal et al. 2001) and a homologue of the apoptosis inhibitor ICP34.5 of herpes simplex

[1]Dpt. Biotecnología, INIA, Ctra. de la Coruña Km7, 28040 Madrid, Spain, e-mail: calonso@inia.es

Progress in Molecular and Subcellular Biology
C. Alonso (Ed.): Viruses and Apoptosis
© Springer-Verlag Berlin Heidelberg 2004

(Zsak et al. 1996). These viral genes function to protect infected cells from apoptosis early in infection and to prevent premature death of the host cell which would impair virus production.

The virus is able to modulate the expression of different antiviral cytokines in infected cells, such as tumor necrosis factor-α, interleukin 8, interferon-α or transforming growth factor-β (Gomez del Moral et al. 1999). Viral genes involved in such modulation are EP204R, a homologue of cellular CD2, a T-cell antigen involved in regulation of cell activation (Rodriguez et al. 1993), EP153R with similarity to the C-type lectin family of adhesion proteins (Neilan et al. 1999), etc. A bifunctional virus-encoded protein, A238L, acts both to inhibit activation of the host transcription factor NFkB and also inhibits activity of the host phosphatase calcineurin thus inhibiting calcineurin-dependent pathways, such as activation of NFAT transcription factor (Miskin et al. 1998, 2000). This protein may therefore inhibit transcriptional activation in infected macrophages of a wide range of host immunomodulatory genes that are dependent on these factors.

2 Virus Interactions with the Host Cell

The pathogenesis of viral infections involves dynamic interactions between the virus and hosts. These can result in cell death, elimination of the virus or latent infection keeping both cells and pathogens alive. The outcome of an infection is often determined by cell signalling. Viruses deliver genomes and proteins with signalling potential into target cells and thereby alter the metabolism of the host. Responses by cells include stimulation of innate and adaptive immunity, growth, proliferation, survival and apoptosis.

2.1 Apoptotic Cell Death in the Macrophage Target Cell

Apoptosis has been considered a defense mechanism to prevent dissemination of viral infection which is counteracted by viral apoptosis suppressor genes (Cuconati and White 2002). Nevertheless, apoptosis could be induced as a consequence of the interaction of viral proteins with determined cell targets. Several structural viral proteins have been reported to induce apoptosis by interaction with the infected cell in an early stage of the infective cycle and in the absence of virus replication (Ramsey-Ewing and Moss 1998; Jan and Griffin 1999; Connolly and Dermody 2002), but the mechanism underlying this trigger in most viral models is still unknown.

The apoptotic signal in the ASFV target cell is triggered without virus replication and before early protein synthesis in a postbinding step

(Carrascosa et al. 2002). It was ruled out that the induction of apoptosis is due to the interaction of virus components with a cellular protein in the plasma membrane, as occurs with Env protein of avian leucosis virus (Brojatsch et al. 1996) or bovine herpes virus1 infection (Hanon et al. 1998), because UV-inactivated ASFV does not induce caspase expression nor apoptosis (Carrascosa et al. 2002). Also, using lysosomotropic drugs it was shown that proper uncoating of ASFV particles is necessary to induce the apoptotic signal in infected cells (Carrascosa et al. 2002). Therefore, apoptosis initiation could be triggered in an early infection step after receptor binding and uncoating, through the virus uptake of key cellular processes such as intracellular transport systems upon internalization. Nevertheless, ASFV-infected macrophages in vitro show late apoptosis, indicating that virus genes negatively regulating the process are acting early after infection (Ramiro-Ibanez et al. 1996).

2.2 Deadly Virus–Cell Interactions

Viruses have evolved unique mechanisms to regulate cellular processes, adapting their proteins to interact with cellular proteins. Some crucial steps for virus replication develop through events that begin to take place after attachment of virus to the cell membrane. Virus binding to the cell membrane may mediate a series of biochemical changes that optimize the intracellular milieu for use of cellular machinery.

The use of cell transport systems during early steps of viral infection enables virus particles to cross the cytosol and gain access to their replication sites. ASFV protein p54 interacts with the light chain of cytoplasmic dynein (DLC), a microtubular-based motor, facilitating virus transport to cytoplasmic factory sites (Alonso et al. 2001). Dyneins are microtubule-based motors that generate a driving force towards the minus-end of microtubules and are involved in several processes. Cytoplasmic dyneins are involved in vesicular transport, movement of endosomes and lysosomes, positioning of the mitotic spindle, etc. Dynein is a multisubunit protein composed of heavy, intermediate and light chains. Light chain cytoplasmic dynein of 8 kDa (DLC8) is the molecule that binds to cargoes and links microtubules to other structures such as transmembrane proteins, scaffold proteins and cross links to other cytoskeletal filaments (King 2000). DLC8 interacts with other molecules of the microtubular complex, links microtubules to other cytoskeletal components and binds to specific cargoes.

Structural ASFV protein p54 interaction with motor protein DLC8 is required for virus intracellular transport (Alonso et al. 2001). A 13 amino acid domain of ASFV protein p54 is sufficient for binding of p54 to DLC8; a SQT motif within this domain being critical for this binding (Alonso et al. 2001). The amino acid residues required for the viral p54–DLC8 interaction differ from those present in other DLC8-binding proteins, such as neuronal nitric oxide synthase (nNOS; Fan et al. 2001; Lo et al. 2001), but it is similar to the binding domain of the proapoptotic protein Bim (Alonso et al. 2001; Rodriguez-Crespo et al. 2001). Based on the sequence homology of their binding site, we investigated whether the interaction of viral p54 with DLC8 could produce changes in binding of Bim to microtubules through DLC8; our results showed that Bim is translocated from its binding to the microtubular complex upon ASFV infection. Proapoptotic Bim translocation triggers apoptosis through the bcl-2 pathway which might initiate the death signal in ASFV-infected cells (Fig. 1; Hernaez et al. 2003).

Bim is a proapoptotic member of the Bcl2 family with a BH3 domain only. BH3-only proteins are mammalian Bad, Bim, Bmf, Bik, Blk, Hrk/DP5, Bid, Noxa and *C. elegans* EGL-1 (Strasser et al. 2000). This group of proteins has been recognized as essential initiators of apoptosis (Huang and Strasser2000). Signals leading to caspase activation are regulated by bcl-2 proteins via regulating mitochondrial pore opening and release of apoptogenic molecules from mitochondria. Initiation of the proteolytic cascade requires the assembly of precursors on a scaffold protein or their translocation to organelles. Proapoptotic Bcl2 proteins Bim and Bmf are localized to distinct cytoskeletal structures and function by sensing intracellular damage (Puthalakath et al. 2001). Under physiological conditions, Bim is sequestered to microtubules by binding to the 8-kDa dynein light chain (Puthalakath et al. 1999, 2001). Apoptotic stimuli result in the release of Bim bound to DLC8 from the dynein motor complex. The freed proapoptotic molecules can translocate to mitochondria where they bind to and neutralize Bcl-2 and its antiapoptotic homologues (Bouillet et al. 2000; Strasser et al. 2000).

The ability of DLC8 to interact with peptides of diverse amino acid sequences implies that this molecule may bridge multiple cellular proteins to molecule motors (Fan et al. 2001). This is possible due to the dimeric DLC8 structure solved by NMR, in which the area which binds to cargo proteins is in a hydrophobic groove on each side of the dimmer (Liang et al. 1999). There are three splice forms of Bim: Bim_S, Bim_L and Bim_{EL} (O'Connor et al. 1998). Single amino acid substitutions in Bim_L that disrupt its interaction with DLC8 increased its killing potency to that of Bim_S (Puthalakath et al. 1999). This fact suggested that DLC8 functions as a physiological regulator of Bim proapoptotic activity.

Subcellular localization of BH3-only proteins plays a critical role in cell death control, enabling the cell to react rapidly and efficiently to death signals. Bim is activated by cytokine withdrawal and calcium flux (Puthalakath et al. 1999; Bouillet et al. 2002), whereas Bmf is activated by loss of cellular attachment-anoikis (Puthalakath et al. 2001). Bim and Bmf function to sense intracellular damage by their localization to distinct cytoskeletal structures. Bim is held inactive through its attachment to the cytoskeleton, and is thus denied access to its apoptotic effector targets. When proapoptotic Bim is displaced from the microtubular complex, it is freed to exert its function at the mitochondrial membrane, increasing mitochondrial permeability (Puthalakath

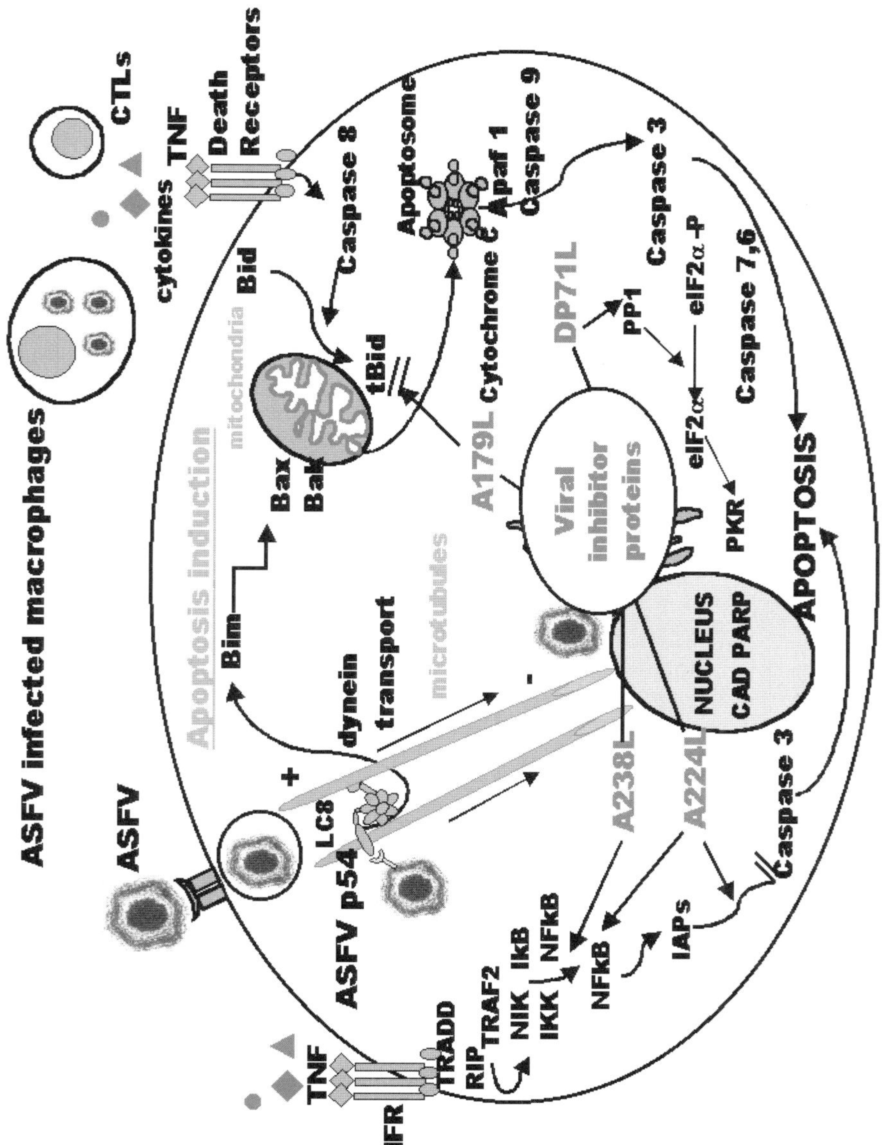

Fig. 1. Proposed model for apoptosis induction and inhibition stimuli in the African swine fever infected cell

et al. 1999; Costantini et al. 2002). This is the sequence of events possibly acting in ASFV-infected cells. The proapoptotic proteins act at the surface of the mitochondrial membrane to increase mitochondrial permeability and promote leakage of apoptosis-inducing factor (AIF) and other apoptosis initiators to the cytosol forming the apoptosome, activating caspase 9. Activated caspase 9 is the central activator of the mitochondrial, intrinsic and also termed bcl-2 regulated apoptotic pathway, which initiates a proteolytic cascade, processing other inactive caspases into their active forms. Caspase 3 activation will ultimately result in apoptosis. Together with Bim translocation to the mitochondria, we found an AIF shift to soluble fractions and caspase 9 and 3 activation in ASFV-infected cell extracts at 12 and 24 h after infection. Also, transient expression of viral p54 alone resulted in effector caspase 3 activation and apoptosis, that was not found with mutant p54 lacking its DLC8 Binding domain.

Other proteins of viral origin have been reported to bind dynein chains (Martinez Moreno et al. 2003). HSV I directly binds to intermediate dynein chains (Ye et al. 2000), and adenovirus binds to light dynein chains through a cellular adapter GTPase to be transported across the cytosol (Suomalainen et al. 1999; Leopold et al. 2000; Lukashok et al. 2000). Two lyssavirus proteins have been reported to bind DLC8 (Jacob et al. 2000; Raux et al. 2000), including P protein of rabies virus; however, it is not yet known if this interaction results in Bim release from microtubules. Interestingly, it has been reported that rabies P protein deletion mutants in the DLC8 binding site maintain their replication in vitro and in vivo in the inoculation site, but are unable to extend to the central nervous system (Mebatsion 2001). The cytoplasmic domain of the poliovirus receptor can interact specifically with the light chain of the dynein motor complex, driving viral invasion of the central nervous system through the retrograde axonal pathway (Mueller et al. 2002).

Chen et al. (2002) reported that Bim mediates secreted HIV-1 Tat-induced apoptosis of bystander lymphocytes. By affinity-purification experiments, they demonstrated that HIV-1 Tat binds to tubulin/microtubules through its conserved core region. Nevertheless, the mechanistic details between Tat alteration of microtubule dynamics and the initiation of the apoptotic signalling pathway through Bim remain to be elucidated. Due to the high specificity of this killing mechanism of uninfected cells, these authors postulated that such death signals could possibly occur in HIV-infected cells, being efficiently blocked by HIV-encoded inhibitors of the bcl2 family in the target cell. Given the fact that diverse viruses use microtubules and microtubule-dependent motors for transport (Whittaker et al. 2000), it will be of interest to determine whether viral-induced apoptosis could be also facilitated by Bim in these model viruses.

Therefore, Bim translocation could represent the early trigger of the cell death cascade that will be subsequently modulated by the viral apoptosis inhibitors which delay the ultimate effectors of nuclear fragmentation. Hence, apoptosis of the target cell will occur in fact relatively late after infection, when several ASFV replication cycles have been completed (24 h post-infection; Ramiro-Ibanez et al. 1996). In the cell it is the relative abundance of the proapoptotic and antiapoptotic Bcl2 family of proteins that determines the susceptibility of the cell to programmed cell death (Cory and Adams 2002; Cuconati and White 2002). Also, viral manipulation of the apoptosis pathways seems to involve both sides of this regulation.

3 Apoptosis in ASFV Pathogenesis

Virus isolates differ in virulence and may produce a variety of disease signs ranging from acute to chronic to unapparent. The acute infection produced by virulent isolates may cause 100% mortality in 5–10 days. Less virulent isolates may produce a mild disease from which a number of infected swine recover and become carriers. ASFV induces a severe immunosuppression in acutely infected animals that is detected after 2–3 days of infection by depletion of lymphoid subpopulations assessed by flow cytometry in peripheral blood (Ramiro-Ibanez et al. 1997). Depletion in lymphoid organ cellularity is due to programmed cell death, with a clear correlation between the degree of apoptosis and the lethality potential of the infecting viral isolate (Ramiro-Ibanez et al. 1996; Oura et al. 1998). The induction of apoptosis in non-infected immunocompetent cells should be considered as an important pathogenic mechanism to subvert host immune defense in ASFV infection.

3.1 Bystander Lymphocyte Apoptosis in ASF

ASFV causes apoptosis of cells of monocyte/macrophage lineage, reticuloendothelial cells and lymphocytes (Ramiro-Ibanez et al. 1996). Given the fact that lymphocytes are not susceptible to the virus, it was suggested that they have a relevant role for soluble mediators to cause lymphoid cell death. Systemic and local release of tumor necrosis factor alpha (TNF-α) is found during infection in mainly affected organs in correlation with viral protein expression. Elevation of TNF-α levels in serum is coincident with the onset of clinical signs (Gomez del Moral et al. 1999). The overproduction of this cytokine may contribute to the major clinical features of acute ASF such as shock, intravascular coagulation and bystander cell apoptosis.

TNF-α is secreted by primary culture macrophages after in vitro infection with virulent isolates and these virus-free TNF-α-containing supernatants induce apoptosis of lymphocytes from healthy donors. This effect is partially abrogated by preincubation with an anti-TNF-α-specific antibody, pointing to a relevant role for this cytokine in the pathogenesis of bystander lymphocyte apoptosis (Gomez del Moral et al. 1999).

4
Survival of the Target Cell

In the context of viral infections, the survival of the target cell in order to allow completion of the replicative cycle becomes a critical issue. However, activation of the programmed cell death pathway during the virus invasion can seriously compromise this survival impairing virus production. Induction of apoptosis in the cell that the virus infects is an important host cell defense mechanism that can be considered either a non-specific response against harmful agents, or the very specific result of the viral interference with specific cellular processes (Hernaez et al. 2003). Moreover, the immune system response with innate effector cells – such as natural killer (NK) cells and dendritic cells, or the adaptive immunity with the antigen-specific cytotoxic T lymphocytes (CTLs) – results in activation of the apoptotic pathway, particularly by members of the tumor necrosis factor cytokine family, as a mechanism to restrict viral replication (Benedict et al. 2002). Therefore, some viruses have incorporated genes that encode antiapoptotic proteins or modulate the expression of cellular regulators of apoptosis.

4.1
ASFV Apoptosis Inhibitor Genes

Several virus models are known to encode homologues of cellular antiapoptotic Bcl2 proteins (cBcl2s) which are central regulators of apoptosis at the mitochondrial level. Viral Bcl2 homologues (vBcl2s) prevent premature death of the host cell, which impairs virus production and contributes to avoid the immune surveillance of the host. Also, inhibition of apoptosis in infected cells can facilitate persistent infection (Cuconati and White 2002). ASFV vBcl2 homologue A179L in the BA71 strain (Yanez et al. 1995) or LMW5-HL in the Malawi strain (Neilan et al. 1993) is a sequence and functional vBcl2 homologue. It is expressed both at early and late times after infection and is an essential gene for virus replication (Brun et al. 1996). When exogenously expressed, it protects cells from apoptosis induced by the interferon-induced p68 kinase (PKR; Brun et al. 1996), by IL-3 withdrawal (Afonso et al. 1996), and the inhibition of macromolecular synthesis (Revilla et al. 1997). Interestingly, A179L is able to inhibit baculovirus-induced apoptosis in Sf9 insect cells (Brun et al. 1998), indicating a low degree of species specificity, as would be required of a viral protein that probably functions in both a mammal and an arthropod. However, the specific mechanism of action of the ASFV Bcl2 remains to be determined.

Immune surveillance by the host against a viral infection would partially entail attacking the infected cell with secreted FasL, TRAIL or TNFα, as is the case in ASF. One possible function of the vBcl2 proteins, therefore, would be to prevent death of the infected cell due to signalling by death cytokines. It has been considered possible that inhibition of apoptosis mediated by death receptors by vBcl2s is an indirect consequence of inhibition of the core apoptotic machinery represented by Bax and Bak, which may block yet another apoptotic pathway (Cuconati and White 2002). Alternatively, it may be so critically important to block apoptosis by death receptor ligands that viruses encode redundant inhibitory mechanisms to ensure survival of the infected cell.

Adenovirus E1B19k interacts with BH3-only proteins Bid, tBid, Bad BH3, but not with Bcl2 and Bcl-xL (Chen et al. 1996; Perez and White 2000). Also, the E1B 19K protein interacts with proapoptotic BH3-only proteins Nbk/Bik and Bnip3 (Boyd et al. 1994, 1995; Han et al. 1996), but the physiological context of these activities still remains to be determined. The ASFV bcl-2 homologue was found to interact with BH3 only proteins specifically with the active forms of Bid which is a central player in the death receptor signalling pathway, being able to inhibit apoptosis induced by overexpression of active or truncated Bid (Fig. 1; Fernandez-Zapatero et al. 2003). Protection conferred by vBcl2 could be a mechanism to shield ASFV-infected cells against secreted cytokines. These cytokines could be deleterious to non-infected immunocompetent cells, contributing to the virus evasion of the immune response.

ASFV encodes other genes with antiapoptotic functions in alternative pathways such as an IκB homologue gene (A228L), an IAP homologue (A224L), a homologue to the herpes simplex apoptotic suppressor proteins ICP 34,5 (DP71L), etc. These redundant functions point out the importance of viral apoptosis regulation during infection of this virus, which expands cell survival to the time needed to replicate its genome (Fig. 1). The inhibitor of apoptosis (IAP), A224L or 4CL, is synthesized late after infection (Chacon et al. 1995) and promotes cell survival by interaction with the catalytic fragment of caspase 3 (Nogal et al. 2001). ASFV IAP is also able to activate NF-κB through enhancing IκB kinase (IKK) basal activity (Rodriguez et al. 2002). ASFV also encodes a homologue of the apoptosis inhibitor γ34.5 of HSV I, ASFV DP71L or NL (Zsak et al. 1996). The HSV protein activates PP1 phosphatase which dephosphorylates eIF-2α and in turn avoids PKR-mediated inhibition of the protein synthesis (He et al. 1998). The ASFV protein could have the same activity given the high sequence homology with the HSV γ34.5. Both long and short forms of the protein are synthesized by all known ASFV isolates and both localize in the nucleus to the nucleolus (Goatley et al. 1999).

The combined action of some of these apoptosis inhibitor genes can facilitate persistent infection. ASFV persistent infection of its natural host, the African wild swine, allows the virus to persist in nature facilitating reinfections of the domestic pig populations in farms. Also, domestic pigs experimentally infected with attenuated isolates that survive a short acute phase develop a lifelong persistent infection (Carrillo et al. 1994). The argasid vector (ticks, genus *Ornithodoros*) also becomes persistently infected, being able to transmit the disease by bite (Kleiboeker et al. 1999).

5 Conclusions

Apoptosis is known to play a central role in virus pathogenesis as the virus induces cell death both in the target cell and immune defense cells. ASFV encodes for a set of homologues of known apoptosis inhibitors. These viral apoptosis suppressor genes function to protect infected cells from apoptosis early in infection and to prevent premature death of the host cell that would impair virus production. A179L interacts with BH3-only proteins. Its interaction with active Bid, inhibits the death receptor pathway in the target cell, which becomes protected from the high TNF-α levels existing in lymphoid organs, that, in contrast, are deleterious to the immunocompetent cells. In spite of this evidence of apoptosis inhibition, the mechanism by which the virus induces apoptosis in the target cell was until recently unknown. Proapoptotic BH3-only protein Bim is the first reported apoptosis initiator underlying ASFV-induced apoptosis and Bim translocation was demonstrated to trigger apoptosis by a bcl-2-dependent pathway in African swine fever virus infection. The elucidation of the complex strategies of virus interference with apoptotic mechanisms can provide further understanding into the host–virus relationship that might help to design new vaccine strategies.

Acknowledgements. This work was supported by EU grant QLK3-2000-362 BMc 2000-1003 and AGL 2002-668 from Plan National I+D Spain.

References

Afonso CL, Neilan JG, Kutish GF, Rock DL (1996) An African swine fever virus Bc1–2 homolog, 5-HL, suppresses apoptotic cell death. J Virol.70:4858–4863

Alonso C, Miskin J, Hernaez B, Fernandez-Zapatero P, Soto L, Canto C, Rodriguez-Crespo I, Dixon L, Escribano JM (2001) African swine fever virus protein p54 interacts with the microtubular motor complex through direct binding to light-chain dynein. J Virol 75:9819–9827

Benedict CA, Norris PS, Ware CF (2002) To kill or be killed: viral evasion of apoptosis. Nat Immunol 3:1013–1018

Bouillet P, Huang DC, O'Reilly LA, Puthalakath H, O'Connor L, Cory S, Adams JM, Strasser A (2000) The role of the pro-apoptotic Bcl-2 family member bim in physiological cell death. Ann NY Acad Sci 926:83–89

Bouillet P, Purton JF, Godfrey DI, Zhang LC, Coultas L, Puthalakath H, Pellegrini M, Cory S, Adams JM, Strasser A (2002) BH3-only Bcl-2 family member Bim is required for apoptosis of autoreactive thymocytes. Nature 415:922–926

Boyd JM, Malstrom S, Subramanian T, Venkatesh LK, Schaeper U, Elangovan B, D'Sa-Eipper C, Chinnadurai G (1994) Adenovirus E1B 19 kDa and Bcl-2 proteins interact with a common set of cellular proteins. Cell 79:341–351

Boyd JM, Gallo GJ, Elangovan B, Houghton AB, Malstrom S, Avery BJ, Ebb RG, Subramanian T, Chittenden T, Lutz RJ, et al (1995) Bik, a novel death-inducing protein, shares a distinct sequence motif with Bcl-2 family proteins and interacts with viral and cellular survival-promoting proteins. Oncogene 11:1921–1928

Brojatsch J, Naughton J, Rolls MM, Zingler K, Young JA (1996) CAR1, a TNFR-related protein, is a cellular receptor for cytopathic avian leukosis-sarcoma viruses and mediates apoptosis. Cell 87:845–855

Brun A, Rivas C, Esteban M, Escribano JM, Alonso C (1996) African swine fever virus gene A179L, a viral homologue of bcl-2, protects cells from programmed cell death. Virology 225:227–230

Brun A, Rodriguez F, Escribano JM, Alonso C (1998) Functionality and cell anchorage dependence of the African swine fever virus gene A179L, a viral bcl-2 homolog, in insect cells. J Virol 72:10227–10233

Carrascosa AL, Bustos MJ, Nogal ML, Gonzalez de Buitrago G, Revilla Y (2002) Apoptosis induced in an early step of African swine fever virus entry into vero cells does not require virus replication. Virology 294:372–382

Carrascosa AL, Sastre I, Vinuela E (1991) African swine fever virus attachment protein. J Virol 65:2283–2289

Carrillo C, Borca MV, Afonso CL, Onisk DV, Rock DL (1994) Long-term persistent infection of swine monocytes/macrophages with African swine fever virus. J Virol 68:580–583

Chacon MR, Almazan F, Nogal ML, Vinuela E, Rodriguez JF (1995) The African swine fever virus IAP homolog is a late structural polypeptide. Virology 214:670–674

Chen D, Wang M, Zhou S, Zhou Q (2002) HIV-1 Tat targets microtubules to induce apoptosis, a process promoted by the pro-apoptotic Bcl-2 relative Bim. EMBO J 21:6801–6810

Chen G, Branton PE, Yang E, Korsmeyer SJ, Shore GC (1996) Adenovirus E1B 19-kDa death suppressor protein interacts with Bax but not with Bad. J Biol Chem 271:24221–24225

Connolly JL, Dermody TS (2002) Virion disassembly is required for apoptosis induced by reovirus. J Virol 76:1632–1641

Cory S, Adams JM (2002) The Bcl2 family: regulators of the cellular life-or-death switch. Nat Rev Cancer 2:647–656

Costantini P, Bruey JM, Castedo M, Metivier D, Loeffler M, Susin SA, Ravagnan L, Zamzami N, Garrido C, Kroemer G (2002) Pre-processed caspase-9 contained in mitochondria participates in apoptosis. Cell Death Differ 9:82–88

Cuconati A, White E (2002) Viral homologs of BCL-2: role of apoptosis in the regulation of virus infection. Genes Dev 16:2465–2478

Dixon LK, Costa JV, Escribano JM, Rock DL, Vinuela E, Wilkinson PJ (2000) Asfarviridae. In: 7th Report of the International Committee on Taxonomy of Viruses. Van Regenmortel MHV, Fauquet CM, Bishop DHL, Cartens E, Estes M, Lemon S, Maniloff J, Mayo MA, McGeoch D, Pringle CR, Wickner RB (eds) Academic Press, San Diego, pp 159–165

Fan J, Zhang Q, Tochio H, Li M, Zhang M (2001) Structural basis of diverse sequence-dependent target recognition by the 8 kDa dynein light chain. J Mol Biol 306:97–108

Fernandez-Zapatero P, Hernaez B, Dixon L, Escribano JM, Alonso C (2003) Viral bcl-2 homolog A179L interacts with Bid to block a death-receptor induced apoptosis pathway in the African swine fever infected cell. In: Proceedings Apoptosis 2003 Meeting, From signalling pathways to therapeutics tools, Luxembourg

Goatley LC, Marron MB, Jacobs SC, Hammond JM, Miskin JE, Abrams CC, Smith GL, Dixon LK (1999) Nuclear and nucleolar localization of an African swine fever virus protein, I14L, that is similar to the herpes simplex virus-encoded virulence factor ICP34.5. J Gen Virol 80:525–535

Gomez del Moral M, Ortuno E, Fernandez-Zapatero P, Alonso F, Alonso C, Ezquerra A, Dominguez J (1999) African swine fever virus infection induces tumor necrosis factor alpha production: implications in pathogenesis. J Virol 73:2173–2180

Gomez-Puertas P, Rodriguez F, Oviedo JM, Brun A, Alonso C, Escribano JM (1998) The African swine fever virus proteins p54 and p30 are involved in two distinct steps of virus attachment and both contribute to the antibody-mediated protective immune response. Virology 243:461–471

Han J, Sabbatini P, White E (1996) Induction of apoptosis by human Nbk/Bik, a BH3-containing protein that interacts with E1B 19 K. Mol Cell Biol 16:5857–5864

Hanon E, Meyer G, Vanderplasschen A, Dessy-Doize C, Thiry E, Pastoret PP (1998) Attachment but not penetration of bovine herpesvirus 1 is necessary to induce apoptosis in target cells. J Virol 72:7638–7641

He B, Gross M, Roizman B (1998) The gamma134.5 protein of herpes simplex virus 1 has the structural and functional attributes of a protein phosphatase 1 regulatory subunit and is present in a high molecular weight complex with the enzyme in infected cells. J Biol Chem 273:20737–20743

Hernaez B, Fernandez-Zapatero P, Garcia-Gallo M, Rodriguez-Crespo I, Alvarez A, Dixon L, Escribano JM, Alonso C (2003) Microtubular motor complex binding by a viral protein through light chain cytoplasmic dynein results in Bim translocation and apoptosis. In: Proceedings Apoptosis 2003 Meeting, From signalling pathways to therapeutics tools, Luxembourg

Huang DC, Strasser A (2000) BH3-only proteins – essential initiators of apoptotic cell death. Cell 103:839–842

Jacob Y, Badrane H, Ceccaldi PE, Tordo N (2000) Cytoplasmic dynein LC8 interacts with lyssavirus phosphoprotein. J Virol 74:10217–10222

Jan JT, Griffin DE (1999) Induction of apoptosis by Sindbis virus occurs at cell entry and does not require virus replication. J Virol 73:10296–10302

King SM (2000) The dynein microtubule motor. Biochim Biophys Acta 1496:60–75

Kleiboeker SB, Scoles GA, Burrage TG, Sur J (1999) African swine fever virus replication in the midgut epithelium is required for infection of *Ornithodoros* ticks. J Virol 73:8587–8598

Leopold PL, Kreitzer G, Miyazawa N, Rempel S, Pfister KK, Rodriguez-Boulan E, Crystal RG (2000) Dynein- and microtubule-mediated translocation of adenovirus serotype 5 occurs after endosomal lysis. Hum Gene Ther 11:151–165

Liang J, Jaffrey SR, Guo W, Snyder SH, Clardy J (1999) Structure of the PIN/LC8 dimer with a bound peptide. Nat Struct Biol 6:735–740

Lo KW, Naisbitt S, Fan JS, Sheng M, Zhang M (2001) The 8-kDa dynein light chain binds to its targets via a conserved (K/R)XTQT motif. J Biol Chem 276:14059–14066

Lukashok SA, Tarassishin L, Li Y, Horwitz MS (2000) An adenovirus inhibitor of tumor necrosis factor alpha-induced apoptosis complexes with dynein and a small GTPase. J Virol 74:4705–4709

Martinez Moreno M, Navarro-Lerida I, Roncal F, Albar JP, Alonso C, Gavilanes F, Rodriguez-Crespo I (2003) Recognition of novel viral sequences that associate to the dynein light chain LC8 identified through a Pepscan technique. FEBS Lett 544:262–267

Mebatsion T (2001) Extensive attenuation of rabies virus by simultaneously modifying the dynein light chain binding site in the P protein and replacing Arg333 in the G protein. J Virol 75:11496–11502

Miskin JE, Abrams CC, Goatley LC, Dixon LK (1998) A viral mechanism for inhibition of the cellular phosphatase calcineurin. Science 281:562–565

Miskin JE, Abrams CC, Dixon LK (2000) African swine fever virus protein A238L interacts with the cellular phosphatase calcineurin via a binding domain similar to that of NFAT. J Virol 74:9412–9420

Mueller S, Cao X, Welker R, Wimmer E (2002) Interaction of the poliovirus receptor CD155 with the dynein light chain Tctex-1 and its implication for poliovirus pathogenesis. J Biol Chem 277:7897–7904

Neilan JG, Lu Z, Afonso CL, Kutish GF, Sussman MD, Rock DL (1993) An African swine fever virus gene with similarity to the proto-oncogene bcl-2 and the Epstein-Barr virus gene BHRF1. J Virol 67:4391–4394

Neilan JG, Borca MV, Lu Z, Kutish GF, Kleiboeker SB, Carrillo C, Zsak L, Rock DL (1999) An African swine fever virus ORF with similarity to C-type lectins is non-essential for growth in swine macrophages in vitro and for virus virulence in domestic swine. J Gen Virol 80:2693–2707

Nogal ML, Gonzalez de Buitrago G, Rodriguez C, Cubelos B, Carrascosa AL, Salas ML, Revilla Y (2001) African swine fever virus IAP homologue inhibits caspase activation and promotes cell survival in mammalian cells. J Virol 75:2535–2543

O'Connor L, Strasser A, O'Reilly LA, Hausmann G, Adams JM, Cory S, Huang DC (1998) Bim: a novel member of the Bcl-2 family that promotes apoptosis. EMBO J 17:384–395
Oura CA, Powell PP, Anderson E, Parkhouse RM (1998a) The pathogenesis of African swine fever in the resistant bushpig. J Gen Virol 79:1439–1443
Oura CA, Powell PP, Parkhouse RM (1998b) African swine fever: a disease characterized by apoptosis. J Gen Virol 79:1427–1438
Perez D, White E (2000) TNF-alpha signals apoptosis through a bid-dependent conformational change in Bax that is inhibited by E1B 19K. Mol Cell 6:53–63
Puthalakath H, Huang DC, O'Reilly LA, King SM, Strasser A (1999) The proapoptotic activity of the Bcl-2 family member Bim is regulated by interaction with the dynein motor complex. Mol Cell 3:287–296
Puthalakath H, Villunger A, O'Reilly LA, Beaumont JG, Coultas L, Cheney RE, Huang DC, Strasser A (2001) Bmf: a proapoptotic BH3-only protein regulated by interaction with the myosin V actin motor complex, activated by anoikis. Science 293:1829–1832
Ramiro-Ibanez F, Ortega A, Brun A, Escribano JM, Alonso C (1996) Apoptosis: a mechanism of cell killing and lymphoid organ impairment during acute African swine fever virus infection. J Gen Virol 77:2209–2219
Ramiro-Ibanez F, Ortega A, Ruiz-Gonzalvo F, Escribano JM, Alonso C (1997) Modulation of immune cell populations and activation markers in the pathogenesis of African swine fever virus infection. Virus Res 47:31–40
Ramsey-Ewing A, Moss B (1998) Apoptosis induced by a postbinding step of vaccinia virus entry into Chinese hamster ovary cells. Virology 242:138–149
Raux H, Flamand A, Blondel D (2000) Interaction of the rabies virus P protein with the LC8 dynein light chain. J Virol 74:10212–10216
Revilla Y, Cebrian A, Baixeras E, Martinez C, Vinuela E, Salas ML (1997) Inhibition of apoptosis by the African swine fever virus Bcl-2 homologue: role of the BH1 domain. Virology 228:400–404
Rodriguez CI, Nogal ML, Carrascosa AL, Salas ML, Fresno M, Revilla Y (2002) African swine fever virus IAP-like protein induces the activation of nuclear factor kappa B. J Virol 76:3936–3942
Rodriguez JM, Yanez RJ, Almazan F, Vinuela E, Rodriguez JF (1993) African swine fever virus encodes a CD2 homolog responsible for the adhesion of erythrocytes to infected cells. J Virol 67:5312–5320
Rodriguez-Crespo I, Yelamos B, Roncal F, Albar JP, Ortiz de Montellano PR, Gavilanes F (2001) Identification of novel cellular proteins that bind to the LC8 dynein light chain using a pepscan technique. FEBS Lett 503:135–141
Strasser A, Puthalakath H, Bouillet P, Huang DC, O'Connor L, O'Reilly LA, Cullen L, Cory S, Adams JM (2000) The role of bim, a proapoptotic BH3-only member of the Bcl-2 family in cell-death control. Ann NY Acad Sci 917:541–548
Suomalainen M, Nakano MY, Keller S, Boucke K, Stidwill RP, Greber UF (1999) Microtubule-dependent plus and minus end-directed motilities are competing processes for nuclear targeting of adenovirus. J Cell Biol 144:657–672
Tulman ER, Rock DL (2001) Novel virulence and host range genes of African swine fever virus. Curr Opin Microbiol 4:456–461
Whittaker GR, Kann M, Helenius A (2000) Viral entry into the nucleus. Annu Rev Cell Dev Biol 16:627–651
Yanez RJ, Rodriguez JM, Nogal ML, Yuste L, Enriquez C, Rodriguez JF, Vinuela E (1995) Analysis of the complete nucleotide sequence of African swine fever virus. Virology 208:249–278
Ye GJ, Vaughan KT, Vallee RB, Roizman B (2000) The herpes simplex virus 1 U(L)34 protein interacts with a cytoplasmic dynein intermediate chain and targets nuclear membrane. J Virol 74:1355–1363
Zsak L, Lu Z, Kutish GF, Neilan JG, Rock DL (1996) An African swine fever virus virulence-associated gene NL-S with similarity to the herpes simplex virus ICP34.5 gene. J Virol 70:8865–8871

Neuronal Apoptosis Pathways in Sindbis Virus Encephalitis

Pablo M. Irusta[1] and J. Marie Hardwick[1]

1 Introduction

In their daily battles against microbial pathogens, infected hosts need to strike a balance between fighting the invaders and minimizing the overall destruction caused by their defensive campaign. Limited and localized responses are crucial to reduce the amount of "collateral damage" that may result in the elimination of vital components of the host essential for the survival of the infected organism. For instance, an excess in prophylactic programmed cell death or an exacerbated cytotoxic T-cell response beyond a certain threshold could cause irreparable harm to fundamental tissues, leading to irreversible sequelae or death. The regulation of the protective response is particularly important in cases like viral infections that cause encephalitis, as most of the target cells belong to non-renewable populations, some of which are essential for life. In the case of Sindbis virus (SV), neurons are the main site of viral replication, and virus-induced neuronal cell death correlates with Sindbis virulence. Specifically, apoptotic mechanisms triggered in infected cells appear to be the main determining factors of the outcome of an infection. In this regard, Sindbis represents a model system to study viruses that induce encephalomyelitis affecting humans, in which decisions taken at the cellular level in infected neurons ultimately dictate the fate of the entire organism. To predict the outcome of infections and to devise strategies to treat encephalitis, it is crucial to identify both host- and virus-related factors that modulate cell death and disease. Ultimately, a better understanding of the molecular mechanisms governing neuronal cell fate under conditions of stress will also help us gain insight into several neurological diseases that affect humans.

2 Sindbis Virus, a Brief Overview

Sindbis virus (SV) is the prototype member of the alphavirus genus within the *Togaviridae* family. This arthropod-borne, single-strand positive-sense RNA

[1]Department of Molecular Microbiology and Immunology, Johns Hopkins Bloomberg School of Public Health, 615 North Wolfe St., Baltimore, Maryland 21205, USA, e-mail: pirusta@jhsph.edu

Progress in Molecular and Subcellular Biology
C. Alonso (Ed.): Viruses and Apoptosis
© Springer-Verlag Berlin Heidelberg 2004

virus forms icosahedral enveloped particles containing primarily three structural proteins, a capsid protein that wraps the 11.7 kb viral genome, and two membrane glycoproteins, E1 and E2, that form protruding spikes on the virion surface (Rice et al. 1982; Lopez et al. 1994; Zhang et al. 2002). SV enters the host cells by receptor-mediated endocytosis and replicates entirely within their cytoplasm (Strauss and Strauss 1994; Griffin and Hardwick 1997). However, the receptors and co-receptors that mediate virus attachment, entry and tropism are not fully characterized.

The natural life cycle of the virus involves vertebrate avian or mammalian hosts and invertebrate mosquito vectors. In the laboratory, SV is able to infect a variety of different cell types, but in vivo its replication is primarily restricted to certain tissues, including brown fat, muscle but particularly the central nervous system in mice. In mosquitoes, SV can be found in gut epithelia, salivary glands and the nervous system. Although laboratory strains of SV are not associated with human disease, SV infections in humans occur worldwide and are known to cause rash and arthritis. SV infections in mice cause an age-dependent acute encephalomyelitis. In contrast, mosquito infection appears to be asymtomatic. At the cellular level, SV infection is productive in many cultured mammalian cell lines resulting primarily in apoptotic death (Levine et al. 1993; Lewis et al. 1996). In mosquito cells, SV causes a persistent infection, although viral persistence in cells of mammalian origin has been reported (Inglot et al. 1973; Igarashi et al. 1977; Levine et al. 1993; Karpf et al. 1997; Karpf and Brown 1998; Sawicki et al. 2003). In the past few years, Sindbis virus has been used as the raw material for the construction of expression vector systems. Typically, genes of interest are inserted into the viral genome under the transcriptional control of an extra copy of the viral subgenomic promoter (Hardwick and Levine 2000). The resulting recombinant viral vectors (double subgenomic SV or dsSV) have typical Sindbis virus tropism, with the advantage of infecting a variety of cell types in vitro, perhaps more importantly primary neurons, as well as of establishing in vivo infections in murine models. Sindbis virus thus provides an efficient way to express exogenous proteins in both isolated cells and whole animal brains and, by the same token, because of the very nature of the virus, to induce programmed cell death and disease in the recipient host. Therefore, these vectors allow us to study the effects of particular gene products in virus-induced apoptosis, and ultimately in encephalitis development and animal survival. These features have made SV a model of choice for studying the molecular mechanisms underlying virus-induced apoptosis during viral infections. Importantly, studies using this system will likely provide relevant information that enhances our understanding of related human arthropod-borne alphavirus encephalitides such as Eastern, Western and Venezuelan equine encephalitis.

3 Pathogenesis: Factors That Determine Sindbis Virus Neurovirulence

3.1 Viral Factors

Several strains of SV have been identified, which vary greatly in their neurovirulence. The wild-type virus, as well as most laboratory strains, are lethal when inoculated into neonatal mice but becomes avirulent in weanling and adult mice. Other strains of SV were generated in the laboratory through serial passage in mouse brains. The resulting neuroadapted SV (NSV) contains mutations in the viral envelope glycoproteins and non-coding regions that render the virus capable of causing hind-limb paralysis and death in adult animals (Jackson et al. 1987, 1988).

Sindbis virus produces only eight proteins during its life cycle. Four non-structural and four structural polypeptides are all that the virus needs to complete the viral cycle. Although determinants of pathogenicity have been mapped primarily to the structural glycoproteins E1 and E2, regions within the non-structural part of the genome also appear to influence virulence (Lustig et al. 1988). At the cellular level, the classical rapid SV-induced apoptosis appears to require the synthesis and transport to the cell surface of the viral envelope glycoproteins. Moreover, expression of these proteins in the absence of other viral factors is sufficient to trigger programmed cell death (PCD; Joe et al. 1998). Major apoptotic determinants appear to lie in the transmembrane domains of E1 and E2, as the expression of these domains in AT3 rat prostate carcinoma cells in the absence of flanking sequences resulted in as much cell death as the expression of their full-length precursors (Joe et al. 1998). A mechanism of action for the viral glycoproteins has been proposed by Jan and Griffin (1999), who described SV-induced apoptosis at the time of virus entry into the cell by a process that requires virus-cell fusion and possibly a receptor-specific pathway. These researchers showed that high doses of UV-inactivated SV (MOI 500) could trigger apoptosis in CHO (Chinese hamster ovary) and N18 (mouse neuroblastoma) cells when the virus was induced to fuse at the plasma membrane by low pH treatment, an event that resembled the natural endosomal route of SV entry. Cell death under these conditions required inhibitors of protein synthesis, occurred in the absence of virus replication, and was inhibited by drugs that prevent endosomal acidification. Notably, the extent of apoptosis induced by UV-inactivated SV in N18 cells depended on the virus strain used, with a strong correlation between increased virulence of the virus and increased amount of cell death. Binding per se of Sindbis virus to N18 neuroblastoma cells was recently reported to be influenced by the composition of the E2 protein in different viral strains (Lee et al. 2002b). The acid-induced fusion process between the viral glycoproteins and the cell membrane is known to require the presence of sphingomyelin in the target membrane, and SV entry has been shown to trigger activation of sphingomyelinases and release of the pro-apoptotic molecule ceramide (Jan et al. 2000). Currently, it is not clear whether the apoptotic pathways observed during virus fusion and those triggered by expression of the envelope glycoproteins or their transmembrane domains are the same. During SV infection, however, cell death has been shown to occur even if the production of the viral glycoproteins is blocked, although at much later times post-infection (Frolov and Schlesinger 1994). This effect may be a consequence of the efficient virus-induced shutoff of host protein synthesis. It seems likely that SV is able to engage diverse signaling pathways at different stages of infection, and a convergence of these pathways may facilitate the ability of the virus to efficiently kill a variety of cell types.

Regardless of their modus operandi, the envelope glycoproteins are important determinants of virulence. This fact is clearly exemplified by two Sindbis virus strains that differ in only one amino acid within the E2 molecule. The Sindbis virus vector dsTE12Q is a non-neurovirulent recombinant virus that causes no detectable disease in 30-day-old mice. In contrast, a derivative neurovirulent strain, dsTE12H, containing a single glutamine to histidine substitution at position 55 in the E2 envelope glycoprotein, efficiently kills 50% of infected animals of the same age (Fig. 1B). Changes in SV lethality correlate with changes in the ability of the virus to induce apoptosis, and viruses containing the glutamine to histidine substitution at position 55 of E2 are such powerful inducers of cell death that they can even overcome the inhibition of apoptosis imposed by Bcl-2 overexpression in certain cell lines like AT3 cells (Ubol et al. 1994). In vivo, neurovirulent viruses containing E2 mutations also achieve higher levels of replication compared with avirulent strains; for instance, mean titers of dsTE12H in infected brains of 10-day-old mice are approximately 100 times higher than those of dsTE12Q (Griffin et al. 1994; Lewis et al. 1996). The correlation between enhanced viral replication and increased induction of apoptosis in mouse brains suggests that the regulation of these processes may be linked in neuronal cells. Further support for this notion comes from experiments showing that enforced expression of anti-apoptotic proteins in mouse brains reduces not only programmed cell death, but also SV replication (Cheng et al. 1996; Levine et al. 1996; Liang et al. 1998). The mechanisms by which mutations in the structural glycoproteins influence viral replication, induction of apoptosis and neurovirulence are currently unknown.

3.2
Age-dependent Host Factors

What are the host factors that protect mice from lethal infection with SV? The most obvious factor is the age of the infected host. As mice age, they acquire

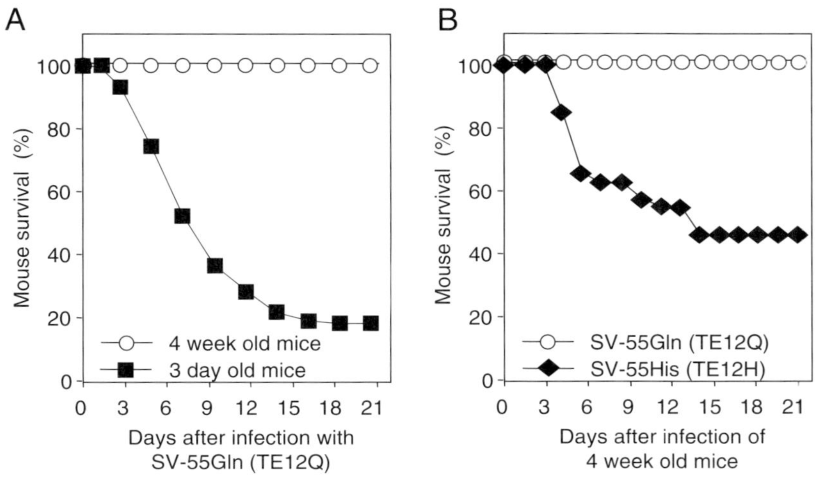

Fig. 1. Examples of viral and host determinants affecting the outcome of a Sindbis virus infection. **A** Survival curves of 3-day-old and 4-week-old CD1 mice infected with an E2-Gln55 Sindbis virus (dsTE12Q) Sindbis virus. **B** Survival curves of 30-day-old CD1 mice infected with either the E2-Gln55 virus (dsTE12Q) or the neurovirulent E2-His55 virus (dsTE12H).

a robust resistance to avirulent strains of Sindbis virus, the molecular basis of which has only recently begun to be unraveled. This age-dependent resistance to lethal infection is not unique to Sindbis virus, and has been observed with other neurotropic viruses, including other alphaviruses, bunyaviruses, flaviviruses, rhabdoviruses and reoviruses (Sigel 1952; Fenner 1968; Griffin et al. 1994; Oliver et al. 1997).

Age-dependent differences in susceptibility to Sindbis virus infection have been specifically associated with neuronal maturation in the infected host. Infections of whole animals as well as primary cultures of dissociated neurons show that the degree of cellular resistance to viral replication and virus-induced apoptosis appears to be affected by the maturation state of the target cells (Levine et al. 1993; Fazakerley and Allsopp 2001; Levine 2002). In an attempt to identify host genes involved in age-dependent resistance, Levine and co-workers undertook a systematic approach using DNA microarray analysis to compare gene expression patterns in the central nervous systems of neonatal versus weanling mice before and after Sindbis virus infection (Labrada et al. 2002). When younger and older animals were compared in the absence of infection, several genes involved in central nervous system development that display transcriptional regulation were identified or confirmed. Among them, apoptosis-related genes such as TRAF-4 and caspase-3 were found to be downregulated in older animals. These findings are consistent with a role of these factors in age-dependent resistance since both proteins are known pro-apoptotic factors. TRAF-4 is a TRAF family member that interacts with p75NTR neurotrophin receptor and caspase-3 is one of the main effector proteases that help execute the apoptotic program. Thus, it is possible that the reduction in the amounts of these pro-death molecules in brain cells as they mature contributes to their resistance to virus-induced apoptosis. Conversely, the neuroprotective chemokine, fractaline, which has been shown to regulate neuronal susceptibility to apoptosis in vitro, and apolipoprotein D, a lipocalin postulated to play a role in neuronal repair, were upregulated during postnatal development in the brain (Montpied et al. 1999; Ranz et al. 1999; Terrisse et al. 1999; Tong et al. 2000; Zujovic et al. 2000; Labrada et al. 2002). Analysis of seven different Bcl-2 family members showed no changes in postnatal developmental expression, indicating that variations in their protein levels are likely unimportant in the generation of resistance. It remains to be determined which if any of the identified differentially expressed genes plays a role in the higher susceptibility of younger animals to Sindbis virus replication, induction of apoptosis and/or fatal outcome of infection.

When a similar analysis was performed in neonatal and weanling mice that were inoculated with a non-neurovirulent SV recombinant strain (dsTE12Q) it was observed that younger mice possessed a larger number of host inflammatory genes with both altered expression and greater changes in expression levels, consistent with a deleterious immune response in the young. Interestingly, only one inflammatory gene, an expressed sequence tag similar to human ISG12, failed to follow this trend and its expression levels were significantly increased in older animals. Furthermore, when 1-day-old mice were inoculated with a recombinant dsTE12Q virus carrying the gene for the putative murine ISG12 orthologue, the animals displayed a modest but significant delay in SV-induced death compared with infections with control virus. Thus, enforced neuronal expression of ISG12 in neonatal mice protected them to some extent from fatal infection, suggesting that this protein may be one of the survival factors acquired by neurons during development (Labrada et al. 2002). When gene expression levels of 1-day-old mice before and after infection were analyzed, it was found that apolipoprotein D was dramatically increased after infection. Since increased expression of this protein was also found in uninfected older mice, it remains possible that apoD plays a role in virus resistance (Labrada et al. 2002). This approach identified novel cellular factors that may directly affect the outcome of Sindbis infection and points out a variety of other candidate regulators of age-dependent viral pathogenesis that merit further investigation.

One problem with this approach is that variations in gene expression probably occur in both neuronal and non-neuronal as well as infected and uninfected cells in the brain, and thus the direct correlation between virus-specific responses and more general secondary consequences of infection cannot be easily separated. In addition, gene regulation at the level of transcription may not necessarily correlate with protein production for all the genes analyzed. Conversely, important changes in protein expression may occur by post-transcriptional modulation or protein activities can be controlled by post-translational modifications, all changes that are missed by this approach. Nevertheless, these are powerful tools with which to explore pathogenesis.

Age-dependent differences in susceptibility to Sindbis virus infection have also been correlated with differences in the immune response mounted against the virus by young and older mice. Younger animals, which sustain higher levels of viral replication than adults, show a robust innate anti-viral immune response, including high interferon (IFN) production and elevated induction of IFN-regulated genes. A detailed analysis of the immune mechanisms that modulate host survival and viral clearance in SV-infected animals has been made by Griffin et al. (1992, 1997, 2001). The role of acquired immune responses in the development of resistance does not appear to be critical, as animals lacking B and T cells still display age-dependent susceptibility (Wesselingh et al. 1994).

Finally, the genetic background of the host can affect profoundly the outcome of infection. For instance, the closely related mouse strains BALB/cj and BALB/cByj are both susceptible to encephalitis induced by the neuroadapted SV (NSV, see above) that contains the E2 Gln55His change as well as other sequence differences. Although both mouse strains support NSV replication to an equal extent, infection of 4.5-week-old BALB/cj mice results in up to 80% mortality whereas only 20% of BALB/cByj mice die under identical experimental circumstances (Tucker et al. 1996). Studies aimed at discovering genetic determinants among mouse strains with different SV susceptibility found that mortality, paralysis and viral RNA loads are related complex traits, and that a yet unidentified quantitative trait locus on chromosome 2 designated *Nsv1* that appears to modulate the outcome of infection (Thach et al. 2001).

3.3
Viral Determinants Affecting Host Factors

Infection of mice of the same age with avirulent or virulent strains of Sindbis virus (SV) results in dramatically different outcomes. Do virulent and avirulent virus infections lead to differential transcriptional regulation of host genes in infected cells, and if so, do any of these gene products play a protective or pathologic role in response to SV infections? In an effort to answer these questions, Levine and colleagues compared the gene expression patterns from brains of adult mice infected with the avirulent dsTE12Q (the double subgenomic virus vector derived from TE12Q) and the neurovirulent dsTE12H strains (Hardwick and Levine 2000; Johnston et al. 2001). Infection with the neurovirulent dsTE12H strain induced increased expression of a larger number of host genes as well as more robust increases compared with the avirulent dsTE12Q strain, whereas the number of genes whose expression was decreased during infection and the levels of reduction were similar for both viruses. The differences in magnitude could be related to the levels of replication attained by the different viruses, since dsTE12H mean titers in mouse brains were significantly higher than those observed for dsTE12Q (Lewis et al. 1996; Johnston et al. 2001). The transcription of a number of genes coding for chemokines as well as proteins involved in IFN signaling were found elevated in brains infected with both SV strains, although the levels were consistently higher for dsTE12H. These proteins include the interferon regulatory factors IRF-1 and IRF-7, and the β-chemokines MCP-1, MCP-3 and MCP-5. Some chemokines, however, were only increased during infection with the neurovirulent strain, mainly RANTES and MIP-1β. The increase in RANTES and MCP-1 expression levels in SV-infected brains is potentially important, since both have been found to be upregulated in other neurotropic virus infections (Sasseville et al. 1996; Asensio and Campbell 1997; Lane et al. 1998; Manchester et al. 1999; Chen et al. 2000). RANTES, a candidate regulator of SV-induced neuropathogenesis, has been implicated in demyelination in the central nervous system in mouse hepatitis virus-infected animals (Lane et al. 2000). IFN-α/β/γ receptor-deficient mice have been reported to have a much higher susceptibility to SV infection than Stat-1-deficient mice, and MCP-1 was one of the few genes induced by IFN in Stat-1 knockout animals (Gil et al. 2001). Thus, MCP-1 could be one of the critical IFN-regulated genes that contribute to resistance to SV.

A few apoptotic mediators were moderately induced only in brains of dsTE12H-infected mice, including Bcl-2 family members like Bfl-1 and Mcl-1 as well as the cell death receptor TNFRp55. The expression of the putative anti-apoptotic factor peripheral benzodiazepine receptor (PBR) was significantly increased in dsTE12Q- and dsTE12H-infected brains (Johnston et al. 2001). The virus-induced upregulation of PBR was only detected in 10-day-old mice but not in newborns, in which PBR expression was undetectable by RT-PCR and immunohistochemistry. PBR is a mitochondrial protein that associates with VDAC (voltage-dependent anion channel) and ANT (adenine nucleotide transporter; McEnery et al. 1992), and displays anti-apoptotic activities in Jurkat cells (Bono et al. 1999). The induction of PBR expression has previously been reported in brains upon neurotoxic insults such as excitotoxic injury, ischemia, and chemical sympathelectomy (Bono et al. 1999). Remarkably, when 1-day-old mice were infected with a recombinant dsTE12Q virus containing the PBR gene, over 50% of the animals survived compared with the typical 100% mortality induced by control viruses. Viruses containing the PBR gene and their respective controls (i.e. those containing a copy of PBR lacking a start codon) replicated to similar levels in these experiments. Thus, enforced PBR expression during SV infection demonstrated that this protein, which is induced in infected adults but not in young animals, was able to mediate a protective host response in neonatal neurons without affecting viral replication. It is possible that the inability of newborn mice to induce the expression of PBR contributes to their increased susceptibility to lethal infection (Johnston et al. 2001).

The identification of age-dependent factors in mice that explain resistance to neurovirulent SV may prove relevant to human diseases, since a variety of neurotropic viruses that cause encephalitis in humans, including La Crosse, Japanese B encephalitis and Western equine encephalitis, also display an age-dependent severity of infection, with higher acute mortality and neurological sequelae in children compared with adults (Roos 1999; Hinson and Tyor 2001).

4 Sindbis Virus Induced Cell Death: Molecular Mechanisms Operating in Infected Cells

4.1 Classical Virus-induced Apoptosis Pathways in Mammalian Cell Lines

The first evidence that SV induces apoptosis in infected cells came from observations in cultured BHK (baby hamster kidney) cells, the cell line of choice to propagate the virus. BHK cells infected with SV display several features that are hallmarks of apoptotic programmed cell death, including active blebbing of the plasma membrane, condensed nuclear chromatin and chromosomal DNA laddering, as well as caspase activation (Levine et al. 1993; Nava et al. 1998). These apoptotic changes were later observed in a variety of cell lines upon SV infection including AT3, N18 and PC12 cells (Levine et al. 1993; Ubol et al. 1994, 1996; Rosen et al. 1995; Cheng et al. 1996; Joe et al. 1996; Nargi-Aizenman et al. 2002). Apoptosis appears to benefit virus production since overexpression of the anti-apoptotic Bcl-2 protein in AT3 cells has been shown to suppress not only cell death, but also SV replication (Levine et al. 1993). The time course of apoptotic events that take place during SV infection has been studied in detail in the neuroblastoma cell line N18. In these cells, virus production starts at approximately 5 h post-infection (h.p.i.), and the first changes in cell morphology, mainly nuclear chromatin condensation, are evident as early as 6.5 h.p.i. At 10 h.p.i., activation of the DNA repair enzyme poly (ADP ribose) polymerase (PARP) can be detected, and 6 h later its inactivation by caspase-3-mediated cleavage becomes apparent, an event that coincides with DNA fragmentation (Griffin and Hardwick 1997). Caspases play a causal role in SV-induced apoptosis as expression of caspase inhibitors CrmA, p35 and dominant-negative caspase-9 from the SV vector protect cultured cells and mice (Nava et al. 1998; Ryan et al. 2002). Fibroblasts lacking Apaf-1 are resistant to cell death triggered by SV infection, further supporting the notion that SV initiates an intrinsic pathway of caspase activation via cytochrome c release (Balachandran et al. 2000). Thus, SV infection of mammalian cells triggers signaling mechanisms that result in classical programmed cell death.

The signaling pathways utilized by SV to initiate death are still under investigation. The available data suggest that the mechanisms used may be different depending on the cellular context (Fig. 2). For instance, in rat prostate adenocarcinoma AT3 cells, addition of κB-specific transcription factor decoys

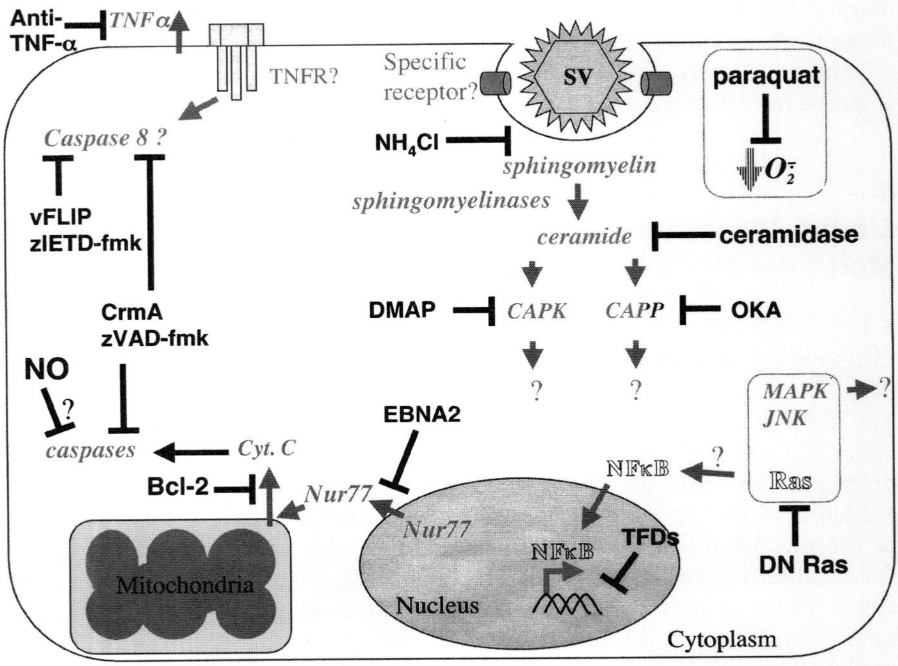

Fig. 2. Summary of molecular events reported in SV-infected cell lines of diverse origins. Note that different mechanisms may operate in different cells (see text) and various combinations of pathways might apply to diverse cellular contexts. Pathways that are confirmed or believed to be induced by translocation or activation during SV infection are shown in gray. Molecules that block SV-activated pathways and impair SV-induced cell death are shown in black. Steady-state activities of NFκB and Ras, rather than those induced by SV, appear to be required for apoptosis are depicted in embossed characters. The bent arrow indicates activation of transcription. The large striped arrow indicates SV-induced decrease of superoxide levels.

(TFDs) before infection efficiently inhibited both the transcription factor NFκB and SV-induced death, indicating that the activation of cellular genes regulated by NFκB was necessary to initiate apoptosis in this cell type (Lin et al. 1995). Steady-state NFκB activity before infection and not that induced during SV infection appears to be critical for the death process, since AT3 cells infected with a recombinant virus expressing a superrepressor form of IκB α, which resulted in complete inhibition of NFκB nuclear functions, showed similar levels of apoptosis to cells infected with control viruses (Lin et al. 1998). In contrast to that observed in AT3 cells, NFκB does not appear to play a role in SV-induced apoptosis in N18 mouse neuroblastoma cells (Lin et al. 1995). When the role of Ca^{2+}, a well-established regulator of programmed cell death, was studied in N18 cells it was found that SV not only did not induce increased intracellular Ca^{2+} levels, but that infection in the absence of extracellular Ca^{2+} resulted in an acceleration of the apoptotic process (Ubol et al. 1996).

In the neuronal cell line PC12, Ras-dependent signaling was shown to be important for SV-induced apoptosis, as expression of a dominant-negative Ras mutant (DN-Ras) delayed the onset of SV-induced apoptosis in these cells (Joe et al. 1996). The effects of DN-Ras could be related to its ability to inhibit cell cycle progression and Ras-induced expression of NFκB-responsive genes (Lin et al. 1998). Activation of p38 mitogen-activated protein kinase (MAPK) and c-Jun NH2-terminal kinases (JNKs) has been reported to precede the onset of apoptosis in SV-infected Vero cells. In this paradigm, induction of the kinases correlated with the phosphorylation of a 27-kDa heat shock protein (HSP27) and its relocalization from the cytoplasm to the perinuclear region (Nakatsue et al. 1998). Recently, Zrachia et al. (2002) have reported that infection of C6 glioma cells with a virulent strain of Sindbis (SVNI) induced a selective translocation of PKCδ to the endoplasmic reticulum and its tyrosine phosphorylation. In addition, the expression of a PKCδ kinase-dead mutant, or one lacking three autophosphorylation sites (tyrosines 52, 64 and 155) as well as preincubation with the specific PKCδ inhibitor, rottlerin, resulted in an increase of virus-induced cell death (Zrachia et al. 2002). These results suggest that PKCδ kinase exerts a negative effect on SV-induced apoptosis in glioma cells through an as yet unidentified mechanism that requires autophosphorylation at specific tyrosine residues.

A newly discovered mediator of SV-induced apoptosis is the protein Nur77, an orphan member of the nuclear hormone receptor superfamily, which localizes to the nucleus of healthy cells where it regulates gene transcription by acting alone or in combination with retinoid-X receptor (Philips et al. 1997; Katagiri et al. 2000; Li et al. 2000; Winoto and Littman 2002). Nur77 appears to have a very different function in apoptotic cells, as a variety of death stimuli induce its synthesis and translocation to the cytoplasm, where Nur77 targets the mitochondria and induces cytochrome c release (Li et al. 2000). In an unexpected finding, Hayward and colleagues discovered that, like other apoptotic stimuli, SV infection also induces Nur77 expression and mitochondrial localization. Moreover, they found that when the EBNA2 protein of Epstein-Barr virus was expressed in NIH3T3 cells during SV infection, Sindbis virus-induced cell death was dramatically reduced (Lee et al. 2002a), an activity that was dependent on EBNA2 binding to Nur77 and its retention in the nucleus of infected cells. Expression of EBNA2 mutants that neither bound nor retained Nur77 in the nucleus was an inefficient inhibitor of SV-induced death. In addition, elimination of Nur77 expression by an anti-sense strategy in NIH3T3 cells was as protective against SV-induced apoptosis as overexpression of the anti-apoptotic Bcl-xL protein (Lee et al. 2002a).

Disruption of cellular redox homeostasis can activate apoptosis. In particular, an imbalance in the redox equilibrium toward oxidant species, a condition known as "oxidative stress", has been shown to cause apoptotic cell death in culture cells whereas antioxidants can prevent it (Larrick and Wright 1990; Lennon et al. 1991; Zhong et al. 1993; Mayer and Noble 1994; Ferrari et al. 1995; Greenlund et al. 1995; Johnston et al. 2001). Free radical species such as superoxide ($O_2^{\cdot}$) and nitric oxide (NO) have been shown to negatively regulate SV-induced apoptosis (Tucker et al. 1996; Lin et al. 1999). Infection of AT3 cells with the Sindbis AR339 strain resulted in a rapid decrease in intracellular superoxide levels, whereas infection with recombinant viruses carrying the human Cu,Zn-SOD gene, an enzyme that catalyzes the conversion of $O_2^{\cdot}$ into H_2O_2, enhanced SV-induced cell death (Lin et al. 1999). Conversely, pretreatment of 3T3 fibroblasts with paraquat, a drug that increases intracellular superoxide levels, significantly inhibited SV-induced death. Recombinant SV carrying the SOD gene or viruses grown in the presence of paraquat replicate with equal efficiency as control viruses, indicating that levels of $O_2^{\cdot}$ did not modulate viral replication. Notably, the reported effects of $O_2^{\cdot}$ do not appear to be mediated by peroxide since addition to cells of pyruvate, a non-enzymatic scavenger of H_2O_2, had no effect in similar experiments (Lin et al. 1999). Thus, decreased intracellular superoxide levels enhanced SV-induced apoptosis independently of H_2O_2 and without affecting viral replication. These experiments suggest that rather than causing oxidative stress, SV infection disturbs the redox equilibrium toward a more reduced state. This type of "reductive stress" due to reduction of reactive oxygen species (ROS) has been reported in dexamethasone-induced apoptosis in thymocytes (Wang et al. 1996). Redox-active agents have been reported to modulate caspase activity (Hampton and Orrenius 1997), and under certain circumstances intracellular $O_2^{\cdot}$ has been shown to function as an endogenous inhibitor of Fas-induced apoptosis. Since SV- and Fas-induced apoptosis are both efficiently blocked by CrmA (a specific inhibitor of caspases 1, 4, 5 and 8) it is plausible that these pathways may share similar molecular components (Nava et al. 1998). Interestingly, vFLIP expression in PC12 cells was reported to block SV-induced cell death (Sarid et al. 2001). V-FLIP, encoded by human herpesvirus-8, is a member of the FLICE-inhibitory family of proteins that possesses potent anti-caspase-8 inhibitory activity. It is not clear if SV triggers the activation of a member of the TNF-receptor family during infection, but a role for the TNF-α pathway in SV virulence has been suggested by several findings. SV-infection induces the expression of TNF-α in PC12 cells, and pretreatment with soluble TNF-α receptor or anti-TNF-α antibodies markedly reduces virus-induced apoptosis. In addition, TNF-α-knockout mice showed elevated resistance to lethal infection (Sarid et al. 2001). However, experiments performed in mouse fibroblasts indicate that SV-induced apoptosis occurs in the absence of FADD, a key signal transducer of the death receptor/caspase-8 pathway (Balachandran et al. 2000). Our unpublished data are consistent with this finding.

Nitric oxide has also been shown to inhibit SV-induced apoptosis in cultured cells and reduce SV-mediated mortality in mice. NO is a gaseous molecule involved in a variety of cellular processes, including neurotransmission, vasodilation and macrophage elimination of microorganisms (Droge 2002). In the brain, NO is synthesized in neurons, microglia, macrophages and astrocytes, and its presence can be neurotoxic or neuroprotective depending on the context. For instance, in cultures of rodent primary neurons, addition of NO is toxic and inhibition of NO production reverses the neurotoxicity developed during cerebral ischemia (Bredt et al. 1991; Dawson et al. 1991). On the other hand, NO is known to block neurotransmission by reacting with the thiol groups of the NMDA receptor and decreasing calcium influx, thereby conferring neuroprotection (Lipton et al. 1993). NO has also been suggested to protect cells from apoptosis through it ability to decrease caspase activity by S-nitrosylation (Tenneti et al. 1997; Kim et al. 2002). Tucker et al. (1996) reported that 4.5-week-old BALB/cByj mice, which are generally resistant to NSV infection, could be rendered susceptible to death without affecting viral replication if treated with L-NAME, a competitive inhibitor of NO synthase. Similar results were obtained in scid/CB17 mice, indicating that L-NAME effects were independent of T- or B-cell immune response against the virus. In vitro, pretreatment of N18 neuroblastoma cells with drugs that produce NO, such as SNAP or SNP, protected cells from SV-induced apoptosis without altering viral replication. Taken together these results suggest that NO is a bona fide modulator of SV-induced cell death. NO is likely to protect mice from fatal SV-induced encephalitis by preventing neuronal cell death, thereby prolonging the survival of essential cells until the immune system initiates the clearing of the virus (Tucker et al. 1996).

A key mediator of SV-induced cell death is ceramide. This molecule has been implicated in a variety of cell death pathways, including those activated by TNF-α, Fas ligand, neurotrophins, ionizing radiation, cytokines, heat shock, UV light, antitumor drugs, and oxidative stress (Cifone et al. 1994; Hannun and Obeid 1995; Santana et al. 1996; Verheij et al. 1996; Dobrowsky and Carter 1998). Like other alphaviruses, Sindbis virus fusion occurs in late endosomes and requires an acid-induced conformational change in the heterodimer formed by the structural glycoproteins E1 and E2, as well as a target membrane containing sphingomyelin (Wahlberg and Garoff 1992; Bron et al. 1993; Nieva et al. 1994; Lu et al. 1999; Smit et al. 1999, 2001; Phinney et al. 2000). NSV infection of N18 cells induces a rapid activation of the membrane-associated acidic sphingomyelinase (aSMase), as well as a delayed activation of cytosolic neutral sphingomyelinase (nSMase; Jan et al. 2000). These events, which can be blocked by the inhibitor of endosomal acidification and viral fusion, NH_4-Cl, result in significant and prolonged increases of intracellular ceramide in infected cells. Interestingly, binding of UV-inactivated NSV to the cell membrane and exposure to low pH also triggered the generation of ceramide. Thus, viral glycoprotein-mediated fusion during viral entry might be sufficient to activate sphingomyelinases. Notably, recombinant Sindbis viruses expressing acid ceramidase, an enzyme that decreased the levels of ceramide, profoundly delayed virus-induced apoptosis in N18 cells. Also, treatment of N18 cells with C2-ceramide, a cell permeable analog of ceramide, mimicked SV-induced cell death, at least in part, by efficiently inducing apoptosis in a zVAD-fmk- and Bcl-2-repressible manner (Jan et al. 2000). Intracellular ceramide transients are known to directly activate the ceramide-activated protein kinase (CAPK) and the ceramide-activated protein phosphatase (CAPP), resulting in broad changes in cellular protein phosphorylation (Dobrowsky and Hannun 1993; Liu et al. 1994). Remarkably, inhibitors of both CAPK and CAPP (DMAP-dimethylaminopurine and okadaic acid, respectively) efficiently block SV-induced apoptosis in N18 cells, suggesting that the signaling cascade that generates ceramide in infected cells requires changes in protein phosphorylation to exert its deadly effects.

In summary, SV infection of cultured cells appears to activate a plethora of signaling pathways depending on the cell line and strain of virus. The cell specificity of many of the reported effects and the sometimes contradictory results obtained using different dividing cell lines illustrate the limitations of extrapolating this information to neurons, the primary in vivo target of viral replication.

4.2 Sindbis Virus-Induced Cell Death in Neurons: Apoptotic Versus Non-apoptotic

Sindbis virus neurovirulence was assumed to be mainly determined by the ability of the virus to induce infected neurons in the central nervous system to undergo programmed cell death (Levine et al. 1996; Lewis et al. 1996). Specific subsets of neurons responsible for mortality in mice have not been clearly identified but presumably include neurons in the brain stem that regulate vital functions. Though there is little doubt that SV-infected neurons die by programmed cell death, the morphologies of different neuron subtypes vary greatly. Apoptotic DNA fragmentation can be readily observed in brain extracts of infected mice a few hours after infection and TUNEL staining is clearly associated with regions of viral replication in infected brain tissue (Lewis et al. 1996). Significantly, endogenous and overexpressed anti-death Bcl-2 family members are very effective at protecting newborn mice from lethal infection with SV, indicating that obstruction of the apoptotic program results in increased survival of the infected host (Levine et al. 1996). The paralysis observed in adult mice infected with neurovirulent strains of SV results from the spread of SV infection to spinal cord motor neurons. These motor neurons degenerate and die following SV infection but apparently by a non-apoptotic (but programmed) mechanism, as these cells do not show characteristic apoptotic morphology in infected animals, consistent with spinal cord morphologies described by others. (Havert et al. 2000). Furthermore, Kerr et al. (2002) showed that when a recombinant neurovirulent Sindbis virus (dsNSV) engineered to express protective Bcl-2 proteins including Bcl-2 was inoculated into 4-week-old mice, these animals survived the lethal infection better than those infected with control viruses, indicating protection of neurons (probably in the brain stem) by Bcl-2. This report demonstrated that the correlation between reduction in neuronal apoptosis and improved survival, as previously observed in newborn mice, could be recapitulated in 4-week-old animals infected with neurovirulent NSV. In contrast, the expression of anti-apoptotic Bcl-2 family proteins during infection failed to prevent degeneration and death of motor neurons in the lumbar spinal cord. Consistent with this finding, Bcl-2 proteins also failed to protect mice from hind-limb paralysis, indicating that these cells die by a distinct non-apoptotic pathway. Although Bcl-2 has been shown to inhibit necrotic death as well as apoptotic cell death, the death pathway in SV-infected spinal cord motor neurons is independent of Bcl-2. A similar outcome was observed for the role of Bax, generally considered to be a pro-death Bcl-2 family member. When Bax knockout mice were infected with the neurovirulent dsNSV, they showed significantly higher mortality rates than their wild-type littermates, indicating that Bax promotes animal survival in this experimental paradigm (see below), while both groups developed paralysis to the same extent indicating that Bax does not protect spinal cord motor neurons (Kerr et al. 2002). It has been postulated that SV infection might trigger an excitatory death pathway that could result in damage not only of infected cells, but also of bystander uninfected neurons, a notion that is supported by the discovery that antagonists of the N-methyl-D-aspartate (NMDA) subtype of glutamate receptors are able to exert a small but significant protection against SV-induced death in neurons (Nargi-Aizenman and Griffin 2001; Kerr et al. 2002). Thus, Sindbis virus infection in mice appears to trigger diverse cell death pathways in different neuronal populations, and, whereas some are clearly apoptotic and affect survival, others are non-apoptotic and remain to be fully understood.

5 Sindbis Virus as a Molecular Trojan Horse to Carry Heterologous Genes into the Killing Fields

Throughout this chapter we have given several examples in which exogenous genes were cloned into the Sindbis virus genome and tested for their potential apoptotic-modulating function. The recombinant double subgenomic Sindbis virus vector system (Hardwick and Levine 2000) has greatly aided the identification of pro-death and pro-survival proteins that affect neuronal cell death and disease, including Bcl-2, CrmA, P35, IAPs, Aven, Beclin, SMN, PBR, Bax, and Bak (Cheng et al. 1996; Liang et al. 1998; Lewis et al. 1999; Chau et al. 2000; Kerr et al. 2000; Johnston et al. 2001; Ryan et al. 2002; Fannjiang et al. 2003). This system also proved useful for mapping apoptotic determinants in known proteins that are important for neuron-specific cell death.

Sindbis-based viral vectors that cause age-dependent neuronal death provide a unique opportunity to study proteins involved in neuronal apoptosis, which are likely to display neuron-specific functions along with the normal alterations of these functions during CNS development. In the case of the Bcl-2 family of apoptotic regulators, discoveries made using the Sindbis system have challenged the "dogmatic view" of this protein family, in which some members were believed to be either entirely pro- or anti-apoptotic. Two examples in which the Sindbis-expression system has unraveled pro-survival functions in "classical" pro-death molecules are given below.

Bax is generally believed to function as a pro-apoptotic protein in diverse cell-death paradigms, although some data suggest that Bax could also have protective functions in some neuronal-derived cells (Middleton et al. 1996; Zhou et al. 1998; Middleton and Davies 2001). Work in our laboratory has shown that infection of newborn mice with control recombinant dsTE12Q viruses carrying the Bax gene cloned in a reverse orientation (anti-sense) resulted in 20% survival, whereas similar infections with viruses carrying the Bax gene cloned in a forward orientation (expression competent) resulted in 80% viability, indicating that expression of Bax during infection dramatically reduced virus-induced mortality (Lewis et al. 1999). The protective effects of Bax were even greater than those observed with Bcl-2, as only 60% of the animals infected with a virus carrying the Bcl-2 gene survived in these experiments. At the cellular level, the ability of Bax to protect neurons from SV-induced death depended on the neuronal subtype analyzed, with Bax displaying anti-apoptotic functions in hippocampal cells of organotypic cultures but pro-apoptotic effects in dissociated dorsal root ganglia neurons. In cultured cell lines, however, Bax-expressing viruses were consistently potent inducers of apoptosis, emphasizing the cell type specificity of the protective effect (Cheng et al. 1996). Recently, it was reported that Bax also protected older mice from lethal infection induced by recombinant neurovirulent Sindbis virus (dsNSV), and here again Bax protected hippocampal neurons but not other subtypes, such as lumbar motor neurons (Kerr et al. 2002). Consequently, Bax was able to increase survival of older animals but did not ameliorate virus-induced paralysis. The protective effects of Bax in this paradigm were confirmed using Bax-deficient mice infected with dsTE12Q or dsNSV, as animals lacking Bax were more susceptible to lethal infection. The use of recombinant Sindbis viruses clearly identified Bax anti-apoptotic functions in a neuronal disease state, demonstrating that this "pro-death" molecule was indeed a potent survival factor in specific subsets of virus-infected neurons under physiologically relevant conditions.

Bak is another classical "killer" Bcl-2 family member, shown to exert pro-apoptotic activities in a variety of cell culture systems (Chittenden et al. 1995; Orth and Dixit 1997; Griffiths et al. 1999; Lee et al. 1999; Wei et al. 2001). However, anti-death activities have also been assigned to Bak under certain circumstances (Kiefer et al. 1995). To investigate the functional role of Bak in pathological neuronal death, our laboratory constructed recombinant dsTE12Q viruses expressing Bak (dsTE12Q-Bak), and tested their lethality in newborn mice. We found that dsTE12Q-Bak viruses were much less efficient at killing newborn mice than control viruses (80 and 40% survival, respectively), indicating that Bak, like Bax, was a potent protective factor in SV-induced death (Fannjiang et al. 2003). dsTE12Q-Bak and control viruses (i.e.

viruses carrying a copy of the Bak gene cloned in an anti-sense orientation) replicated to equal extents in these assays. When similar experiments were performed in tissues prepared from newborn mice, Bak was very efficient at protecting hippocampal and spinal cord neurons from virus-induced apoptosis. However, viruses expressing Bak were more powerful killers than controls when tested in several immortalized cell lines, demonstrating that, as in the case of Bax, the protective effects of Bak were also cell-type-specific. The results obtained in the animal model were confirmed by using Sindbis virus dsTE12Q to infect Bak knockout mice, which showed an increased mortality compared with similarly infected wild-type littermates. These results were also recapitulated in vitro, where hippocampal and spinal cord neurons derived from Bak knockout mice were more susceptible to SV-induced cell death than wild-type cultures. Bak deficiency was shown to be the sole contributor to the reduced survival in these experiments, as the observed effects were reverted when Bak knockout mice or cultures derived from these animals were infected with a recombinant dsTE12Q virus carrying a copy of wild-type Bak. In contrast, infection of older mice with a recombinant neuroadapted dsNSV expressing Bak resulted in higher mortality than infection with control dsNSV viruses, indicating that, as animals aged, Bak switched to a pro-apoptotic function. In vitro, Bak protected hippocampal neurons of older mice but increased cell death in spinal cord neurons. These results indicate that, during postnatal development, Bak is converted from an anti-apoptotic to a pro-death factor in spinal cord neurons, but not in hippocampal neurons.

The discoveries made with Bax and Bak exemplify the advantages of the Sindbis virus system to study neuronal cell death at different stages of mammalian development. The understanding of these differences displayed by Bcl-2 family members during virus-induced apoptosis in neurons of diverse origin and age may prove critical if in the future we are to consider treating diseases like encephalitis through pharmacological manipulation.

6 Summary

Sindbis virus infects neurons of the brain and spinal cord leading to neuronal apoptosis and encephalitis in mice. During postnatal development, neurons of mice remain susceptible to infection but become refractory to SV-induced programmed cell death. Failure to undergo programmed cell death results in a persistent infection. However, some neurovirulent strains of Sindbis virus overcome the age-dependent protective function in neurons, leading to enhanced apoptotic cell death in the central nervous system and higher mortality rates. Sindbis virus infections can also cause hind-limb paralysis due to the death of infected spinal cord motor neurons. However, spinal cord neuron death in older mice appears to occur by mechanisms that differ from classical apoptosis observed in newborn mice based on the morphology of dying neurons at these two sites. Sindbis virus infections of mosquitoes and some mosquito cell lines, on the other hand, do not induce cell death but persistent infections, a phenomenon also observed occasionally in cultured mammalian cells as well as in brains of infected mice surviving lethal infections. Thus, both viral and cellular factors contribute to the varied outcomes of infection. The molecular mechanisms that govern the susceptibility or resistance of particular cell types to SV-induced cell death are not well understood. Furthermore, the cellular execution machinery that produces the characteristic morphological distinctions between brain and spinal cord (i.e. apoptotic versus non-apoptotic) remain to be discovered.

Acknowledgements. The work from our laboratory described here was supported by the D'Arbeloff Fellowship in Biological Sciences (PI) and the National Institutes of Health grant NS34175 (JMH).

References

Asensio VC, Campbell IL (1997) Chemokine gene expression in the brains of mice with lymphocytic choriomeningitis. J Virol 71:7832–7840

Balachandran S, Roberts PC, Kipperman T, Bhalla KN, Compans RW, Archer DR, Barber GN (2000) Alpha/beta interferons potentiate virus-induced apoptosis through activation of the FADD/caspase-8 death signaling pathway. J Virol 74:1513–1523

Bono F, Lamarche I, Prabonnaud V, Le Fur G, Herbert JM (1999) Peripheral benzodiazepine receptor agonists exhibit potent antiapoptotic activities. Biochem Biophys Res Commun 265:457–461

Bredt DS, Glatt CE, Hwang PM, Fotuhi M, Dawson TM, Snyder SH (1991) Nitric oxide synthase protein and mRNA are discretely localized in neuronal populations of the mammalian CNS together with NADPH diaphorase. Neuron 7:615–624

Bron R, Wahlberg JM, Garoff H, Wilschut J (1993) Membrane fusion of Semliki Forest virus in a model system: correlation between fusion kinetics and structural changes in the envelope glycoprotein. EMBO J 12:693–701

Chau BN, Cheng EH, Kerr DA, Hardwick JM (2000) Aven, a novel inhibitor of caspase activation, binds Bcl-xL and Apaf-1. Mol Cell 6:31–40

Chen CJ, Liao SL, Kuo MD, Wang YM (2000) Astrocytic alteration induced by Japanese encephalitis virus infection. Neuroreport 11:1933–1937

Cheng EH, Levine B, Boise LH, Thompson CB, Hardwick JM (1996) Bax-independent inhibition of apoptosis by Bcl-XL. Nature 379:554–556

Chittenden T, Harrington EA, O'Connor R, Flemington C, Lutz RJ, Evan GI, Guild BC (1995) Induction of apoptosis by the Bcl-2 homologue Bak. Nature 374:733–736

Cifone MG, De Maria R, Roncaioli P, Rippo MR, Azuma M, Lanier LL, Santoni A, Testi R (1994) Apoptotic signaling through CD95 (Fas/Apo-1) activates an acidic sphingomyelinase. J Exp Med 180:1547–1552

Dawson VL, Dawson TM, London ED, Bredt DS, Snyder SH (1991) Nitric oxide mediates glutamate neurotoxicity in primary cortical cultures. Proc Natl Acad Sci USA 88:6368–6371

Dobrowsky RT, Carter BD (1998) Coupling of the p75 neurotrophin receptor to sphingolipid signaling. Ann NY Acad Sci 845:32–45

Dobrowsky RT, Hannun YA (1993) Ceramide-activated protein phosphatase: partial purification and relationship to protein phosphatase 2A. Adv Lipid Res 25:91–104

Droge W (2002) Free radicals in the physiological control of cell function. Physiol Rev 82:47–95

Fannjiang Y, Kim C-H, Huganir RL, Zou S, Lindsten T, Thompson BB, Mito T, Traystman RJ, Larsen T, Griffin DE, Mandir AS, Dawson TM, Dike S, Sappington AL, Kerr DA, Jonas EA, Kaczmarek LK, Hardwick JM (2003) Bak alters neuronal excitability and can switch from anti- to pro-death function during postnatal development. Dev Cell 4:575–585

Fazakerley JK, Allsopp TE (2001) Programmed cell death in virus infections of the nervous system. Curr Top Microbiol Immunol 253:95–119

Fenner FJ (1968) The pathogenesis of viral infections: the influence of age on resistance to viral infections, vol 2. Academic Press, New York

Ferrari G, Yan CY, Greene LA (1995) N-Acetylcysteine (D- and L-stereoisomers) prevents apoptotic death of neuronal cells. J Neurosci 15:2857–2866

Frolov I, Schlesinger S (1994) Comparison of the effects of Sindbis virus and Sindbis virus replicons on host cell protein synthesis and cytopathogenicity in BHK cells. J Virol 68:1721–1727

Gil MP, Bohn E, O'Guin AK, Ramana CV, Levine B, Stark GR, Virgin HW, Schreiber RD (2001) Biologic consequences of Stat1-independent IFN signaling. Proc Natl Acad Sci USA 98:6680–6685

Greenlund LJ, Deckwerth TL, Johnson EM Jr (1995) Superoxide dismutase delays neuronal apoptosis: a role for reactive oxygen species in programmed neuronal death. Neuron 14:303–315

Griffin DE, Hardwick JM (1997) Regulators of apoptosis on the road to persistent alphavirus infection. Annu Rev Microbiol 51:565–592

Griffin DE, Levine B, Tyor WR, Irani DN (1992) The immune response in viral encephalitis. Semin Immunol 4:111–119

Griffin DE, Levine B, Tyor WR, Tucker PC, Hardwick JM (1994) Age-dependent susceptibility to fatal encephalitis: alphavirus infection of neurons. Arch Virol Suppl 9:31–39

Griffin DE, Levine B, Tyor W, Ubol S, Despres P (1997) The role of antibody in recovery from alphavirus encephalitis. Immunol Rev 159:155–161

Griffin DE, Ubol S, Despres P, Kimura T, Byrnes A (2001) Role of antibodies in controlling alphavirus infection of neurons. Curr Top Microbiol Immunol 260:191–200

Griffiths GJ, Dubrez L, Morgan CP, Jones NA, Whitehouse J, Corfe BM, Dive C, Hickman JA (1999) Cell damage-induced conformational changes of the pro-apoptotic protein Bak in vivo precede the onset of apoptosis. J Cell Biol 144:903–914

Hampton MB, Orrenius S (1997) Dual regulation of caspase activity by hydrogen peroxide: implications for apoptosis. FEBS Lett 414:552–556

Hannun YA, Obeid LM (1995) Ceramide: an intracellular signal for apoptosis. Trends Biochem Sci 20:73–77

Hardwick JM, Levine B (2000) Sindbis virus vector system for functional analysis of apoptosis regulators. Methods Enzymol 322:492–508

Havert MB, Schofield B, Griffin DE, Irani DN (2000) Activation of divergent neuronal cell death pathways in different target cell populations during neuroadapted sindbis virus infection of mice. J Virol 74:5352–5356

Hinson VK, Tyor WR (2001) Update on viral encephalitis. Curr Opin Neurol 14:369–374

Igarashi A, Koo R, Stollar V (1977) Evolution and properties of *Aedes albopictus* cell cultures persistently infected with sindbis virus. Virology 82:69–83

Inglot AD, Albin M, Chudzio T (1973) Persistent infection of mouse cells with Sindbis virus: role of virulence of strains, auto-interfering particles and interferon. J Gen Virol 20:105–110

Jackson AC, Moench TR, Griffin DE, Johnson RT (1987) The pathogenesis of spinal cord involvement in the encephalomyelitis of mice caused by neuroadapted Sindbis virus infection. Lab Invest 56:418–423

Jackson AC, Moench TR, Trapp BD, Griffin DE (1988) Basis of neurovirulence in Sindbis virus encephalomyelitis of mice. Lab Invest 58:503–509

Jan JT, Griffin DE (1999) Induction of apoptosis by Sindbis virus occurs at cell entry and does not require virus replication. J Virol 73:10296–10302

Jan JT, Chatterjee S, Griffin DE (2000) Sindbis virus entry into cells triggers apoptosis by activating sphingomyelinase, leading to the release of ceramide. J Virol 74:6425–6432

Joe AK, Ferrari G, Jiang HH, Liang XH, Levine B (1996) Dominant inhibitory Ras delays Sindbis virus-induced apoptosis in neuronal cells. J Virol 70:7744–7751
Joe AK, Foo HH, Kleeman L, Levine B (1998) The transmembrane domains of Sindbis virus envelope glycoproteins induce cell death. J Virol 72:3935–3943
Johnston C, Jiang W, Chu T, Levine B (2001) Identification of genes involved in the host response to neurovirulent alphavirus infection. J Virol 75:10431–10445
Karpf AR, Blake JM, Brown DT (1997) Characterization of the infection of *Aedes albopictus* cell clones by Sindbis virus. Virus Res 50:1–13
Karpf AR, Brown DT (1998) Comparison of Sindbis virus-induced pathology in mosquito and vertebrate cell cultures. Virology 240:193–201
Katagiri Y, Takeda K, Yu ZX, Ferrans VJ, Ozato K, Guroff G (2000) Modulation of retinoid signalling through NGF-induced nuclear export of NGFI-B. Nat Cell Biol 2:435–440
Kerr DA, Nery JP, Traystman RJ, Chau BN, Hardwick JM (2000) Survival motor neuron protein modulates neuron-specific apoptosis. Proc Natl Acad Sci USA 97:13312–13317
Kerr DA, Larsen T, Cook SH, Fannjiang YR, Choi E, Griffin DE, Hardwick JM, Irani DN (2002) BCL-2 and Bax protect adult mice from lethal Sindbis virus infection but do not protect spinal cord motor neurons or prevent paralysis. J Virol 76:10393–10400
Kiefer MC, Brauer MJ, Powers VC, Wu JJ, Umansky SR, Tomei LD, Barr PJ (1995) Modulation of apoptosis by the widely distributed Bcl-2 homologue Bak. Nature 374:736–739
Kim PK, Kwon YG, Chung HT, Kim YM (2002) Regulation of caspases by nitric oxide. Ann NY Acad Sci 962:42–52
Labrada L, Liang XH, Zheng W, Johnston C, Levine B (2002) Age-dependent resistance to lethal alphavirus encephalitis in mice: analysis of gene expression in the central nervous system and identification of a novel interferon-inducible protective gene, mouse ISG12. J Virol 76:11688–11703
Lane TE, Asensio VC, Yu N, Paoletti AD, Campbell IL, Buchmeier MJ (1998) Dynamic regulation of alpha- and beta-chemokine expression in the central nervous system during mouse hepatitis virus-induced demyelinating disease. J Immunol 160:970–978
Lane TE, Liu MT, Chen BP, Asensio VC, Samawi RM, Paoletti AD, Campbell IL, Kunkel SL, Fox HS, Buchmeier MJ (2000) A central role for CD4(+) T cells and RANTES in virus-induced central nervous system inflammation and demyelination. J Virol 74:1415–1424
Larrick JW, Wright SC (1990) Cytotoxic mechanism of tumor necrosis factor-alpha. FASEB J 4:3215–3223
Lee JM, Zipfel GJ, Choi DW (1999) The changing landscape of ischaemic brain injury mechanisms. Nature 399:A7–14
Lee JM, Lee KH, Weidner M, Osborne BA, Hayward SD (2002a) Epstein-Barr virus EBNA2 blocks Nur77-mediated apoptosis. Proc Natl Acad Sci USA 99:11878–11883
Lee P, Knight R, Smit JM, Wilschut J, Griffin DE (2002b) A single mutation in the E2 glycoprotein important for neurovirulence influences binding of Sindbis virus to neuroblastoma cells. J Virol 76:6302–6310
Lennon SV, Martin SJ, Cotter TG (1991) Dose-dependent induction of apoptosis in human tumour cell lines by widely diverging stimuli. Cell Prolif 24:203–214
Levine B (2002) Apoptosis in viral infections of neurons: a protective or pathologic host response? Curr Top Microbiol Immunol 265:95–118
Levine B, Huang Q, Isaacs JT, Reed JC, Griffin DE, Hardwick JM (1993) Conversion of lytic to persistent alphavirus infection by the bcl-2 cellular oncogene. Nature 361:739–742
Levine B, Goldman JE, Jiang HH, Griffin DE, Hardwick JM (1996) Bc1-2 protects mice against fatal alphavirus encephalitis. Proc Natl Acad Sci USA 93:4810–4815
Lewis J, Wesselingh SL, Griffin DE, Hardwick JM (1996) Alphavirus-induced apoptosis in mouse brains correlates with neurovirulence. J Virol 70:1828–1835
Lewis J, Oyler GA, Ueno K, Fannjiang YR, Chau BN, Vornov J, Korsmeyer SJ, Zou S, Hardwick JM (1999) Inhibition of virus-induced neuronal apoptosis by Bax. Nat Med 5:832–835

Li H, Kolluri SK, Gu J, Dawson MI, Cao X, Hobbs PD, Lin B, Chen G, Lu J, Lin F, Xie Z, Fontana JA, Reed JC, Zhang X (2000) Cytochrome c release and apoptosis induced by mitochondrial targeting of nuclear orphan receptor TR3. Science 289:1159–1164

Liang XH, Kleeman LK, Jiang HH, Gordon G, Goldman JE, Berry G, Herman B, Levine B (1998) Protection against fatal Sindbis virus encephalitis by beclin, a novel Bcl-2-interacting protein. J Virol 72:8586–8596

Lin KI, Lee SH, Narayanan R, Baraban JM, Hardwick JM, Ratan RR (1995) Thiol agents and Bcl-2 identify an alphavirus-induced apoptotic pathway that requires activation of the transcription factor NF-kappa B. J Cell Biol 131:1149–1161

Lin KI, DiDonato JA, Hoffmann A, Hardwick JM, Ratan RR (1998) Suppression of steady-state, but not stimulus-induced, NF-kappaB activity inhibits alphavirus-induced apoptosis. J Cell Biol 141:1479–1487

Lin KI, Pasinelli P, Brown RH, Hardwick JM, Ratan RR (1999) Decreased intracellular superoxide levels activate Sindbis virus-induced apoptosis. J Biol Chem 274:13650–13655

Lipton SA, Choi YB, Pan ZH, Lei SZ, Chen HS, Sucher NJ, Loscalzo J, Singel DJ, Stamler JS (1993) A redox-based mechanism for the neuroprotective and neurodestructive effects of nitric oxide and related nitroso-compounds. Nature 364:626–632

Liu J, Mathias S, Yang Z, Kolesnick RN (1994) Renaturation and tumor necrosis factor-alpha stimulation of a 97-kDa ceramide-activated protein kinase. J Biol Chem 269:3047–3052

Lopez S, Yao JS, Kuhn RJ, Strauss EG, Strauss JH (1994) Nucleocapsid-glycoprotein interactions required for assembly of alphaviruses. J Virol 68:1316–1323

Lu YE, Cassese T, Kielian M (1999) The cholesterol requirement for sindbis virus entry and exit and characterization of a spike protein region involved in cholesterol dependence. J Virol 73:4272–4278

Lustig S, Jackson AC, Hahn CS, Griffin DE, Strauss EG, Strauss JH (1988) Molecular basis of Sindbis virus neurovirulence in mice. J Virol 62:2329–2336

Manchester M, Eto DS, Oldstone MB (1999) Characterization of the inflammatory response during acute measles encephalitis in NSE-CD46 transgenic mice. J Neuroimmunol 96:207–217

Mayer M, Noble M (1994) N-Acetyl-L-cysteine is a pluripotent protector against cell death and enhancer of trophic factor-mediated cell survival in vitro. Proc Natl Acad Sci USA 91:7496–7500

McEnery MW, Snowman AM, Trifiletti RR, Snyder SH (1992) Isolation of the mitochondrial benzodiazepine receptor: association with the voltage-dependent anion channel and the adenine nucleotide carrier. Proc Natl Acad Sci USA 89:3170–3174

Middleton G, Davies AM (2001) Populations of NGF-dependent neurones differ in their requirement for Bax to undergo apoptosis in the absence of NGF/TrkA signalling in vivo. Development 128:4715–4728

Middleton G, Nunez G, Davies AM (1996) Bax promotes neuronal survival and antagonises the survival effects of neurotrophic factors. Development 122:695–701

Montpied P, deBock F, Lerner-Natoli M, Bockaert J, Rondouin G (1999) Hippocampal alterations of apolipoprotein E and D mRNA levels in vivo and in vitro following kainate excitotoxicity. Epilepsy Res 35:135–146

Nakatsue T, Katoh I, Nakamura S, Takahashi Y, Ikawa Y, Yoshinaka Y (1998) Acute infection of Sindbis virus induces phosphorylation and intracellular translocation of small heat shock protein HSP27 and activation of p38 MAP kinase signaling pathway. Biochem Biophys Res Commun 253:59–64

Nargi-Aizenman JL, Griffin DE (2001) Sindbis virus-induced neuronal death is both necrotic and apoptotic and is ameliorated by N-methyl-D-aspartate receptor antagonists. J Virol 75:7114–7121

Nargi-Aizenman JL, Simbulan-Rosenthal CM, Kelly TA, Smulson ME, Griffin DE (2002) Rapid activation of poly(ADP-ribose) polymerase contributes to Sindbis virus and staurosporine-induced apoptotic cell death. Virology 293:164–171

Nava VE, Rosen A, Veliuona MA, Clem RJ, Levine B, Hardwick JM (1998) Sindbis virus induces apoptosis through a caspase-dependent, CrmA-sensitive pathway. J Virol 72:452–459
Nieva JL, Bron R, Corver J, Wilschut J (1994) Membrane fusion of Semliki Forest virus requires sphingolipids in the target membrane. EMBO J 13:2797–2804
Oliver KR, Scallan MF, Dyson H, Fazakerley JK (1997) Susceptibility to a neurotropic virus and its changing distribution in the developing brain is a function of CNS maturity. J Neurovirol 3:38–48
Orth K, Dixit VM (1997) Bik and Bak induce apoptosis downstream of CrmA but upstream of inhibitor of apoptosis. J Biol Chem 272:8841–8844
Philips A, Lesage S, Gingras R, Maira MH, Gauthier Y, Hugo P, Drouin J (1997) Novel dimeric Nur77 signaling mechanism in endocrine and lymphoid cells. Mol Cell Biol 17: 5946–5951
Phinney BS, Blackburn K, Brown DT (2000) The surface conformation of Sindbis virus glycoproteins E1 and E2 at neutral and low pH, as determined by mass spectrometry-based mapping. J Virol 74:5667–5678
Ranz G, Reindl M, Beer SCPR, Unterrichter I, Berger T, Schmutzhaard E, Poewe W, Kampfl A (1999) Increased expression of apolipoprotein D following experimental traumatic brain injury. J Neurochem 73:1615–1625
Rice CM, Bell JR, Hunkapiller MW, Strauss EG, Strauss JH (1982) Isolation and characterization of the hydrophobic COOH-terminal domains of the sindbis virion glycoproteins. J Mol Biol 154:355–378
Roos KL (1999) Encephalitis. Neurol Clin 17:813–833
Rosen A, Casciola-Rosen L, Ahearn J (1995) Novel packages of viral and self-antigens are generated during apoptosis. J Exp Med 181:1557–1561
Ryan CA, Stennicke HR, Nava VE, Burch JB, Hardwick JM, Salvesen GS (2002) Inhibitor specificity of recombinant and endogenous caspase-9. Biochem J 366:595–601
Santana P, Pena LA, Haimovitz-Friedman A, Martin S, Green D, McLoughlin M, Cordon-Cardo C, Schuchman EH, Fuks Z, Kolesnick R (1996) Acid sphingomyelinase-deficient human lymphoblasts and mice are defective in radiation-induced apoptosis. Cell 86:189–199
Sarid R, Ben-Moshe T, Kazimirsky G, Weisberg S, Appel E, Kobiler D, Lustig S, Brodie C (2001) vFLIP protects PC-12 cells from apoptosis induced by Sindbis virus: implications for the role of TNF-alpha. Cell Death Differ 8:1224–1231
Sasseville VG, Smith MM, Mackay CR, Pauley DR, Mansfield KG, Ringler DJ, Lackner AA (1996) Chemokine expression in simian immunodeficiency virus-induced AIDS encephalitis. Am J Pathol 149:1459–1467
Sawicki DL, Silverman RH, Williams BR, Sawicki SG (2003) Alphavirus minus-strand synthesis and persistence in mouse embryo fibroblasts derived from mice lacking RNase L and protein kinase R. J Virol 77:1801–1811
Sigel MM (1952) Influence of age on susceptibility to virus infections with particular reference to laboratory animals. Annu Rev Microbiol 6:247–280
Smit JM, Bittman R, Wilschut J (1999) Low-pH-dependent fusion of Sindbis virus with receptor-free cholesterol- and sphingolipid-containing liposomes. J Virol 73:8476–8484
Smit JM, Klimstra WB, Ryman KD, Bittman R, Johnston RE, Wilschut J (2001) PE2 cleavage mutants of Sindbis virus: correlation between viral infectivity and pH-dependent membrane fusion activation of the spike heterodimer. J Virol 75:11196–11204
Strauss JH, Strauss EG (1994) The alphaviruses: gene expression, replication, and evolution. Microbiol Rev 58:491–562
Tenneti L, D'Emilia DM, Lipton SA (1997) Suppression of neuronal apoptosis by S-nitrosylation of caspases. Neurosci Lett 236:139–142
Terrisse L, Sequin D, Bertrand P, Poirier J, Milne R, Rassart E (1999) Modulation of apolipoprotein D and apolipoprotein E expression in rat hippocampus after entorhinal cortex lesion. Brain Res Mol Brain Res 70:26–35
Thach DC, Kleeberger SR, Tucker PC, Griffin DE (2001) Genetic control of neuroadapted sindbis virus replication in female mice maps to chromosome 2 and associates with paralysis and mortality. J Virol 75:8674–8680

Tong N, Perry SW, Zhang Q, James HJ, Guo H, Brooks A, Bal H, Kinnear SA, Fine S, Epstein LG, Dairaghi D, Schall TJ, Gendelman HE, Dewhurst S, Sharer LR, Gelbard HA (2000) Neuronal fractalkine expression in HIV-1 encephalitis: roles for macrophage recruitment and neuroprotein in the central nervous system. J Immunol 164:1333–1339

Tucker PC, Griffin DE, Choi S, Bui N, Wesselingh S (1996) Inhibition of nitric oxide synthesis increases mortality in Sindbis virus encephalitis. J Virol 70:3972–3977

Ubol S, Tucker PC, Griffin DE, Hardwick JM (1994) Neurovirulent strains of alphavirus induce apoptosis in bcl-2-expressing cells: role of a single amino acid change in the E2 glycoprotein. Proc Natl Acad Sci USA 91:5202–5206

Ubol S, Park S, Budihardjo I, Desnoyers S, Montrose MH, Poirier GG, Kaufmann SH, Griffin DE (1996) Temporal changes in chromatin, intracellular calcium, and poly(ADP-ribose) polymerase during Sindbis virus-induced apoptosis of neuroblastoma cells. J Virol 70:2215–2220

Verheij M, Bose R, Lin XH, Yao B, Jarvis WD, Grant S, Birrer MJ, Szabo E, Zon LI, Kyriakis JM, Haimovitz-Friedman A, Fuks Z, Kolesnick RN (1996) Requirement for ceramide-initiated SAPK/JNK signalling in stress-induced apoptosis. Nature 380:75–79

Wahlberg JM, Garoff H (1992) Membrane fusion process of Semliki Forest virus. I. Low pH-induced rearrangement in spike protein quaternary structure precedes virus penetration into cells. J Cell Biol 116:339–348

Wang JF, Jerrells TR, Spitzer JJ (1996) Decreased production of reactive oxygen intermediates is an early event during in vitro apoptosis of rat thymocytes. Free Radic Biol Med 20:533–542

Wei MC, Zong WX, Cheng EH, Lindsten T, Panoutsakopoulou V, Ross AJ, Roth KA, MacGregor GR, Thompson CB, Korsmeyer SJ (2001) Proapoptotic Bax and Bak: a requisite gateway to mitochondrial dysfunction and death. Science 292:727–730

Wesselingh SL, Levine B, Fox RJ, Choi S, Griffin DE (1994) Intracerebral cytokine mRNA expression during fatal and nonfatal alphavirus encephalitis suggests a predominant type 2 T cell response. J Immunol 152:1289–1297

Winoto A, Littman DR (2002) Nuclear hormone receptors in T lymphocytes. Cell 109:S57–66

Zhang W, Mukhopadhyay S, Pletnev SV, Baker TS, Kuhn RJ, Rossmann MG (2002) Placement of the structural proteins in sindbis virus. J Virol 76:11645–11658

Zhong LT, Sarafian T, Kane DJ, Charles AC, Mah SP, Edwards RH, Bredesen DE (1993) bcl-2 inhibits death of central neural cells induced by multiple agents. Proc Natl Acad Sci USA 90:4533–4537

Zhou M, Demo SD, McClure TN, Crea R, Bitler CM (1998) A novel splice variant of the cell death-promoting protein Bax. J Biol Chem 273:11930–11936

Zrachia A, Dobroslav M, Blass M, Kazimirsky G, Kronfeld I, Blumberg PM, Kobiler D, Lustig S, Brodie C (2002) Infection of glioma cells with Sindbis virus induces selective activation and tyrosine phosphorylation of protein kinase C delta. Implications for Sindbis virus-induced apoptosis. J Biol Chem 277:23693–23701

Zujovic V, Benavides J, Vige X, Carter C, Taupin V (2000) Fractalkine modulated TNF-α secretion and neurotoxicity induced by microglial activation. Glia 294:305–315

Apoptotic Pathways Triggered By HIV and Consequences on T Cell Homeostasis and HIV-Specific Immunity

M.-L. Gougeon[1]

1 Introduction

HIV infects cells of the immune system and the hallmark of HIV infection is the progressive destruction of the CD4 T lymphocyte subset. Shortly after this syndrome was reported, it was discovered that the HIV envelope could bind to the CD4 receptor, that HIV could replicate within human CD4 T cells in vitro and kill them, and that circulating CD4 T cells decreased in number as disease progressed. Mature CD4 T helper cells are key effectors in coordinating cellular and humoral immunity against pathogens, since they synthesize cytokines required for the induction of innate immunity involved in early killing of virus-infected cells, the maturation of B lymphocyte producers of neutralizing antibodies, and the differentiation of virus-specific $CD8^+$ killer T cells (CTL). In addition, they are a source of chemokines which control the migration of lymphocytes to the site of infection and also inhibit HIV entry into CD4-expressing targets. Therefore, HIV-dependent destruction of CD4 T lymphocytes is responsible for inefficient immune control of HIV replication and for the development of severe immune deficiency that leads to opportunistic infections, neurological impairment, malignancies, and ultimately death (Levy 1998).

Mechanisms responsible for the disappearance of CD4 T cells in HIV-infected persons are still a matter of debate (McCune 2001; Grossman et al. 2002). Because CD4 T cells are the targets of HIV, it was initially proposed that the progressive decrease of this subset in infected subjects is the direct consequence of their killing by HIV. However, this intuitive assumption proved to be too simplistic faced with recent insights into the pathogenic mechanisms of CD4 T cell loss. In vivo analysis of both virus turnover and T cell homeostasis in HIV-infected humans or SIV-infected monkeys has revealed that a number of mechanisms are involved in peripheral CD4 T loss. They include increased turnover and destruction of T cells linked to a high rate of HIV replication in blood and lymph nodes, homing of virus-specific T cells in lymphoid tissues, accelerated destruction by apoptosis of both infected and

[1]Antiviral Immunity, Biotherapy and Vaccine Unit, Molecular Medicine Department, Institut Pasteur, 28 rue du Dr. Roux, 75724 Paris cedex 15, France, e-mail: mlgougeo@pasteur.fr

Progress in Molecular and Subcellular Biology
C. Alonso (Ed.): Viruses and Apoptosis
© Springer-Verlag Berlin Heidelberg 2004

uninfected lymphocytes, not compensated by central regeneration because of an impaired production of T cells by the thymus. Although our understanding of the gradual depletion of the CD4 T cell pool is still incomplete and controversial, a growing body of evidence points to HIV-driven lymphocyte apoptosis as an important contributor to $CD4^+$ T cell depletion and disease progression. This review will summarize recent advances in the mechanisms by which HIV triggers the apoptotic program in target cells, the functional consequences of this viral strategy to escape immune attack, and the effect of antiviral therapies on this process.

2
Why Are CD4 T Cells Depleted in HIV Infection?

2.1
Accelerated T Cell Destruction

For about a decade, HIV infection was supposed to be static, with a long asymptomatic phase, considered as quiescent with regard to viral replication and defined as "clinical latency". However, from 1990 onwards, the development of quantitative assays for HIV RNA in plasma and the use of antiviral drugs to perturb quasi steady-state levels of this RNA allowed kinetics studies of HIV-1 viral load and T cell dynamics. It was thus discovered that, in asymptomatic individuals, the virus is continuously replicating in blood and tissues with a high turnover rate, driving a rapid lymphocyte turnover and contributing to the destruction of mature T cells (Ho et al. 1995; Wei et al. 1995).

To better understand host lymphocyte dynamics in response to chronic virus replication, in vivo studies of DNA labeling with either 5-bromodeoxyuridine (BrdU) or deuterated glucose to tag proliferating cells have been performed in infected humans or macaques. BrdU administration in chronically SIV-infected or uninfected macaques showed that all lymphocyte subsets from infected animals had increased proliferation and death rates, and it concerned CD4 and CD8 T lymphocytes, memory and naive cells, natural killer (NK) and B cells, suggesting that chronic SIV replication provokes a state of generalized immune activation, inducing entry of CD4 and CD8 T cells into a rapid proliferating pool leading to increased cell turnover. Similar in vivo DNA labeling studies were performed in humans and they confirmed the enhanced proliferation rate in both CD4 and CD8 T cell subsets from HIV-infected patients (Mohri et al. 1998, 2001; Hellerstein et al. 1999; Lempicki et al. 2000; Kovacs et al. 2001). Another way to determine the extent of virus-driven lymphocyte proliferation is the ex vivo detection of the nuclear antigen Ki67, which is expressed during mitosis. Increased proportions of proliferating (Ki67+) CD4 and CD8 T cells were detected in peripheral blood T cells from infected patients, and reduction of plasma HIV-RNA load by highly active antiviral therapy (HAART) was accompanied by a marked decrease in peripheral T cell proliferation and turnover rate (Fleury et al. 2000; Hazenberg et al. 2000a; Lempicki et al. 2000; Douek 2001; Mohri et al. 2001). Therefore, HIV infection leads to a sustained increase in lymphocyte proliferation and death rates, that is not limited to the CD4 T cell pool, and may be driven by infection-induced immune activation, and the question of why HIV infection leads to the preferential disappearance of CD4 T cells is still unsolved (Hazenberg et al. 2000b; McCune 2001; Grossman et al. 2002). A recent study has attempted to address this question using a mathematical model to analyze D-glucose labeling of in vivo dividing T cells in HIV-infected patients, and it reported that, in CD4 T cells, the per cell death rate is increased compared with control donors, whereas, in the CD8 T cell pool, it is the fraction of proliferating cells that is increased (Ribeiro et al. 2002). These results suggest that CD4 lymphocyte depletion in AIDS is primarily a consequence of increased cellular destruction. In fact, the rapid proliferation and loss of T cells, detected by the above studies, largely reflects the response of the immune system to persistently high and continuous virion production, inducing a state of chronic activation and continuous destruction.

Several non-exclusive mechanisms are thought to contribute to HIV-dependent CD4 T cell death (as detailed below). It can be directly mediated by HIV as a consequence of viral gene expression and cytopathic effect in infected cells, or indirectly through either the killing of bystander cells by proapoptotic viral proteins released by infected cells, or as a consequence of persistent immune activation leading to the triggering of activation-induced death in various uninfected immune cells (Badley et al. 2000; Gougeon and Montagnier 2000; Selliah and Finkel 2001). In addition to these pathways, a complementary cytopathic effect is probably provided by the immune system, since infected cells may be killed by HIV-specific CTL or antibody-dependent cell-mediated cytotoxicity (ADCC).

2.2 Impaired T Cell Regeneration

CD4 T cell lymphopenia may be linked to interference of HIV with T cell renewal, preventing appropriate replacement of prematurely destroyed mature T cells. Evidence for suppression of hematopoiesis by HIV is supported by a number of studies. The generation of multiple hematopoietic lineages from stem cells is blocked in vivo by HIV infection in humanized SCID-Hu mice and bone marrow $CD34^+$ progenitors from HIV-infected patients show a reduced capacity to differentiate into erythrocyte, granulocyte and T lymphocyte lineages in vitro (Hazenberg et al. 2000b). Interference of HIV at the level of progenitor cells was also shown in fetal thymus organ cultures (FTOC) in which murine fetal thymus is repopulated with human $CD34^+$ cells. Patients progressing to AIDS have a dramatic loss in T cell development capacity of progenitors shortly after seroconversion in contrast to long-term non-progressors who retain progenitor capacity 8 years after seroconversion (Clark et al. 2000). Interestingly, the majority of patients investigated had an improvement in T cell development capacity after receiving HAART (Clark et al. 2000). HIV infection of the thymus may also drive the selection of dysfunctional thymocytes.

Recent evidence that the adult thymus is a functional organ (Douek et al. 1999), housing many CD4 T cells in varying stages of maturation, encouraged studies on the impact of HIV infection on thymic-dependent T cell renewal. The input of naive T cells from the thymus appears to be substantially reduced in HIV infection (Douek et al. 1999), either as a result of virus-induced destruction of thymocytes or because of inhibition of their production. Computed tomography of thymic tissue showed that HIV-infected patients with abundant thymic tissue had higher circulating levels of naive ($CD45RA^{+}CD62L^{+}$) T cells and a faster restoration of naive T cell numbers during HAART, suggesting that thymic output may play an important role in T cell homeostasis in HIV disease (Franco et al. 2002). More direct measurement of thymic function can be achieved by using a polymerase chain reaction method that identifies recent thymic emigrants (naive T cells) by detecting circular excision products, formed during T cell rearrangement in the thymus (Spits 2002). The number of T cell receptor excision circles (TRECs), as measured in peripheral blood mononuclear cells or in purified CD4 and CD8 T cells, is thought to correlate with thymic function (Poulin et al. 1999). TREC content of blood cells is reduced within a few months after HIV infection (Douek et al. 1999), and it was found to have pronostic value, such that individuals with low numbers of TRECs progress to AIDS at a faster pace (Hatzakis et al. 2000). Considering that TREC content in a cell population is affected both by changes in thymic output and by the division history of the cells (Hazenberg et al. 2000a), interpretation of TRECs data can be tricky and the conclusion of diminished thymic function after HIV infection because of decreased TRECs content is not supported by some authors (Hazenberg et al. 2000a,b). However, the essential role of the thymus for T cell renewal has been suggested at least in early HIV-infected humans in whom the loss of TRECs within naive T cells occurs in the absence of increased proliferation of this subset (Douek 2001), and also in adult and pediatric patients on HAART showing concomitant increased thymic volume, increased TREC content and enhanced numbers of naive T cells (Douek et al. 2000; Franco et al. 2002; Ometto et al. 2002).

2.3
Cytokines Regulate T Cell Homeostasis

Considering that accelerated destruction of mature CD4 T cells might stimulate de novo production of T cells, thymic rebound represents compensatory feedback adaptation. Such thymic rebound has been ascribed to increased peripheral secretion of a positive regulator and, among them, IL-7 is thought to mediate the homeostasis of naive and memory T cells in vivo (Schluns et al. 2000) and may contribute to T cell homeostasis in HIV disease. Thus, in cross-sectional and longitudinal studies of HIV-infected individuals at varying stages of the disease, falling CD4 T cell counts are associated with increased circulating levels of IL-7 and, reciprocally, IL-7 levels fall when CD4 T cell counts increase after antiretroviral therapy (Napolitano et al. 2001). The positive influence of IL-7 on T cell regeneration is also suggested in a thymic organ culture system in which exogenous IL-7 increases TREC frequency in fetal as well as infant thymus (Okamoto et al. 2002) and, in synergy with the chemokine SDF-1, IL-7 enhances the viability of $CD34^+$ T cell precursors (by modulating the expression of Bxl-2 and Bax) and stimulates their proliferation (Hernandez-Lopez et al. 2002). However, IL-7 may contribute to the destruction of $CXCR4^+CD4^+$ progenitors since it was found to favor HIV replication in thymocytes by inducing expansion of mature $CD27^+$ thymocytes expressing the CXCR4 co-receptor. In addition, IL-7 pretreatment of peripheral naive T cells mediates their expansion and enhances their susceptibility to primary isolates of HIV. These data have to be considered in evaluating IL-7 as an immunomodulator for HIV-infected patients.

3 Apoptosis as an HIV Strategy To Escape the Immune System

3.1 The Apoptotic Pathways

Apoptosis is a highly regulated form of cell death that is essential for the maintenance of a constant lymphocyte population size in the face of a continuous influx of new lymphocytes and the homeostatic proliferation of existing cells, and it is required during an immune response to foreign antigen to eliminate the majority of activated antigen-specific T lymphocytes in order to prevent autoimmunity (Khaled and Durum 2002). Cell death occurs via two main pathways: activation-induced cell death (AICD) via death receptors (extrinsic pathway), and activated T cell autonomous death (ACAD) via Bcl-2-related proteins (intrinsic pathway). The extrinsic pathway is initiated by tumor necrosis factor (TNF) family death receptor ligands binding to their cognate death receptors, and the intrinsic pathway is initiated by internal sensors which propagate signals to the mitochondria (Krammer 2000; Martinou and Green 2001; Fig. 1).

The extrinsic apoptotic pathway is initiated by the specific binding of ligands to the death receptors of the TNF family, such as CD95 (Fas/Apo-1) or TNF-Rs. Via their death domains (DDs), multimerized receptors interact with DDs of adaptor proteins such as FAAD (Fas-associated via death domain) or

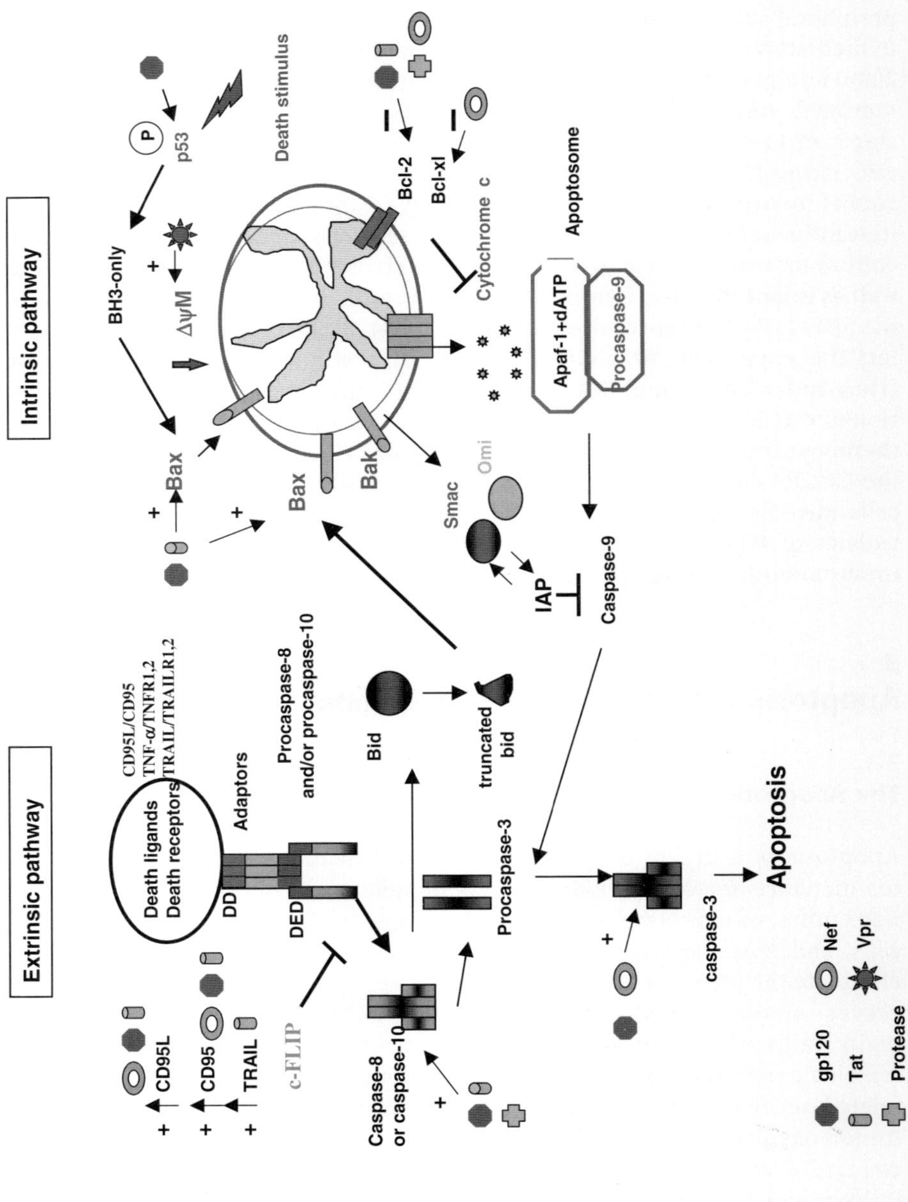
Extrinsic pathway
Intrinsic pathway
Death ligands
Death receptors
CD95L/CD95
TNF-α/TNFR1,2
TRAIL/TRAILR1,2
Adaptors
DD
DED
CD95L
CD95
TRAIL
c-FLIP
Caspase-8
or caspase-10
Procaspase-8
and/or procaspase-10
Bid
truncated
bid
Procaspase-3
caspase-3
Apoptosis
BH3-only
P
p53
Death stimulus
ΔψM
Bax
Bax
Bak
Smac
Omi
Bcl-2
Bcl-xl
Cytochrome c
Apoptosome
Apaf-1+dATP
Procaspase-9
IAP
Caspase-9
gp120
Tat
Protease
Nef
Vpr

TRADD (TNFR-associated via death domain). These adaptor proteins also contain DEDs (death effector domains) that facilitate their binding to pro-caspase-8 and/or pro-caspase-10 to form the DISC (death-inducing signaling complex). As part of the DISC, the pro-caspases are cleaved into their active forms and initiate the intrinsic pathway of apoptosis by cleaving Bid into tBid and activate the effector caspase cascade, which includes caspase-2, -3, -6 and -7. Death-receptor induced apoptosis can be blocked by the cellular FLICE inhibitory protein (cFLIP), which is recruited in the DISC and inhibits proteolytic processing of caspase-8 (FLICE). The intrinsic apoptotic pathway is initiated by internal sensors, such as p53, which propagate signals to the mitochondria via activation of BH3 domain-only members of the Bcl-2 family, which mediate the assembly of pro-apoptotic members of the Bcl-2 family (Bax, Bak, Bok, Bcl-rambo) into heterooligomeric pores in the outer membrane of the mitochondria. This results in the release of factors such as cytochrome c, Smac (second mitochondria-derived activator of caspase) and Omi (also known as PRSS25, protease serine 25) into the cytoplasm. This is associated with the loss of mitochondrial membrane potential. Alteration of mitochondrial membrane integrity can be blocked by the anti-apoptotic Bcl-2 family members (Bcl-2, Bcl-xL, Bcl-w, Mcl-l and Bcl-B). Release of cytochrome c promotes the formation of the apoptosome which includes Apaf-1 and pro-caspase-9. Autocatalytic activation of caspase-9 initiates the effector caspase cascade. Caspase activation is negatively regulated by inhibitors of apoptosis (IAPs), which are counterbalanced by pro-apoptotic Smac and Omi.

3.2
HIV Gene Products Involved in the Control of Cell Death

In vitro infection of activated CD4 T lymphocytes by HIV is associated with a cytopathic effect manifested by ballooning of cells, formation of syncytia and apoptosis of both productively infected and bystander uninfected cells which depends on viral replication. The envelope glycoprotein complex (Env, gp120/gp41), which causes apoptosis of both infected and uninfected cells, appears to be one of the dominant apoptosis-inducing molecules encoded by the HIV-1 genome (Ferri et al. 2000). Env expressed on the plasma membrane of infected cells can interact with the CD4 molecule and a suitable co-receptor to

---

Fig. 1. Apoptotic pathways triggered by HIV. Both the extrinsic and the intrinsic pathways are triggered by HIV proteins. gp120, Nef and Tat upregulate CD95 and CD95L expression, Tat upregulates TRAIL expression, gp120, Tat and protease upregulate caspase-8 and gp120 and Nef increase caspase-3 activity. The mitochondrial pathway is triggered by gp120, which induces mTOR-mediated phosphorylation of p53, Tat and gp120 which promote Bax insertion into mitochondrial membrane and subsequent release of cytochrome c, Vpr which has a direct effect on the mitochondrial permeability transition pore, gp120, Tat and Nef which inhibit Bcl-2 expression while the protease cleaves it, and Nef which inhibits Bcl-xL expression

trigger cell-to-cell fusion; the resulting syncytia subsequently undergo apoptosis. Syncytial apoptosis is not mediated by the death receptor pathway (Ohnimus and Jassoy 1997) but HIV-infected T cells become more susceptible to CD95-induced apoptosis. This increased susceptibility to CD95-induced apoptosis is induced by the HIV gene product Vpu (Casella et al. 1999). Syncytia arising from the fusion of cells expressing the HIV envelope protein with cells expressing the CD4/CXCR4 complex undergo apoptosis through a mitochondrion-dependent pathway, initiated by upregulation of cyclin B-Cdk-1, nuclear translocation of mammalian target of rapamycin (mTOR), mTOR-mediated phosphorylation of p53 on serine 15 (p53S15), p53-dependent upregulation of Bax, and activation of the mitochondrial pathway, subsequent release of cytochrome c, caspase activation and apoptosis (Castedo et al. 2002). The relevance of fusion-induced apoptosis to AIDS pathogenesis is suggested by the positive correlation between infection by syncytium-inducing HIV isolates and the decline in $CD4^+$ T cell numbers in vivo (Blaak et al. 2000). Moreover, the in vivo association of syncytia and apoptosis is suggested by the observation that, in lymph nodes from HIV^+ individuals, syncytia express markers of early apoptosis such as tissue transglutaminase (Amendola et al. 1996), and exhibit an increase in cyclin B and mTOR expression, correlating with viral load (Castedo et al. 2002).

In addition to being induced in the context of syncytia, apoptosis of host cells can be triggered exogenously by several HIV proteins which modulate the extrinsic and intrinsic pathways (Fig. 1). For example, the death receptor pathway is activated by gp120, Nef and Tat, which upregulate CD95 and CD95L expression, by Tat which upregulates TRAIL expression, gp120, Tat and protease which upregulate caspase-8, and gp120 and Nef which increase caspase-3 activity. Several HIV proteins trigger the mitochondrion-controlled pathway, such as gp120, which induces mTOR-mediated phosphorylation of p53, Tat and gp120 which promote Bax insertion into mitochondrial membrane and subsequent release of cytochrome c, Vpr which has a direct effect on the mitochondrial permeability transition pore, gp120, Tat and Nef which inhibit Bcl-2 expression while the protease cleaves it, and Nef which inhibits Bcl-xL expression.

Like many other viruses (Benedict et al. 2002), HIV has also developed strategies to inhibit cellular apoptosis, at least until high levels of progeny virus are produced. Indeed, this strategy is needed for efficient production of virions by host cells and rapid virus dissemination into the organism. Several HIV gene products have been shown to exhibit anti-apoptotic activity. Nef, gp120 and Vpu contribute to the downregulation of CD4 receptor on infected cells, preventing subsequent gp120-CD4-mediated apoptosis (Crise et al. 1990; Willey et al. 1992). Nef downmodulates expression of MHC class I molecules and upregulates CD95L expression on infected cells, a strategy that may function to protect infected cells from cytolysis by CTL or NK cells (Schwartz et al. 1996; Collins et al. 1998). Tat decreases transcription of p53, so promoting cell cycle progression, and allowing the cells to increase virus production (Li et al. 1995),

Table 1. In vivo and in vitro effects of HIV therapies on cell death

	Effects of anti-HIV drugs	References
In vivo	Decreased apoptosis in lymph node and blood lymphocytes	Badley et al. (1999); Gougeon et al. (1999); Sloand et al. (1999); Ledru et al. (2000); De Oliveira Pinto et al. (2002a)
	Upregulation of Bcl-2 in T lymphocytes	Ledru et al. (2000)
	Decreased susceptibility to CD95- and TNFR-induced apoptosis	Badley et al. (1999); Gougeon et al. (1999); Sloand et al. (1999); De Oliveira Pinto et al. (2002a,b)
	Suppression of physiological apoptosis in blood TNF-α T cell producers	Ledru et al. 2000)
	Decreased caspase-1 expression in T lymphocytes	Sloand et al. (1999)
	Apoptosis of peripheral adipocytes	Bastard et al. (2002)
In vitro	Inhibition of CD95-induced apoptosis, caspase-1 and caspase-3 activation in PBMC from healthy donors and HIV-infected patients	Sloand et al. (1999); De Oliveira Pinto et al. (2002b)
	Decreased HIV-mediated CD95 expression in HIV-infected CD4 T cells	Estaquier et al. (2002)
	Enhanced survival and improved TcR-induced proliferation in T cells from HIV^+ patients	Lu and Andrieu (2000)
	Activation of CD95-induced apoptosis in PBMC from control donors	Oliveira Pinto et al. (2002b)
	Inhibition of cytochrome c release by mitochondria	Phenix and Badley (2002)
	Inhibition of mitochondrial permeability, transition pore channel opening induced by HIV-Vpr peptide	Phenix and Badley (2002)
	Inhibition of apoptosis in bone marrow CD34+ stem cells and increase in colony formation	Sloand et al. (2000)
	Inhibition of activation, proliferation and NF-κB transcription by TNF-α and HIV-Tat in human endothelial cells	Pati et al. (2002)
	Inhibition of adipocyte differentiation	Meng et al. (2001); Bastard et al. (2002)

and low-level constitutive expression of Vpr inhibits apoptosis by causing the upregulation of Bcl-2 and downmodulation of Bax (Conti et al. 1998). The in vivo relevance of these in vitro observations is suggested by studies performed on lymph node biopsies from HIV-infected individuals and SIV-infected monkeys, showing that productively infected cells are not apoptotic, while apoptosis is detected predominantly in uninfected bystander cells (Finkel et al. 1995). These findings argue for HIV-driven indirect mechanisms of cell destruction, confirmed by the in vivo detection of apoptosis in cells which are not targets of HIV, such as $CD8^+$ T cells and B cells (Muro-Cacho et al. 1995; Gougeon et al. 1996). Therefore, HIV may ensure viral survival by manipulating the apoptotic machinery to its advantage in infected cells before destroying the immune system through the activation of apoptotic programs in uninfected virus-specific lymphocytes.

3.3 Activation-Induced Cell Death and Activated T Cell Autonomous Death

Activation-induced cell death (AICD) is required during an immune response to foreign antigen to eliminate the majority of activated antigen-specific T lymphocytes in order to prevent autoimmunity. This mechanism of cell elimination is mediated by the death receptor pathway and it is normally induced at the clonal level during an antigenic stimulation (Ju et al. 1995). Because HIV is chronically expressed in the host, the immune system is permanently activated by HIV proteins and, consequently, host T cells are abnormally susceptible to AICD. This is suggested by the following observations: (1) ex vivo activation of T cells from HIV^+ individuals by T cell receptor (TcR) ligands consistently results in enhanced levels of apoptosis compared with T cells from control donors (Groux et al. 1992; Meyaard et al. 1992; Gougeon and Montagnier 1993; Gougeon et al. 1993); (2) the in vivo involvement of the CD95 pathway is suggested by a concomitant increased expression of CD95 on both $CD4^+$ and $CD8^+$ T cell subsets from HIV^+ individuals and increased susceptibility to CD95-induced apoptosis, which is positively correlated with disease progression (Katsikis et al. 1995; Estaquier et al. 1996; Gougeon et al. 1997; Sloand et al. 1997); (3) elevated levels of soluble CD95 are detected in the plasma of HIV^+ individuals and this can be used as a predictive marker for progression to AIDS (Medrano et al. 1998); (4) upregulation of CD95L is also observed in patients' $CD4^+$ and $CD8^+$ T cells, and a significant increase in macrophage-associated CD95L has been detected in lymphoid tissue from HIV^+ subjects, which is correlated with the degree of tissue apoptosis (Badley et al. 1996; Dockrell et al. 1998; Silvetris et al. 1998); (5) in addition to the CD95 pathway, the TNFR death pathway is also triggered in HIV^+ individuals since both $CD4^+$ and $CD8^+$ T cells from HIV^+ individuals are susceptible to TNFR1- and TNFR2-induced apoptosis and this priming for TNFR-induced apoptosis is associated with the in vivo downregulation of Bcl-2 protein and the expression of active forms of caspase-8 and -3 (De Oliveira Pinto et al. 2002a); and (6) another member of the TNF family, tumor necrosis factor-related apoptosis-inducing ligand (TRAIL/Apo-2L), is involved in HIV-associated T cell apoptosis, since T cells from HIV^+ individuals are susceptible to TRAIL-mediated killing, in contrast to cells from control donors (Jeremias et al. 1998), and AICD is partially inhibited by antagonistic TRAIL specific antibodies. The possible contribution of AICD to AIDS pathogenesis is suggested by the positive correlation between TcR-dependent apoptosis and CD4 T cell depletion (Gougeon et al. 1996), the very low level of CD95 expression and AICD in non-pathogenic

HIV-1-infection in chimpanzees (Gougeon et al. 1997), and the in vivo observation, in hu-PBL-NOD-SCID mice infected with HIV, that apoptosis occurs predominantly in uninfected bystander CD4 T cells, including those which harbor latent proviral HIV DNA and constitute reservoirs for HIV (Lum et al. 2001; Miura et al. 2001).

Continuous triggering of the CD95/CD95L pathway on T cell subsets make them possible effectors in killing activated CD95-expressing cells, which can be found in high proportions in HIV^+ individuals. This is corroborated by the demonstration that activated $CD4^+$ T lymphocytes expressing CD95L can kill CD95-expressing $CD8^+$ T lymphocytes (Piazza et al. 1997) and CD95L-expressing macrophages are potential killers of CD95-sensitive target T cells in an MHC-unrestricted and CD95/TNFR-dependent manner (Badley et al. 1997). HIV-specific CTL are also potential killers of CD95-expressing activated lymphocytes since an anti-Nef HLA class I restricted CTL clone, derived from an HIV^+ subject, is able to mediate both perforin- and CD95-dependent cytotoxic activities on either Nef-presenting target cells or CD95-expressing cells, respectively (Garcia et al. 1997). Therefore, HIV-activated T lymphocytes may be deleterious to the immune system of an HIV^+ individual through the CD95L-dependent destruction of uninfected CD95-expressing T cells, induced by persistent HIV-driven immune stimulation.

Activated T cell autonomous death (ACAD) depends on the mitochondrial pathway and it is induced in activated lymphocytes during the termination of an immune response as the consequence of decreased growth factor availability. ACAD can be observed ex vivo in T cells from HIV-infected persons following short-time incubation in culture medium. Indeed, patients' T cells undergo spontaneous apoptosis at a greater rate than cells from healthy subjects when cultured in the absence of exogenous stimulus (Meyaard et al. 1992; Gougeon et al. 1993). Phenotypic characterization of cells primed for spontaneous apoptosis indicates that they include all lymphocyte subsets, i.e. $CD4^+$, $CD8^+$ T cells, B cells and NK cells (Gougeon et al. 1996), as detected in vivo in lymph nodes from HIV^+ individuals (Muro-Cacho et al. 1995); they exhibit an activated phenotype (Gougeon et al. 1996) and they downregulate the Bcl-2 molecule (Adachi et al. 1996; Boudet et al. 1996). Therefore, ACAD is primarily the consequence of immune activation induced by repetitive viral antigen exposure that tips the balance in favor of the pro-apoptotic Bcl-2 family members, a mechanism known to contribute to the physiological homeostasis of activated T cells in vivo (Hildeman et al. 2000). Interestingly, the non-pathogenicity of HIV-1 infection in chimpanzees is associated with the lack of chronic immune activation, a very low level of spontaneous T cell apoptosis and normal expression of Bcl-2 (Gougeon et al. 1997). ACAD in patients' T cells can be prevented by cytokines, particularly IL-2 and IL-15, which upregulate the expression of Bcl-2 (Adachi et al. 1996; Naora et al. 1999).

In conclusion, the normal processes of elimination of activated cells following activation by pathogens, described above, might be detrimental for the immune system in the case of a chronic infection such as that induced by HIV.

Indeed, continuous production of viral proteins by infected cells leads to an unbalanced immune activation and to the triggering of apoptotic programs (death receptor- and mitochondria-dependent), leading to the destruction of healthy non-infected immune cells.

3.4 Impaired Immunity and Disease Evolution

Like any other virus, HIV stimulates strong cytotoxic T lymphocyte (CTL) responses in infected people. In the acute phase of HIV infection, the CTL response initially follows the rise in HIV in the blood and when that response reaches a peak the virus levels fall. In the chronic phase of the infection, there is an inverse relationship between CTL response and virus load (Ogg et al. 1998). However, this CTL response fails ultimately to control HIV infection, in contrast to other persistent human virus infections, such as Epstein-Barr virus (EBV) and cytomegalovirus (CMV; McMichael and Rowland-Jones 2001). Recent studies have shown functional defects in HIV-specific CTL, since they express low levels of perforin and consequently poorly kill appropriate target cells ex vivo (McMichael and Rowland-Jones 2001; Migueles et al. 2002). This may be related to a defect in CTL maturation, as suggested by their phenotype, which corresponds to an intermediate-differentiated subset, in contrast to CMV-specific effectors in the same donors (Appay et al. 2002). Altered differentiation of CTL might be linked to increased apoptosis since, in lymph nodes and blood of individuals chronically infected with HIV, an important fraction of CD8 T cells with the phenotype of CTL express low levels of Bcl-2 and show apoptotic characteristics (Bofill et al. 1995; Boudet et al. 1996). Bcl-2 downregulation and CTL apoptosis may be induced by high doses of HIV during the primary infection, causing the rapid deletion of high avidity HIV-specific CTL, as suggested in a murine model (Alexander-Miller et al. 1998). Therefore, AICD may contribute to differentiation defects in CTL preventing the elimination of the virus in the infected host.

Defective help from CD4 T cells may also contribute to altered differentiation of CTL. Naive Th cells can differentiate into at least two functional classes of cells during an immune response – T_h1 cells, which secrete IFN-γ, and T_h2 cells, which secrete IL-4 (Murphy and Reiner 2002). T_h1 immunity is thought to be crucial for appropriate antiviral CTL response and HIV infection induces an altered pattern of cytokine responses to a variety of stimuli. So, progression of HIV infection is accompanied by a reduced production of IL-12, IFNγ and IL-2 by PBMCs in response to antigenic simulation in vitro and an increased production of IL-4 and IL-10, and this pattern is predictive for the level of $CD4^+$ T cell loss and disease evolution (Clerici and Shearer 1994). The pattern of cytokines produced by a subset can be studied using a single cell analysis method by flow cytometry, which allows determination of the phenotype, frequency, and priming for apoptosis of cytokine producers following their stimulation (Lecoeur et al. 1998). Ex vivo single cell quantification of the frequency of T_h1/T_h2 subsets derived from peripheral T cells stimulated in short-term cultures showed that HIV infection is associated with a differential alteration in the frequency of T_h1 subsets, and a significant decrease in the frequency of T cells primed for IL-2, with a preserved frequency of IFNγ-producing T cells, was observed throughout HIV infection and found to be a good indicator of disease progression (Ledru et al. 1998). It is noteworthy that the progressive decrease in the proportion of IL-2-producing T cells is correlated with their susceptibility to apoptosis, suggesting an important relationship between exacerbated activation-induced cell death in peripheral T helper cells and the impairment of HIV-specific immunity in HIV disease (Fig. 2).

4 HIV Therapy: Restoration of the Immune System and Metabolic Complications

4.1 Phases of Immune Restoration

The availability of potent combined antiretroviral therapies [combinations of HIV nucleosidic reverse transcriptase inhibitors (NRTI) and protease inhibitors (PI)] (HAART) that reduce viral load to undetectable levels and concomitantly increase CD4 T cell counts has considerably challenged the question of the mechanisms controlling the homeostasis and reconstitution of the immune system in treated-HIV-infected persons. Recent studies have pointed out the complexity of the mechanisms involved. The increase in CD4 T cells after HAART results from a combination of both release of sequestered cells from lymphoid compartments to peripheral blood (Pakker et al. 1998) and increased production rate of circulating T lymphocytes due to peripheral homeostatic T cell expansion (Hellerstein et al. 1999), suppression of apoptosis (Gougeon et al. 1999; Sloand et al. 1999) and production of new T cells by the thymus (Douek et al. 1999), suggesting that the mechanisms of renewal of CD4 T cells are still operational. Rapidly after initiation of HAART, an important drop in spontaneous, activation-induced and CD95-triggered apoptosis is observed in both CD4 and CD8 T cells from all treated patients. This occurs before the decrease in immune activation, and the resistance to CD95-induced apoptosis precedes the downregulation of CD95 expression (Badley et al. 1999; Gougeon et al. 1999). Thus, suppression of the plasmatic viral load is associated with normalization of cell survival in many infected patients, associated with quantitative restoration of CD4 T cell numbers. However, high levels of lymphocyte apoptosis may persist in some individuals despite effective therapy, and this is associated with lamivudin-(3TC)containing regimen in vivo; this nucleoside

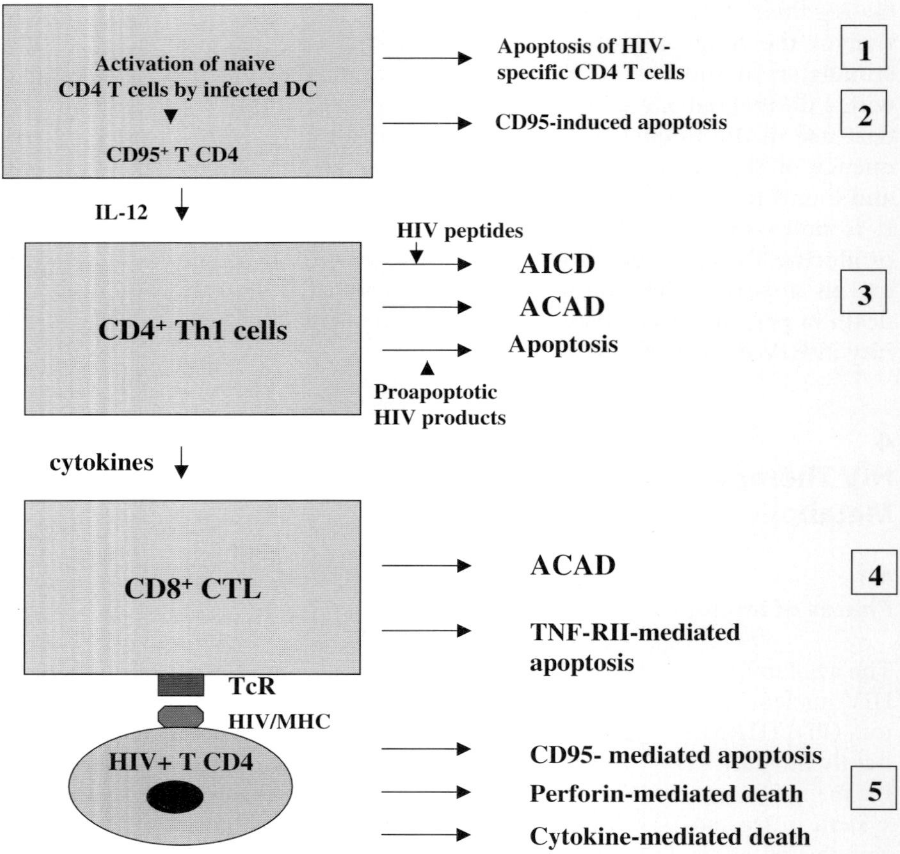

Fig. 2. Apoptosis of HIV-specific immune effectors. HIV-specific T helper cells are primed by dendritic cells (DC) presenting HIV-peptides on their surface. HIV can directly infect DC and infected DC may in turn infect and subsequently induce apoptosis of naive CD4 T cells, contributing to the early loss of HIV-specific T CD4 (*1*). Once primed by DC, activated CD95-expressing HIV-specific CD4 T cells may be killed by fratricide apoptosis mediated by CD95L expressed on activated killers (T CD4, CTL, NK; *2*). Following their differentiation into T helper cells (Th1), CD4 T cells may be killed by AICD, ACAD or under the influence of pro-apoptotic HIV proteins, as detailed in the text (*3*). CD4 Th1 cells help maintain memory CD8 T cells and contribute to the maturation of cytotoxic CD8 T cells (CTL). Because of the loss of CD4 T cells, defects in maturation and function occur in CD8 T cells which may die, either as a consequence of low Bcl-2 expression (ACAD) or because of the activation of the TNF-R pathway (*4*). Infected CD4 T cells can be killed by HIV-specific CTL which use several means of target destruction (CD95L/ CD95, perforin/granzyme pathway or cytokines; *5*)

reverse transcriptase inhibitor exhibiting a pro-apoptotic activity in vitro (De Oliveira Pinto et al. 2002b). Suppression of activation-induced apoptosis under HAART is likely to be a multifactorial process, including the decreased expression of pro-apoptotic HIV proteins and the downregulation of the in vivo immune activation state. In addition, we cannot exclude a direct immunomodulatory effect of HIV PIs on T cells. Indeed a number of in vitro effects of PIs on cell survival and differentiation have been reported: PIs suppress in vitro physiological apoptosis by inhibition of cellular proteases and mitochondrial apoptosis (Sloand et al. 1999; Estaquier et al. 2002; Phenix and Badley 2002), they decrease apoptosis in stem cells and promote hematopoiesis (Sloand et al. 2000), they inhibit activation, proliferation and cytokine synthesis by lymphocytes from control donors or HIV-infected persons (Lu and Andrieu 2000) and they exert antitumorigenic effect on Kaposi sarcoma cells (Pati et al. 2002; Table 1).

4.2 HAART-Associated Metabolic Complications. Contribution of Apoptosis

Long-term HAART has been associated with serious metabolic complications, including a syndrome of lipodystrophy, characterized by dyslipidaemia, changes in body fat distribution (peripheral loss of adipose tissue and visceral abdominal fat accumulation), lactic acidosis, insulin resistance and cardiovascular disease (Carr et al. 1998). The mechanisms involved are poorly understood but some anti-HIV drugs and cytokines were shown to induce functional alteration and apoptosis of peripheral adipocytes and lymphocytes, possibly contributing to these complications Several pathophysiological mechanisms have been proposed to explain the appearance of this syndrome (Gougeon et al. 2003): (1) The adipose tissue may be a target of PI-mediated effects. Some PIs induce in vitro an insulin-resistant state in differentiated adipocytes and a decreased level of sterol regulatory element binding transcription factor 1 (SREBP1), peroxisome proliferator activated receptor γ (PPARγ) and CCAAT/enhancer binding protein α (C/EBPα), leading to inhibition of adipocyte differentiation and apoptosis, as found in vivo in atrophic adipose tissue of individuals with lipodystrophy (Bastard et al. 2002). (2) NRTI may contribute to lipoatrophy through their mitochondrial toxicity on adipocytes since they inhibit DNA polymerase γ, leading to depletion of cellular mtDNA content and apoptosis (Phenix and Badley 2002), and depletion of mtDNA has been demonstrated in subcutaneous fat samples taken from patients with lipodystrophy. (3) Adipocytes are targets of inflammatory cytokines, including TNF-α, IL-1, IL-6 and IFN-α, which inhibit preadipocyte differentiation in vitro and might play a role in atrophy of adipose tissue and insulin resistance in antiretroviral-treated HIV^{+} patients. TNF-α inhibits adipogenesis in preadipocytes by blocking induction of PPARγ and C/EBPα, and it stimulates lipolysis and promotes insulin resistance (Meng et al. 2001). Altered homeostasis of TNF-α-producing

T cells has been observed in HIV^+ patients with lipodystrophy, resulting in the accumulation of $CD4^+$ and $CD8^+$ T cells polarized to TNFα synthesis, which is correlated with increased serum levels of triglycerides, and cholesterol (Ledru et al. 2000). This polarization of T cells is related to their escape from physiological apoptosis, possibly due to the anti-apoptotic effects of PIs (Ledru et al. 2000). In addition, TNF-α expression is upregulated in patient's adipose tissue, and this is negatively correlated with the expression of SREBP1c (Bastard et al. 2002). Dyslipidaemia in patients under HAART is also related to increased production of IFNα, which regulates lipid metabolism, and to hormonal perturbations (increased cortisol/DHEA ratio), and positive correlations are found between both serum IFN-α level and cortisol/DHEA ratio and atherogenic lipid levels (Christeff et al. 2002). Therefore, alterations in the apoptosis process may contribute to metabolic complications associated with antiretroviral therapies through the PI-dependent inhibition of adipocyte differentiation, NRTI-dependent mitochondrial toxicity and increased TNF-α synthesis, resulting in adipocyte death, lipoatrophy and possibly insulin resistance.

5 Conclusions

Continuous production of viral proteins by HIV-infected cells leads to an unbalanced immune activation and to the triggering of apoptotic programs, turning mononuclear cells, including CD4 T cells, CD8 T cells and APC, into effectors of apoptosis and leading to the destruction of healthy non-infected cells. This viral strategy could constitute an important mechanism of immune evasion, since apoptotic cells, through phagocytosis and cross-presentation by dendritic cells, are believed to induce tolerance rather than immunity. The progressive destruction of virus-specific effectors, such as $CD4^+$ T helper cells, either through their direct infection during cognate interaction with infected dendritic cells, or through a bystander process involving upregulation of death receptors and their ligands and downregulation of Bcl-2 family survival factors, leads to impaired immunity and subsequently to the lack of control of HIV replication. HAART limits HIV replication and retards disease progression but drug toxicity and the emergence of drug-resistant variants preclude long-term control in infected persons. Several strategies are under investigation to help the host fighting against HIV, including structured treatment interruption (STI) and immune-based therapies. STI is aimed at inducing "auto-vaccination" (a concept based on the theory that increased exposure to autologous virus through STI could stimulate HIV-specific immune responses and attenuate viral rebound) while reducing dependence on antiretroviral drugs (Havlir 2002). Use of therapeutic vaccines to increase the strength and breadth of HIV-specific cellular immune responses, resulting in clinical benefit, might also be successful. This has been recently shown in a monkey model in which a dramatic reduction in the level of AIDS virus in the blood and an increase in $CD4^+$ T cells was induced by immunization of monkeys with dendritic cells pre-incubated with chemically inactivated SIV (Lu et al. 2003). This strategy offers new hopes of the possibility of therapeutic vaccines to modulate the course of HIV disease.

Acknowledgements. This work was supported by grants from the Agence Nationale de Recherche sur le SIDA (ANRS), Ensemble contre le SIDA (Sidaction), the CNRS, the Pasteur Institute and the European Union.

References

Adachi Y, Oyaizu N, Than S, McCloskey TW, Pahwa S (1996) IL-2 rescues in vitro lymphocyte apoptosis in patients with HIV infection: correlation with its ability to block culture-induced down-modulation of Bcl-2. J Immunol 157:4184–4193

Alexander-Miller MA, Derby MA, Sarin A, Henkart PA, Berzofsky JA (1998) Supraoptimal peptide-major histocompatibility complex causes a decrease in Bcl-2 levels and allows tumor necrosis factor α receptor II-mediated apoptosis of cytotoxic T lymphocytes. J Exp Med 188:1391–1401

Amendola A, Gougeon M-L, Poccia F, Bondurand A, Fesus L, Piacentini M (1996) Induction of tissue transglutaminase in HIV pathogenesis. Evidence for a high rate of apoptosis of CD4 T lymphocytes and accessory cells in lymphoid tissues. Proc Natl Acad Sci USA 93:11057

Appay V, Dunbar PR, Callan M, et al. (2002) Memory CD8+ T cells vary in differentiation phenotype in different persistent virus infections. Nat Med 8:379–385

Badley A, McElhinny JA, Leibson PJ, Lynch DH, Alderson MR, Paya CV (1996) Upregulation of Fas ligand expression by human immunodeficiency virus in human macrophages mediates apoptosis of uninfected T lymphocytes. J Virol 70:199–206

Badley AD, Dockrell D, Simpson M, et al. (1997) Macrophage dependent apoptosis of CD4+ T lymphocytes from HIV-infected individuals is mediated by FasL and tumor necrosis factor. J Exp Med 185:55–64

Badley AD, Parato K, Cameron DW, et al. (1999) Dynamic correlation of apoptosis and immune activation during treatment of HIV infection. Cell Death Differ 6:420–432

Badley AD, Pilon AA, Landay A, Lynch DH (2000) Mechanisms of HIV-associated lymphocyte apoptosis. Blood 96:2951–2964

Bastard JP, Caron M, Vidal H, et al. (2002) Association between altered expression of adipogenic factor SREBP1 in lipoatrophic adipose tissue from HIV-1-infected patients and abnormal adipocyte differentiation and insulin resistance. Lancet 359:1026–31

Benedict CA, Norris PS, Ware CF (2002) To kill or be killed: viral evasion of apoptosis. Nat Immunol 3:1013–1018

Blaak H, van't Woot AB, Brouwer M, et al. (2000) In vivo HIV-1 infection of CD45RA+CD4+ T cells is established primarily by syncytium-inducing variants and correlates with the rate of CD4+ T cell decline. Proc Natl Acad Sci USA 97:1269–1274

Bofill M, Gombert W, Borthwick NJ, et al. (1995) Presence of $CD3^+CD8^+Bcl\text{-}2^{low}$ lymphocytes undergoing apoptosis and activated macrophages in lymph nodes of $HIV\text{-}1^+$ patients. Am J Pathol 146:1542–1550

Boudet F, Lecoeur H, Gougeon M-L (1996) Apoptosis associated with ex vivo down-regulation of bcl-2 and up-regulation of Fas in potential cytotoxic CD8+ T lymphocytes during HIV infection. J Immunol 156:2282

Carr A, Samaras K, Burton S, et al. (1998) A syndrome of lipodystrophy, hyperlipidaemia and insulin resistance in patients receiving HIV protease inhibitors. AIDS 12:F51–F58

Casella CR, Rapaport EL, Finkel TH (1999) Vpu increases susceptibility of human immunodeficiency virus type-1-infected cells to Fas killing. J Virol 73:92–100
Castedo M, Roumier T, Blanco J, et al. (2002) Sequential involvement of Cdk1, mTOR and p53 in apoptosis induced by the HIV-1 envelope. EMBO J 21:4070–4080
Christeff N, de Truchis P, Melchior JC, Perronne C, Gougeon M-L (2002) Longitudinal evolution of HIV-1 associated lipodystrophy is correlated to serum cortisol/DHEA ratio and interferon-α. Eur J Clin Invest 32:775–784
Clark DR, Repping S, Pakker NG, et al. (2000) T-cell progenitor function during progressive human immunodeficiency virus-1 infection and after antiretroviral therapy. Blood 96:242–249
Clerici M, Shearer GM (1994) The Th1/Th2 hypothesis of HIV infection: new insights. Immunol Today 15:575–580
Collins KL, Chen BK, Kalams SA, Walker BD, Baltimore D (1998) HIV-1 Nef protein protects infected primary cells against killing by cytotoxic lymphocytes. Nature 391:397–401
Conti L, Rainaldi G, Matarrese P, et al. (1998) The HIV-1 vpr protein acts as a negative regulator of apoptosis in a human lymphoblastoid T cell line: possible implications for the pathogenesis of AIDS. J Exp Med 187:403–413
Crise B, Buonocore L, Rose JK (1990) CD4 is retained in the endoplasmic reticulum by the human immunodeficiency virus type 1 glycoprotein precursor. J Virol 64:5585–5593
De Oliveira Pinto L, Garcia S, Lecoeur H, Rapp C, Gougeon ML (2002a) Increased sensitivity of T lymphocytes to tumor necrosis factor receptor 1 (TNFR1)- and TNFR2-mediated apoptosis in HIV infection: relation to expression of Bcl-2 and active caspase-3 and caspase-3. Blood 99:1666–1675
De Oliveira Pinto L-M, Lecoeur H, Ledru E, Rapp C, Patey O, Gougeon M-L (2002b) Lack of control of T cell apoptosis under HAART. Influence of therapy regimen in vivo and in vitro. AIDS 16:329–339
Dockrell DH, Badley AD, Villacian JS, et al. (1998) The expression of Fas ligand by macrophages and its upregulation by HIV infection. J Clin Invest 101:2394–2405
Douek DC (2001) Evidence for increased T cell turnover and decreased thymic output in HIV infection. J Immunol 167:6663–6668
Douek DC, McFarland RD, Keiser PH, et al. (1999) Changes in thymic function with age and during the treatment of HIV infection. Nature 396:690–695
Douek DC, Koup RA, McFarland RD, Sullivan JL, Luzuriaga K (2000) Effects of HIV on thymic function before and after antiretroviral therapy in children. J Infect Dis 181:1478–1482
Estaquier J, Tanaka M, Suda T, et al. (1996) Fas-mediated apoptosis of CD4+ and CD8+ T cells from human immunodeficiency virus-infected persons: differential in vitro preventive effect of cytokines and protease antagonists. Blood 87:4959–4966
Estaquier J, Lelievre JD, Petit F, et al. (2002) Effects of antiretroviral drugs on human immunodeficiency virus type 1-induced CD4(+) T-cell death. J Virol 7:5966–5973
Ferri KF, Jacotot E, Blanco J, et al. (2000) Apoptosis control in syncytia induced by the HIV-1 envelope glycoprotein complex: role of mitochondria and caspases. J Exp Med 192:1081–1092
Finkel TH, Tudor-Williams G, Banda NK, et al. (1995) Apoptosis occurs predominantly in bystander cells and not in productive cells of HIV- and SIV-infected lymph nodes. Nat Med 1:129–133
Fleury S, Rizzardi GP, Chapuis A, Tambussi G, Knabenhans C, Simeoni E, Meuwly JY, Corpataux JM, Lazzarin A, Miedema F, Pantaleo G (2000) Long-term kinetics of T cell production in HIV-infected subjects treated with highly active antiretroviral therapy. Proc Natl Acad Sci USA 97:5393–82
Franco JM, Rubia M, Martinez-Moya M, et al. (2002) T-cell repopulation and thymic volume in HIV-1-infected adult patients after highly active antiretroviral therapy. Blood 99:3702–3706
Garcia S, Fevrier M, Dadaglio G, Lecoeur H, Riviere Y, Gougeon M-L (1997) Potential deleterious effect of anti-viral cytotoxic lymphocytes through the CD95 (Fas/APO-1)-mediated pathway during chronic HIV infection. Immunol Lett 57:53

Gougeon M-L, Montagnier L (1993) Apoptosis in AIDS. Science 260:1269–1270
Gougeon M-L, Montagnier L (2000) Programmed cell death as a mechanism of CD4 and CD8 T cells deletion in AIDS. Molecular control and effect of highly active antiretroviral therapy. Ann NY Acad Sci 887:199–212
Gougeon M-L, Garcia S, Heeney J, et al. (1993) Programmed cell death of T lymphocytes in AIDS related HIV and SIV infections. AIDS Res Hum Retrov 9:553–563
Gougeon M-L, Lecoeur H, Dulioust A, et al. (1996) Programmed cell death in peripheral lymphocytes from HIV-infected persons: the increased susceptibility to apoptosis of CD4 and CD8 T cells correlates with lymphocyte activation and with disease progression. J Immunol 156:3509–3520
Gougeon M-L, Lecoeur H, Boudet F, et al. (1997) Lack of chronic immune activation in HIV-infected chimpanzees correlates with the resistance of T cells to Fas/Apo-1 (CD95)-induced apoptosis and preservation of a Th1 phenotype. J Immunol 158:2964
Gougeon M-L, Lecoeur H, Sasaki Y (1999) The CD95 system in HIV infection. Impact of HAART. Immunol Lett 66:97–103
Gougeon M-L, Penicaud L, Fromenty B, et al. (2003) Adipocytes targets and actors in the pathogenesis of HIV-associated lipodystrophy syndrome and metabolic alterations. Antiviral Ther (in press)
Grossman Z, Meier-Schellersheim M, Sousa AE, et al. (2002) CD4+ T-cell depletion in HIV infection: are we closer to understanding the cause? Nat Med 8:319–323
Groux H, Torpier G, Monte D, et al. (1992) Activation-induced death by apoptosis in lymphocytes from human immunodeficiency virus infected asymptomatic individuals. J Exp Med 175:331–340
Hatzakis A, Touloumi G, Karanicolas R, et al. (2000) Effect of recent thymic emigrants on progression of HIV-1 disease. Lancet 355:599–604
Havlir DV (2002) Structured intermittent treatment for HIV disease: necessary concession or premature compromise? Proc Natl Acad Sci USA 99:4–6
Hazenberg MD, Otto SA, Cohen Stuart JW, et al. (2000a) Increased cell division but not thymic dysfunction rapidly affects the T-cell receptor excision circle content of the naive T cell population in HIV-1 infection. Nat Med 6:1036–1042
Hazenberg MD, Hamann D, Schuitemaker H, Miedema F (2000b) T cell depletion in HIV-1 infection: how CD4+ T cells go out of stock. Nat Immunol 1:285–289
Hellerstein M, Hanley MB, Cesar D, et al. (1999) Directly measured kinetics of circulating T lymphocytes in normal and HIV-1 infected humans. Nat Med 5:83–89
Hernandez-Lopez C, Varas A, Sacedon R, et al. (2002) Stromal cell-derived factor 1/CXCR4 signaling is critical for early human T-cell development. Blood 99:546–554
Hildeman DA, Zhu Y, Mitchell TC, Kappler J, Marrack P (2000) Molecular mechanisms of activated T cell death in vivo. Curr Opin Immunol 14:354–359
Ho D, Neumann AU, Perelson AS, et al. (1995) Rapid turnover of plasma virions and CD4 lymphocytes in HIV-1 infection. Nature 373:123–126
Jeremias I, Herr I, Boehler T, Debatin K-M (1998) TRAIL/Apo-2 ligand-induced apoptosis in human T cells. Eur J Immunol 28:143–152
Ju ST, et al. (1995) Fas (CD95) FasL interactions required for programmed cell death after T-cell activation. Nature 373:444–448
Katsikis PD, Wunderlich ES, Smith CA, Herzenberg LA, Herzenberg LA (1995) Fas antigen stimulation induces marked apoptosis of T lymphocytes in human immunodeficiency virus-infected individuals. J Exp Med 181:2029–2037
Khaled AR, Durum SK (2002) Lymphocide: cytokines and the control of lymphoid homeostasis. Nat Rev Immunol 2:817–830
Kovacs JA, Lempicki RA, Sidorov IA, et al. (2001) Identification of dynamically distinct subpopulations of T lymphocytes that are differentially affected by HIV. J Exp Med 194:1731–1741
Krammer PH (2000) CD95's deadly mission in the immune system. Nature 407:789–795

Lecoeur H, Ledru E, Gougeon M-L (1998) A cytofluorometric method for the simultaneous detection of both intracellular and surface antigens on apoptotic peripheral lymphocytes. J Immunol Methods 217:11

Ledru E, Lecoeur H, Garcia S, Debord T, Gougeon M-L (1998) Differential susceptibility to activation-induced apoptosis among peripheral Th1 subsets: correlation with Bcl-2 expression and consequences for AIDS pathogenesis. J Immunol 160:3194–3206

Ledru E, Christeff N, Patey O, Melchior J-C, Gougeon M-L (2000) Alteration of TNF-a T cell homeostasis following potent antiretroviral therapy. Contribution to the development of HIV-associated lipodytrophy syndrome. Blood 95:1391–3198

Lempicki RA, Kovacs JA, Baseler MW, et al. (2000) Impact of HIV-1 infection and highly active antiretroviral therapy on the kinetics of CD4+ and CD8+ T cell turnover in HIV-infected patients. Proc Natl Acad Sci USA 97:13778–13783

Levy JA (1998) HIV and the pathogenesis of AIDS. American Society for Microbiology, Washington, DC

Li CJ, Wang C, Friedman DJ, Pardee AB (1995) Reciprocal modulation between p53 and Tat of human immunodeficiency virus type 1. Proc Natl Acad Sci 92:5461–5464

Lu W, Andrieu JM (2000) HIV protease inhibitors restore impaired T-cell proliferative response in vivo and in vitro: a viral-suppression-independent mechanism. Blood 96:250–258

Lu W, Wu X, Lu Y, Guo W, Andrieu JM, et al. (2003) Therapeutic dendritic cell vaccine for simian AIDS. Nat Med 9:27–32

Lum JJ, Pilon AA, Sanchez-Dardon J, et al. (2001) Induction of cell death in human immunodeficiency virus-infected macrophages and resting memory CD4 T cells by TRAIL/Apo2L. J Virol 75:11128–11136

Martinou J-C, Green DR (2001) Breaking the mirochondrial barrier. Nat Rev Mol Cell Biol 2:63–67

McCune JM (2001) The dynamics of CD4+ T cell depletion in HIV disease. Nature 410:974–979

McMichael AJ, Rowland-Jones S (2001) Cellular immune responses to HIV. Nature 410:980–987

Medrano FJ, Leal M, Arienti D, et al. (1998) TNFβ and soluble APO-1/Fas independently predict progression to AIDS in HIV-seropositive patients. AIDS Res Hum Retrov 14:835–843

Meng L, Zhou J, Sasano H, et al. (2001) Tumor necrosis factor alpha and interleukin 11 secreted by malignant breast epithelial cells inhibit adipocyte differentiation by selectively down-regulating CCAAT/enhancer binding protein alpha and peroxisome proliferator-activated receptor gamma: mechanism of desmoplastic reaction. Cancer Res 61:2250–2255

Meyaard L, Otto SA, Jonker RR, Mijnster MJ, Keet R, Miedema F (1992) Programmed death of T cells in HIV-1 infection. Science 257:217–219

Migueles SA, Laborico AC, Shupert WL, et al. (2002) HIV-specific CD8+ T cell proliferation is coupled to perforin expression and is maintained in nonprogressors. Nat Immunol 3:1061–1068

Miura Y, Misawa N, Maeda N, et al. (2001) Critical contribution of tumor necrosis factor-related apoptosis inducing ligand (TRAIL) to apoptosis of human CD4+ T cells in HIV-1 infected hu-PBL-NOD-SCID mice. J Exp Med 193:651–660

Mohri H, Bonhoeffer S, Monard S, Perelson A, Ho D (1998) Rapid turnover of T lymphocytes in SIV-infected rhesus macaques. Science 279:1223–1227

Mohri H, Perelson AS, Tung K, et al. (2001) Increased turn over of T lymphocytes in HIV-1 infection and its reduction by antiretroviral therapy. J Exp Med 194:1277–1287

Muro-Cacho CA, Pantaleo G, Fauci A (1995) Analysis of apoptosis in lymph nodes of HIV-infected persons. Intensity of apoptosis correlates with the general state of activation of the lymphoid tissue and not with the stage of disease or viral burden. J Immunol 154:5555–5561

Murphy KM, Reiner SL (2002) The lineage decisions of helper T cells. Nat Rev Immunol 2:933–947

Naora H, Gougeon M-L (1999) Interleukin-15 is a potent survival factor in the prevention of spontaneous but not CD95-induced apoptosis in CD4 and CD8 T lymphocytes of HIV-infected individuals. Correlation with its ability to increase Bcl-2 expression. Cell Death Diff 6:1002–1011

Napolitano LA, Grant RM, Deek SG, et al. (2001) Increased production of IL-7 accompanies HIV-1-mediated T-cell depletion: implications for T-cell homeostasis. Nat Med 7:73–79

Ogg GS, Jin X, Bonhoeffer S, et al. (1998) Quantitation of HIV-specific cytotoxic T lymphocytes and plasma viral RNA load. Science 279:2103–2106

Ohnimus H, Jassoy C (1997) Apoptotic cell death upon contact of CD4 T lymphocytes with HIV glycoprotein expressing cells is mediated by caspases but bypasses CD95 (Fas/Apo-1) and TNF receptor-1. J Immunol 159:5246–5252

Okamoto Y, Douek DC, McFarland RD, Koup RA (2002) Effects of exogenous interleukin-7 on human thymus function. Blood 99:2851–2858

Ometto L, De Forni D, Patiri F, et al. (2002) Immune reconstitution in HIV-1-infected children on antiretroviral therapy: role of thymic output and viral fitness. AIDS 16:839–842

Pakker NG, Notermans DW, de Boer RJ, et al. (1998) Biphasic kinetics of peripheral blood T cells after triple combination therapy in HIV-1 infection: a composite redistribution and proliferation. Nat Med 4:208–214

Pati S, Pelser CB, Dufraine J, et al. (2002) Antitumorigenic effects of HIV protease inhibitor ritonavir: inhibition of Kaposi sarcoma. Blood 99:3771–3779

Phenix BN, Badley AD (2002) Influence of mitochondrial control of apoptosis on the pathogenesis, complications and treatment of HIV infection. Biochimie 84:251–264

Piazza C, Montani MS, Gilardini S, et al. (1997) CD4+ T cells kill CD8+ T cells via Fas/Fas ligand-mediated apoptosis. J Immunol 158:1503–1511

Poulin J-F, Viswanathan MN, Harris JM, et al. (1999) Direct evidence for thymic function in adult humans. J Exp Med 190:479–486

Ribeiro RM, Mohri H, Ho DD, Perelson AS (2002) In vivo dynamics of T cell activation, proliferation, and death in HIV-1 infection: why are CD4+ but not CD8+ T cells depleted? Proc Natl Acad Sci USA 99:15572–15577

Schluns KS, Kieper WC, Jameson SC, Lefrançois L, et al. (2000) Interleukin-7 mediates the homeostasis of naive and memory CD8 T cells in vivo. Nat Immunol 1:426–432

Schwartz O, Marechal V, Le Gall S, Lemonnier F, Heard JM (1996) Endocytosis of major histocompatibility class I molecules is induced by the HIV-1 Nef protein. Nat Med 2:338–342

Selliah N, Finkel TH (2001) Biochemical mechanisms of HIV induced T cell apoptosis. Cell Death Diff 8:127–136

Silvetris F, Camarda G, Cafforio P, Dammacco F (1998) Upregulation of Fas ligand secretion in non-lymphopenic stages of HIV-1 infection. AIDS 12:1103–1118

Sloand EM, Young NS, Kumar P, Weichold FF, Sato T, Maciejewski JP (1997) Role of Fas ligand and receptor in the mechanism of T-cell depletion in AIDS: effect on $CD4^+$ lymphocyte depletion and HIV replication. Blood 89:1357–1363

Sloand EM, Kumar PN, Kim S, Chaudhuri A, Weichold FF, Young NS (1999) HIV-1 protease inhibitor modulates activation of peripheral blood CD4+ T cells and decreases their susceptibility to apoptosis in vitro and in vivo. Blood 94:1021–1030

Sloand EM, Maciejewski JP, Kumar P, et al. (2000) Protease inhibitors stimulate hematopoiesis and decrease apoptosis and ICE expression in CD34(+) cells. Blood 96:2735–2739

Spits H (2002) Development of $\alpha\beta$ T cells in the human thymus. Nat Rev Immunol 2:760–772

Wei X, Ghosh SK, Taylor ME, et al. (1995) Viral dynamics in human immunodeficiency virus type 1 infection. Nature 373:117–120

Willey RL, Maldarelli F, Martin MA, Strebel K (1992) Human immunodeficiency virus type 1Vpu protein induces rapid degradation of CD4. J Virol 66:7193–7200

HIV and Apoptosis: a Complex Interaction Between Cell Death and Virus Survival

J. Gil[2], M. Bermejo[1] and J. Alcamí[1]

1 Introduction

1.1 Apoptosis and the Immune System

Apoptosis, or programmed cell death, is a physiological mechanism by which the cell fragments its DNA and "commits suicide" in a controlled way (Coultas and Strasser 2000). The mechanisms of apoptosis are natural and even protect against uncontrolled cellular growth as well as playing a very important role in all development systems: embryogenesis, hematopoietic differentiation and proliferation, control of tumor proliferation, and regulation of immune activation (Holtzman et al. 2000). In the particular context of the immune system, apoptosis is a highly regulated form of cell death that is essential for maintaining a constant lymphocyte population size in the face of the continuous influx of new lymphocytes and homeostatic proliferation of existing cells (Khaled and Durum 2002). In addition, it is required during an immune response to foreign antigens in order to eliminate most activated antigen-specific T-cells and thereby prevent autoimmunity (Hilderman et al. 2002). Cell death occurs through two main pathways: activation-induced cell death, which is initiated by the binding of tumour-necrosis factor family death-receptor ligands to their cognate death receptors (the extrinsic pathway), and activated T-cell autonomous death, which is mediated by Bcl2-related proteins (the intrinsic pathway) and is initiated by internal sensors that transmit signals to the mitochondria (Kramer 2000; Martinou and Green 2001).

1.2 Virus-Induced Apoptosis

Infection with pathogenic viruses results in the death of target cells due to direct toxicity of viral products, compromised biochemical capacity of the cell

[1]AIDS Immunophathogenesis Unit, Centro Nacional de Microbiología, Ctra. Majadahonda a Pozuelo, 28220 Majadahonda, Spain, e-mail: ppalcami@isciii.es
[2]Wolfson Institute for Biomedical Research, University College, London, UK,

Progress in Molecular and Subcellular Biology
C. Alonso (Ed.): Viruses and Apoptosis
© Springer-Verlag Berlin Heidelberg 2004

or the antiviral immune mechanisms elicited by the host (Roulston et al. 1999; Benedict et al. 2002). In many of these processes, cell death is mediated by induction of the apoptotic machinery of the cell (Hay and Kannourakis 2002). Nevertheless, viruses are parasitic life forms that depend on the cell factory for entire life cycle. Viral products (including regulatory proteins) alone cannot achieve a complete viral cycle and cellular factors are necessary at virtually all stages of viral replication. Therefore, in many cases, viral infection modifies the cell environment at both transcriptional and post-transcriptional levels to allow or enhance its own replication. These viral counter-strategies include the control of key cell regulatory elements involved in both intrinsic and extrinsic apoptotic pathways: inactivation of p53, mimicking or induction of anti-apoptotic proteins, caspase inhibition or interference with death-receptor expression and the biochemical pathways involved in TNF-induced apoptosis. The final goal of this process is to allow viral replication despite the strong toxicity of viral products and the immune response elicited by viral infection (Hay and Kannourakis 2002).

1.3 HIV and Apoptosis

Many of the mechanisms described above are found in HIV infection (Badley et al. 2000), but original features of this virus deserve special consideration. First, the immune system, and in particular CD4 lymphocytes, are the main target of HIV. However, HIV-induced immunodeficiency is not exclusively related to direct CD4 killing. Other mechanisms including aberrant cell activation, impaired T-cell homeostasis and destruction of non-infected bystander cells by viral proteins have been described (Gougeon and Montagnier 2000; Gougeon 2003). In all these pathogenic mechanisms, viral-induced apoptosis plays a major role. Second, in the CD4 T-cell environment, HIV can enter a latency phase (Persaud et al. 2003). By means of this Trojan-horse mechanism, HIV persists in reservoirs (Blackson et al. 2002) and escapes from immune response but becomes absolutely dependent on cell activation through the T-cell receptor, cytokines or other immune stimuli to replicate in the infected cell. HIV reactivation from latency requires the activation of the Rel/NFκB family of transcription factors that initiate viral transcription and cooperate with the Tat protein in driving full RNA elongation (Alcami et al. 1995). Interestingly, NFκB is also a key element in regulating cell apoptosis and it has been reported that this factor protects from apoptosis (Barkett and Gilmore 1999). Even though contradictory data regarding the anti-apoptotic capacities of NFκB have been communicated, it is important to emphasise that the same factor that initiates HIV transcription promotes cell survival. Third, it has been reported that gene expression and cell machinery are modified by HIV. In particular, it has been demonstrated that HIV replication can induce the transcription factors required for viral expression, such as NFκB (Bachelerie et al. 1991), activate genes involved in signalling regulation (Geiss et al. 2000; van't

Wout et al. 2003) and modify the cell cycle (Somasundaran et al. 2002). These changes in cell physiology are directly linked to alteration of pro- and anti-apoptotic programs. These aspects of HIV infection draw an extremely complex pathogenic picture in which the interplay between viral-induced apoptosis and cell survival is mediated through an accurate balance between pro- and anti-apoptotic signals triggered by cellular and viral proteins.

In this chapter, we analyse the role of HIV proteins, in particular Env, Nef, Vpr Tat and protease, in virus-induced apoptosis. We will also review the functions of NFκB, a key transcription factor involved in both HIV transcription and apoptosis regulation.

2 The Virus Cycle

2.1 Interaction with Receptors and Virus Entry

The entry of HIV into the cell takes place via interaction with two types of receptors. The CD4 molecule, which is specific and common to all HIV subtypes, and a series of HIV co-receptors which correspond to different chemokine receptors (Weiss 1996). Although it has been reported that several chemokine receptors can act as viral receptors, CCR5 and CXCR4 are probably the most important in vivo (Berger et al. 1999). The CCR5 molecule binds the CC-chemokines RANTES, MIP-1α and MIP-1β, and is the main receptor of R5 HIV strains. The receptor CXCR4 has the chemokine SDF1/CXCL12 as a natural ligand and is the main receptor of the so-called X4 HIV strains. In addition to the viruses with a strict tropism by CCR5 or CXCR4, there have been reports of viral variants capable of entering the cell by way of other receptors or multiple co-receptors (dual or extended tropism strains). The chemokines which bind to CCR5 and CXCR4, especially RANTES and SDF, are capable of inhibiting HIV infection due to a phenomenon of interference with HIV at the level of binding with their co-receptors, and promoting their internalisation. In all probability, these mechanisms constitute a potent antiviral effector and modify the evolution of HIV infection in vivo (Kinter et al. 2000). Once interaction between gp120 and its receptors has taken place, there is a process of fusion between the viral and cellular membrane in which gp41 participates, and which allows the internalisation of the viral nucleocapsid and uncoating of the viral genome (Wyatt and Sodroski 1998) (Fig. 1).

2.2 Retrotranscription and Integration

Once the nucleocapsid enters the cell, retrotranscription of one of the strands of viral RNA takes place by means of the reverse transcriptase (RT) enzyme

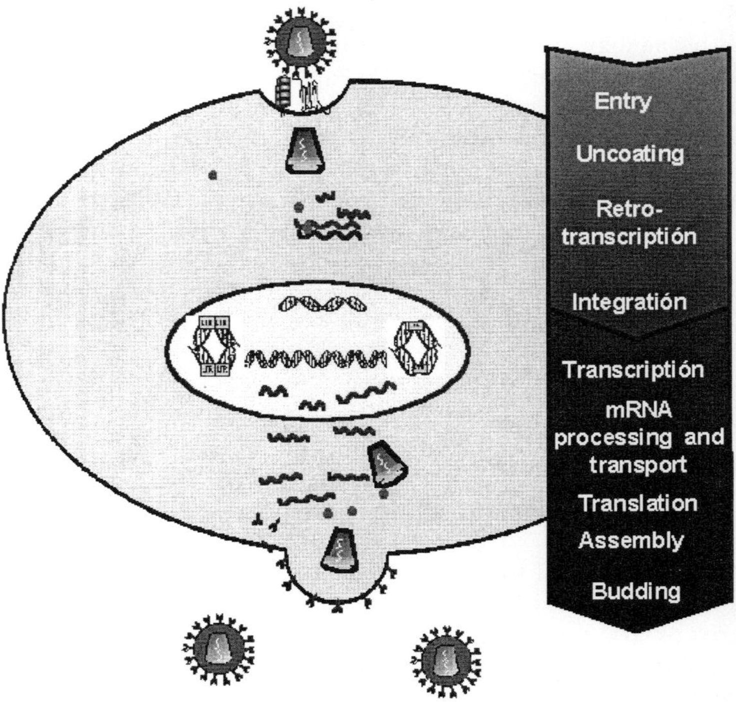

Fig. 1. The HIV cycle

which is transported in the virion itself. Retrotranscription is a complex process carried out in the cell cytoplasm via which RT generates a double strand of DNA that duplicates the LTR situated at both extremes of the proviral genome. Once the DNA is synthesised, it is transported to the nucleus and is integrated in the cellular genome via the action of a viral "integrase" which makes up what is known as an integrated "provirus". The transport process involves the active participation of viral proteins such as *Vpr* and the viral matrix protein p17 (Greene and Peterlin 2002). Similarly, it has also been shown that the process of retrotranscription and integration depends not only on viral factors, but also on cellular factors induced during the processes of cell activation. In resting CD4 lymphocytes, once the viral genome has been internalised, it is retrotranscribed incompletely and retrotranscription and integration do not finalise unless the cell is activated (Zack et al. 1990). These non-integrated proviral forms can remain in cytosol for up to 2 weeks and constitute a potential reservoir of HIV (Chun et al. 1997).

2.3 Latency and Replication

Once HIV is integrated in the genome of the infected cell, the viral genome can behave in different ways: it can remain latent, replicate at low level or undergo a massive replication with a consequent cytopathic effect on the infected cell. HIV replication is a sequential process which depends on the action of viral and cellular factors and which can be broken down into the following steps:

The initiation of transcription represents the beginning of HIV messenger RNA synthesis from proviral DNA in the cell genome. The move from a situation of transcriptional "silence" to one of transcriptional "activity" does not depend on viral proteins, but on cellular factors which interact with regulatory sequences located in the viral LTR. These factors, which act at the level of the enhancer and promoter genetic regulation sequences of HIV, allow the formation of the primary transcriptional complex (RNA polymerase II and associated factors), responsible for gene transcription. Among these factors, NFκB represents the main regulatory element in the transcription of HIV in CD4 lymphocytes from their state of latency (Alcami et al. 1995). Complete transcription of the viral genome requires the participation of the viral protein *tat*, which acts by increasing the transcription rate of the HIV genome 10^2–10^3-fold and allows the synthesis of all the viral RNA. *Tat* acts as a direct transactivator in direct cooperation with other cellular factors, in particular NFκB, and allows complete elongation of the viral messenger RNA (Cullen 1993). In the absence of *Tat*, there is no transcription of complete viral RNA. HIV mRNA is synthesised as a single transcript which must be transported to the cytosol and processed in transcripts of different sizes. Both steps, processing and transport, are essentially carried out by another viral protein, *Rev*, which tends to be found in the nucleus. In the absence of *Rev*, HIV mRNA accumulates in the nucleus and is not processed in its different transcripts. Similarly, *Rev* participates in the process of assembling the RNA messengers using the protein synthesis machinery and accelerates the synthesis of viral proteins by polysomes.

Once synthesised, the viral proteins must be processed after translation before being assembled in what will constitute the mature viral particles. Different viral proteins take part in this process, the most notable being *Vif*, *Vpu* and viral protease (Emerman and Malim 1998). The product of the *vif* gene is not essential for viral replication, but its deletion reduces infectivity by between 100 and 1000 times. The *vpu* gene is not essential for viral replication either, but, in its absence, proteins accumulate in the cytoplasm and fewer virions are produced. Viral protease plays a central role in the production of viral particles. It processes the pre-protein precursors p55-gag and gag-pol in the proteins of the nucleocapsid, the reverse transcriptase of the virus itself and the viral protease. Processing of gp160 in gp41 and gp120 is via a cellular protease. Final maturation of the virions and correct assembly of viral proteins take place at the last point in the infective cycle, during the process of budding of the virus via the cell membrane, and allow a mature viral particle to be constituted.

3 Mechanisms of HIV-Induced Lymphopenia

Loss of CD4 T cells and a progressive immune deficiency represent the hallmark of HIV infection but this phenomenon is not produced by a single cause. In addition to a direct cytopathic effect, other hypotheses to explain CD4 destruction have been proposed including killing of infected cells by CTL, aberrant activation of the immune system, apoptosis of bystander cells by HIV products and impaired production and altered homeostasis of CD4 cells. There is a fierce controversy surrounding the mechanisms involved in HIV-induced immunosuppression, especially with regard to the relative importance of these mechanisms in vivo. Nevertheless, these hypotheses are not exclusive and probably all the mechanisms reported contribute to the quantitative and qualitative alterations of the different lymphocyte subpopulations in HIV infection.

3.1 Cytopathic Effect

Given the aggressive kinetics of viral replication, killing of CD4 lymphocytes by direct cytopathic effect is considered to be the most important cause of lymphocyte destruction. The mathematical models developed from modifications in levels of viral load and CD4 lymphocytes after highly active antiretroviral therapy calculate that HIV destroys around 10^8 CD4 lymphocytes daily by direct cytopathic effect (Perelson et al. 1996).

3.2 Trapping and Redistribution

The cytopathic effect model referred to above is not accurate since it presupposes a homogeneous traffic and distribution of lymphocytes between peripheral blood (which contains only 1% of total lymphocytes) and the lymphoid organs. However, the CD4 lymphopenia observed in HIV infection not only reflects the direct destruction of CD4 lymphocytes, but it is secondary, in part, to a process of lymphocyte redistribution. It has been shown that the accumulation of the virus in lymph nodes leads to a phenomenon of trapping of lymphocytes in the lymphoid organs around the virion-covered dendritic cells (Haase et al. 1996). The existence of this hijacking in the lymph nodes has been confirmed by the observation that the increase in CD4 produced during the first weeks after the initiation of antiretroviral treatment is, in fact, a redistribution of lymphocytes from lymph nodes to peripheral blood when viral load falls drastically in the lymphoid organs (Parker et al. 1998). Therefore, the mathematical models that do not consider this component of redistribution probably overestimate the number of lymphocytes infected and destroyed by the direct viral mechanism.

3.3 Immune-Mediated Destruction

As CD4 lymphocytes are the main type of cell infected by HIV, they become a target cell of the immune system itself. Therefore, the infected CD4 cells are destroyed by cytotoxic lymphocytes in a specific HLA context, as was demonstrated in models where the infusion of CD8 lymphocytes activated against HIV led to a reduction in the number of infected CD4 (Brodie et al. 1999). Similarly, it has been observed that, during primary infection, there is a correlation between the fall in CD4 and the expansion of antiviral CD8 clones (Ogg et al. 1998). These clinical and experimental data suggest that the anti-HIV CTL response is a mechanism which contributes to the destruction of CD4.

3.4 Apoptosis

Increasing evidence points to HIV-driven lymphocyte apoptosis as an important contributor to the destruction of the immune system. The many mechanisms that contribute to HIV-associated lymphocyte apoptosis include chronic immunologic activation, interaction of envelope proteins with viral receptors, toxicity of viral proteins, enhanced expression of cytotoxic ligands and cytokine production by lymphocytes and monocytes. The role of premature lymphocyte apoptosis in CD4+ T cell loss during HIV infection was indicated by early reports (Ameisen and Capron 1991; Groux et al. 1992) suggesting that apoptosis could be a mechanism of destruction of CD4 lymphocytes in HIV infection which would affect not only the infected cells, but which could also lead to the destruction of non-infected lymphocytes. According to this hypothesis, it has been demonstrated that in the lymph nodes of infected patients, there are, in vivo, a majority of apoptotic cells which are not infected and a minority of cells which actively replicate the virus and do not present signs of apoptosis (Finkel et al. 1995). These data together with many different in vitro experiments suggest that apoptosis may be a indirect mechanism of lymphocytic destruction in HIV infection (Jaworovski and Crowe 1999; Basanez and Zimmerberg 2001; Cloyd et al. 2001).

3.5 Persistent Immune Activation

Chronic uncontrolled infections provide continuous antigenic stimulation that causes persistent immune activation. Given that the number of cycles of cell division is limited some authors have proposed that the reduction in the number of CD4 and CD8 along HIV infection may be secondary to the exhaustion of the cell response when the system undergoes chronic and massive antigenic pressure (Hazenberg et al. 2000a; van Baarle et al. 2002). In these circumstances, CD8 lymphocytes with antiviral activity would be continuously generating effector cells and the CD4 pool would also undergo an increased turnover to compensate continuous lymphocytic destruction by viral replication (Perelson et al. 1996).

3.6 Blockade in Cell Regeneration

The models of viral replication and lymphocyte destruction by direct cytopathic effect calculate that around 10^8 lymphocytes are destroyed every day. This implies that a similar quantity of cells must be produced in order to maintain the number of CD4 and to reach a state of equilibrium in which the number of destroyed cells is compensated by an abnormal state of CD4 proliferation which would submit the system to a severe immunological "stress" (Perelson et al. 1996). However, when it has been possible, using metabolic labelling or cell division parameters, to measure lymphocyte kinetics in seropositive patients before beginning treatment, it has been observed that the percentage of CD4 lymphocytes undergoing cell proliferation is reduced in seropositive patients with respect to non-infected patients. Contrary to expectations, antiviral treatment and control of viremia do not reduce the number of proliferating CD4 lymphocytes 4, but instead this percentage increases and becomes similar to that observed in non-infected patients. In global terms, these data support the hypothesis that viral replication leads to a blockade in the kinetics of CD4 lymphocyte regeneration both at central level (thymus and perhaps bone marrow) and peripheral level (lymph nodes; McCune 2001). Other authors criticise this hypothesis and defend the experimental results obtained as not being due to a phenomenon of thymic dysfunction but secondary to changes in the activation of the immune system and to the redistribution of lymphoid subpopulations which occurs in the phases of active replication and control of viremia (Hazenberg et al. 2000a,b, 2003).

4
HIV-Induced Apoptosis

As previously described, CD4 lymphopenia represents the end-point of different processes produced by HIV infection which share apoptosis as a common pathway to induce lymphocytic death. In this section, the apoptotic and anti-apoptotic mechanisms triggered by viral products that lead to killing of infected or bystander non-infected cells are examined. Particularly, the role of Tat, Vpr, Nef, Vpu and the viral protease will be considered. Furthermore, the role of NFκB as a key transcription factor involved in both apoptosis control and regulation of the HIV cycle will be analysed. It is important to emphasise that apoptosis in HIV infection is not only involved in the destruction of CD4 lymphocytes, since it has been reported that other subpopulations such as CD8, monocytes or nervous system cells activate cell death programs after contact with HIV proteins (Takahashi et al. 2002; Peterlin and Trono 2003).

4.1
Envelope–Receptor Interactions

It has been shown that blockade of the CD4 molecule by monoclonal antibodies before activation of the antigen receptor induces apoptosis of the activated cell (Oyaizu et al. 1993). Therefore, incomplete or asynchronic activation of CD4 lymphocytes produces an abnormal activation of the cell that, in turn, leads to an apoptotic death programme instead of inducing an adequate immune response. In preliminary experiments, it was proven that contact between viral particles or the gp120 HIV glycoprotein with CD4 lymphocytes has a similar effect to that reported using anti-CD4 antibodies and can induce apoptosis if the cells are later activated (Banda et al. 1992). The molecular mechanisms involved in apoptosis induction through CD4–receptor interactions have been extensively investigated. It has been shown that cross-linking of CD4 T cells by gp120 activates the CD95-CD95L pathway and enhances susceptibility to Fas-mediated killing (Banda et al. 1992; Oyaizu et al. 1994). Moreover, in previously activated cells, gp120 cross-linking induces downregulation of BCL-2 (Hashimoto et al. 1997) and activates caspase 3 (Cicala et al. 2000). The apoptotic response to gp120 is almost completely inhibited by soluble CD4 and by anti-gp120 antibodies (Laurent-Crawford et al. 1993). Finally, this interaction must also involve CD4 signalling because deletion or mutation of the intracytoplasmic domain of CD4 also abrogated the apoptotic response (Guillerm et al. 1998; Moutouh et al. 1998). In addition to CD4, CXCR4–gp120 interaction has been reported to be important in mediating envelope-induced apoptosis. Indeed, bystander killing in CD4 T cells is completely prevented by the addition of the CXCR4 inhibitor AMD3100 or an anti-CXCR4 monoclonal antibody, neither of which interferes with the engagement of gp120 with CD4 (Blanco et al. 2000). Interestingly, not only can apoptosis be mediated through interaction between gp120 and HIV receptors (Basanez and Zimmerberg 2001) but, in the process of fusion mediated by gp41, transduction pathways which induce cell death are also activated (Blanco et al. 2003).

It was initially suggested that circulating immune complexes and replication-incompetent viruses that contain gp120 can induce death in a similar manner (Aceituno et al. 1997; Kameoka et al. 1997) but further work performed with lymphoid tissue ex vivo has shown that inactivated virions cannot deplete CD4 lymphocytes (Sylwester et al. 1998; LaBonte et al. 2003).

In vitro experiments show that as well as in soluble form, the viral envelope expressed on the plasma membrane of infected cells can interact with viral receptors on uninfected cells and trigger apoptosis (Sodroski et al. 1986; Ferri et al. 2000). Cells expressing the HIV-1 envelope glycoprotein complex (gp120/gp41, Env) induce the death of target cells either after cell-to-cell fusion or after cell-to-cell contact in a fusion-independent fashion and both processes require gp120 and gp41 function (Blanco et al. 2003). Syncytia that arise from the fusion of Env-expressing cells with cells that express the CD4-CXCR4 complex undergo apoptosis through a mitochondrion-dependent pathway. This pathway is initiated by the upregulation of the cyclin B-CDK1 (cyclin dependent kinase 1) pathway and nuclear translocation of the mammalian target of rapamycin (MTOR). This leads to MTOR-mediated phosphorylation of p53, p53-dependent upregulation of expression of Bax (BCL-2 associated X protein) and activation of the mitochondrial pathway. Interestingly, peripheral-blood and lymph-node cells from HIV-positive individuals have increased expression of cyclin B and MTOR that correlates with p53Ser15 phosphorylation and viral load (Castedo et al. 2002). Whether these mechanisms elicited by env-dependent cell fusion in vitro are relevant in vivo remains a matter of debate because the formation of syncytia is rarely observed in tissue biopsies from HIV-infected patients.

Given that in HIV-infected patients the emergence of X4 or R5X4 strains correlate with rapid CD4 decrease and progression to AIDS (Glushakova et al. 1998), it has been proposed that X4-tropic envelopes could be more cytopathic than R5 strains. In agreement with this, in human lymphoid tissues ex vivo, apoptosis of uninfected bystander CD4 T cells is a major mechanism of lymphoid depletion caused by X4 and R5X4 HIV-1 strains but is only a minor mechanism of depletion by R5 strains (Glushakova et al. 1999; Penn et al. 1999; Jekle et al. 2003). However, both CXCR4- and CCR5-tropic envelopes are equally cytopathic in vitro (Grivel and Margolis 1999; LaBonte et al. 2003) and the observed differences are probably due to the different expression of CCR5 and CXCR4 receptors in lymphocytes (Bermejo et al. 1998; Grivel and Margolis 1999).

4.2
Viral Protein R (Vpr)

Vpr is an accessory protein of HIV that is dispensable for replication in lymphocytes but not in non-dividing cells such as macrophages. Vpr is involved in many different functions. Regarding the viral cycle, Vpr is incorporated in the preintegration complex and collaborates in its transport to the nucleus. It is a heterologous transactivator of the HIV-LTR and other viral promoters and increases expression from unintegrated DNA. Vpr also reduces CD4 proliferation through G2 cell cycle arrest which favours viral integration (Ayavoo et al. 1997; Somasundaran et al. 2002). Importantly, Vpr is found in the virion and can be secreted from infected cells and detected in culture supernatant and plasma from HIV-infected patients but it is not known if these soluble forms are biologically active. In vitro extracellular Vpr is extremely neurotoxic and can induce apoptosis and G2 arrest in bystander cells thus contributing to cell depletion in lymphoid tissues and CNS damage. Although prolonged G2 arrest can lead to apoptosis and during this phase of the cell cycle expression of apoptosis-involved proteins such as Wee-1 kinase (Yuan et al. 2003) and Survivin (Zhu et al. 2003) is inhibited, apoptosis is not directly related with G2 arrest. In agreement with this hypothesis, protein determinants responsible for G2 cell cycle arrest and apoptosis are different in the Vpr molecule (Somasundaran et al. 2002). Vpr-induced apoptosis is mainly mediated through activation of the intrinsic pathway (Brenner and Kroemer 2003; Muthumani et al. 2003). Although different mechanisms such as caspase activation (Shostak et al. 1999; Stewart et al. 2000; Muthumani et al. 2002) have been described, a direct interaction between Vpr and the adenine nucleotide translocator (ANT) in the inner membrane of the mitochondria seems to be the mechanism that triggers apoptosis-mediated Vpr (Jacotot et al. 2000). Recently, a Vpr mutant that is unable to interact with ANT has been described in long-term surviving HIV-infected patients (Lum et al. 2003), thus supporting a pathogenic role for Vpr-induced apoptosis in vivo.

4.3
Transcription Anti-terminator (Tat)

Tat is an essential protein in the HIV cycle that interacts with different cellular factors to drive viral transcription. Tat can be secreted by infected cells. Although a small number of articles have described anti-apoptotic properties for Tat (McCloskey et al. 1997; Gibellini et al. 2001) there is solid evidence showing that Tat is a pro-apoptotic factor in different cell types including endothelial cells (Kim et al. 2003), neurons (Haughey and Mattson 2002) and T lymphocytes (Li et al. 1995; Westendorp et al. 1995) and can be secreted from infected cells and mediate its apoptotic effects on bystander cells. However, it is not known whether the concentration and activity of soluble Tat in vivo are able to mediate its apoptotic effect.

Different mechanisms have been proposed to explain Tat-mediated apoptosis. In cells from the CNS calcium deregulation and release of NO (Haughey and Mattson 2002; Kim et al. 2003) trigger apoptosis in a TNF-independent manner. In T lymphocytes Tat induces spontaneous and activation-dependent apoptosis through different pathways including an increase in CD95L (Yang et al. 2002), upregulation of caspases (Bartz and Emerman 1999) and interaction with the microtubule network (Chen et al. 2002). Interestingly, HIV infection or treatment with soluble Tat upregulates Bcl-2 expression in monocytes which become protected from apoptosis (Zhang et al. 2002). However, both infected and Tat-treated monocytes secrete TRAIL (TNF-related apoptosis-induced ligand) that kills uninfected CD4 lymphocytes (Yang et al. 2003).

4.4 Vpu

Vpu is an accessory protein that increases the release of viral particles. In addition, Vpu retains CD4 in the reticulum (Speth and Dierich 1999) and inhibits degradation of cellular proteins through its interaction with the ubiquitin complex that targets proteins for degradation by the proteasome (Akari et al. 2001). Vpu increases susceptibility to Fas killing (Casella et al. 1999) through this latter mechanism. Vpu blocks IκB degradation in the proteasome thus decreasing NFκB activity and subsequent transcription of NFκB-dependent anti-apoptotic genes such as TRAF and Bcl-xL (Akari et al. 2001).

4.5 Nef

Nef is an important protein involved in the pathogenesis of AIDS although some of its mechanisms are not fully understood (reviewed in Fackler and Baur 2002). Nef induces HIV replication and enhances infectivity in some experimental systems but, more importantly, permits viral evasion from the immune response of the host through downregulation of MHC-I and CD4 molecules. Nef not only protects the infected cell from CTL killing by reducing MHC-I expression, but also interacts with the T-cell receptor ξ chain and induces CD95L in infected cells (Zauli et al. 1999). CD95L expression stimulates apoptosis on the surrounding viral-specific CTL through the cross-linking of Fas/CD95 in these cells (Xu et al. 1997). Together with these mechanisms mediating immune evasion, Nef directly protects the infected cell from apoptosis by repressing pro-apoptotic signalling. It has been reported that Nef interacts with apoptosis signal regulating kinase-1 (ASK1), an intermediary in Fas-dependent apoptosis, and inhibits its activity (Geleziunas et al. 2001).

Moreover, Nef activates PAK and PI3 kinases, which results in Bad inactivation and increases levels of the anti-apoptotic protein Bcl-2 (Wolf et al. 2001). Finally, Nef interacts with p53, a critical regulatory element in the intrinsic apoptotic pathway, and could inhibit its apoptotic function (Greenway et al. 2002). In summary, Nef protects from apoptosis triggered both by the external (Fas-dependent) and internal pathways in experimental models and could enhance survival of infected cells (Xu and Screaton 2001). Generation of HIV replication from latency is an extremely rapid process and takes between 6 and 8 h to generate an infectious progeny (Bermejo et al., submitted). Even if the anti-apoptotic effect of Nef delays cell death for only a few hours, this can be critical to allow full HIV replication and viral propagation. As it has been cleverly expressed, the main function of Nef could be to provide both the armour and the sword through apoptosis subversion (Ameisen 2001).

4.6 Viral Protease

It has been reported that, in addition to their antiviral effects, protease inhibitors (PI) could improve immune reconstitution through apoptosis inhibition (for a review, see Phenix et al. 2002). The anti-apoptotic effect of protease inhibitors has been related to inhibition of TNF-mediated cell death (Wolf et al. 2003). In agreement with this hypothesis, protease inhibitors decrease immune activation and CD95 expression in vitro (Estaquier et al. 2002) and in vivo (Jimenez et al. 2002). On the other hand, a pro-apoptotic effect for PI through changes in mitochondrial membrane potential has also been reported (Estaquier et al. 2002; Matarrese et al. 2003). The relevance of these effects in vivo deserves further study.

5 The NFκB System and Its Role in HIV-Induced Apoptosis

Initially, NFκB was identified as a nuclear factor interacting with the immunoglobulin enhancer sequence (Sen and Baltimore 1986). Only when it was cloned, did it become clear that NFκB was not only a nuclear factor, but also a family of transcription factors, with an intricate regulation. NFκB regulates a wide set of genes involved in the regulation of immune and inflammatory responses (Pahl 1999). These genes include receptors (MHC-I, MHC-II, TCR, β2-microglobulin), cell adhesion molecules (ICAM-I, VCAM-I), cytokines (IFN-γ, GM-CSF, IL-2, IL-6, IL-8, TNF-α), transcription factors (NFκB, c-myc, IRF-1), and molecules regulating apoptosis (Fas, FasL, XIAP, TRAFs). Through the upregulation of all these genes, NFκB controls both innate and adaptative immune responses, and its deregulation is associated with different inflammatory diseases.

NFκB is also a key controller of the antiviral response, for example, through IFN regulation (Stark et al. 1998). In addition, viruses have exploited NFκB induction for transcribing their own genomes, as is the case with HIV-1 and HTLV-1, for stimulating proliferation and survival of lymphocytes.

5.1 The NFκB Family of Transcription Factors

There are five members of the NFκB family of transcription factors in mammals: Rel (c-Rel), RelA (p65), RelB, NFκB1 (p50 and its precursor p105), and NFκB2 (p52 and its precursor p100; Fig. 2). Structurally, the proteins have a conserved Rel homology region (RHR), spanning around 300 aas, which is involved in dimerisation, DNA binding and interaction with IκB proteins. There is also a nuclear localisation signal (NLS) located in the same region.

Proteins of the NFκB family can combine to form homodimeric or heterodimeric transcription complexes that activate a varied set of genes in response to proinflammatory cytokines, viral and bacterial infections and a huge range of cellular stresses (Ghosh et al. 1998). Different combinations of NFκB transcription factors present different transactivation or even repression activities, and also distinct affinities for kB sites located in promoters of different genes (Perkins et al. 1992, May and Ghosh 1997).

In resting cells, NFκB stays in the cytoplasm as an inactive form (Baeuerle and Baltimore 1988). This first report established the central dogma of NFκB

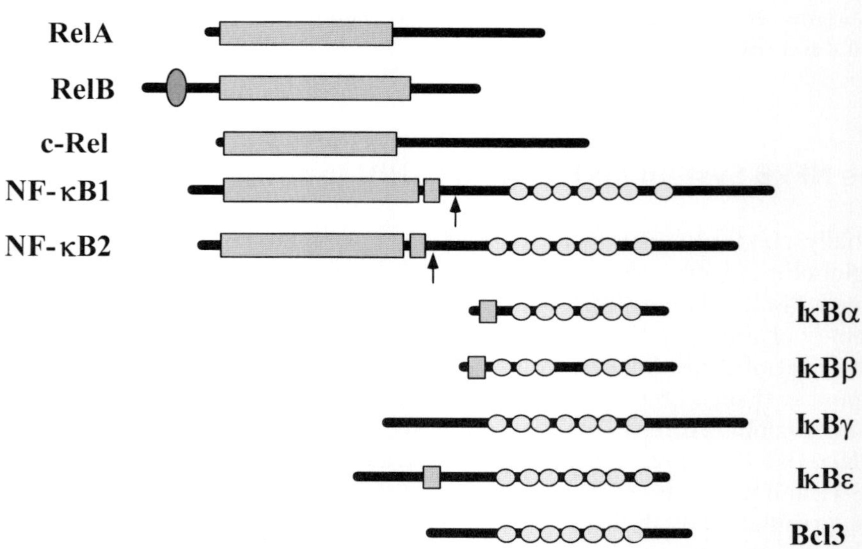

Fig. 2. The NF-κB and IκB families

regulation by the translocation of activated NFκB from the cytoplasm to the nucleus, and remains basically true up to the present. The factors responsible for maintaining NFκB inactive in the cytoplasm are proteins of the IκB family.

5.2 The IκB Family

This family of inhibitory proteins includes IκBα, IκBβ, IκBγ, IκBε, Bcl3, p100 and p105 (Fig. 2; Gosh et al. 1998). Structurally, they contain six or seven ankyrin repeats that are involved in binding to the RHR of NFκB family proteins, thus hiding their NLS. IκBα, IκBβ and IκBγ contain the N-terminal regulatory regions necessary for their controlled degradation in response to different stimuli. The different IκB proteins show preferences for binding to specific forms of NFκB. IκBs also play an important role in ending NFκB activation. Hence, newly synthesised IκBα enters the nucleus and binds NFκB, thus enhancing its dissociation from the DNA and relocating it to the cytoplasm by means of the IκBα nuclear export sequence (NES; Arenzana-Seisdedos et al. 1997).

5.3 Mechanisms of NFκB Activation

Although initially NFκB was thought to be a B-cell-specific transcription factor, it was soon shown that NFκB activity could be induced in a wide range of cell types by a plethora of stimuli (Fig. 3). These stimuli include PMA, TNF-α, IL-1, bacterial endotoxin, dsRNA, viruses or viral proteins such as Tax protein from HTLV-1 (Gosh et al. 1998). Upon stimulation, NFκB translocates to the nucleus where it binds to the promoter region of its target genes, thus regulating their transcription. NFκB localisation to the nucleus is preceded by degradation of IκB proteins, which constitutes the key controlled event in NFκB activation.

Potent NFκB activators induce complete IκB degradation in minutes. This process is mediated by the 26S proteasome (de Martino and Slaughter 1999) and is concomitant to IκB phosphorylation in two conserved adjacent serine residues (serines 32 and 36 in human IκBα) located in the N-terminal regulatory domain of IκB family proteins (DiDonato et al. 1996). Upon phosphorylation, IκB suffers a second post-translational modification; the addition by an SCF family ubiquitin ligase of multiple ubiquitin molecules in two lysines (lysines 21 and 22 in IκBα; Ben Neriah 2002). Thus, the strictly regulated step, essential for NFκB activation, is the inducible phosphorylation of IκB in response to stimuli, given the fact that other components of this complex activation pathway, such as the enzymes involved in ubiquitination or the proteasome, are constitutively active.

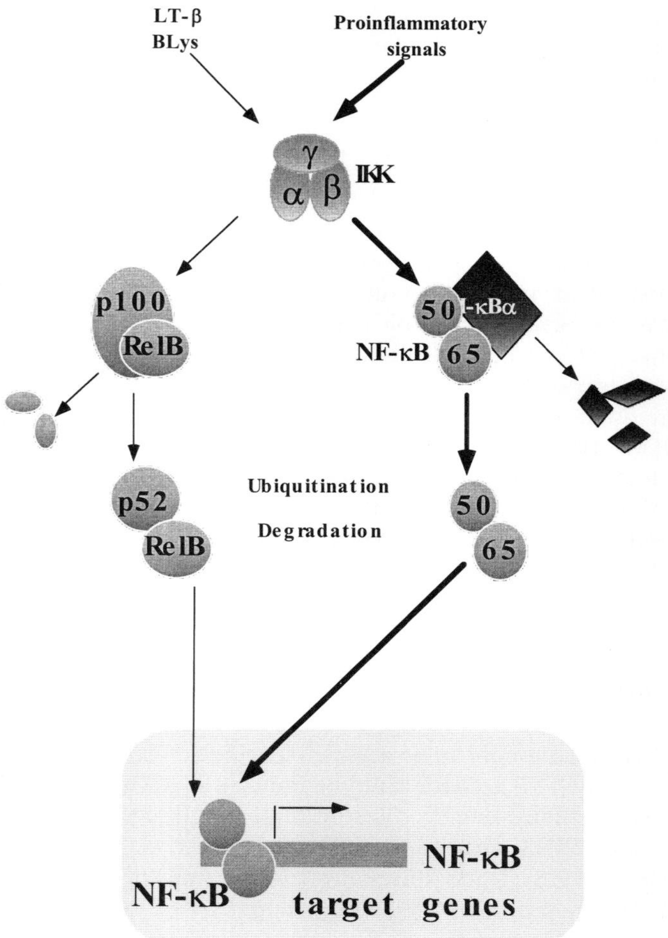

Fig. 3. NF-κB activation

5.4
IκB Kinase: the Key Regulator of the NF-κB Pathway

IκB is phosphorylated by a high molecular weight complex, formed by two different protein kinases (termed IKKα and IKKβ) and an essential adapter molecule, known as IKKγ or NEMO (Karin 1999). Structurally, IKKα and IKKβ are highly similar. They are composed of a kinase domain, followed by a leucine zipper, and a helix-loop-helix motif (HLH). IKKγ presents three regions with a helix structure, including a leucine zipper. Biochemical fractioning and characterisation of an IKK complex purified from mammalian cells indicate that they are composed of IKKα homodimers or IKKαIKKβ heterodimers bound to an undetermined number of IKKγ subunits (Rothwarf et al. 1998). IKK becomes activated in response to pro-inflammatory stimuli in a way dependent on the IKKα phosphorylation on an array of serine residues located in its activation loop. Other serines located in the C-terminal region of IKKα are involved in the deactivation process when phosphorylated. This explains the transient nature of IKK activation (Delhase et al. 1999).

In addition, both IKKα and IKKβ are required for NFκB activation triggered by pro-inflammatory stimuli including bacterial and viral infections, dsRNA and antigens (Chu et al. 1999). All these stimuli lead to NFκB activation whose kinetics are similar to those observed for IKK activation. Thus, a key question is how all these different stimuli converge in activating the IKK complex. Basically, there are two main hypotheses. The first proposes that a scaffolding protein could bring the IKK complex close to the receptor, resulting in a conformational change and causing IKK autophosphorylation. The other option is that an upstream kinase is involved in phosphorylating the IKK complex. Several candidates have been proposed, including MEKK1, MEKK2, MEKK3, NIK, TAK, TBK1 or PKCζ, but no definitive evidence exists, especially as knockouts of these kinases failed to impair NFκB activation (Li and Verma 2002).

5.5
From Death Receptors to NFκB

In addition to surviving signals acting as a reassurance mechanism for not being eliminated by the apoptotic machinery, mammalian cells have developed other pathways to signal for self-destruction of certain cells. This kind of apoptotic signalling has striking importance for maintaining homeostasis of the immune system (Boise and Thompson 1996). The key molecules involved in initiating apoptosis under these conditions are receptors of the TNFR superfamily (reviewed in Ashkenazi and Dixit 1998; Yeh et al. 1999), also termed death receptors (DRs). Upon binding to their respective ligands, DRs become activated and signal the cell to initiate certain genetic programmes, including activation of NFκB and triggering of programmed cell death.

5.5.1
Family of Death Receptors

The TNFR superfamily of transmembrane proteins includes two gene families encompassing 18 ligands and 28 receptors. Their best-characterised members are Fas (or CD95), whose ligand is FasL (CD95L), and TNFR-1, whose ligand is TNF-α. Other family members are DR3, DR4, DR5, and DR6. Apo3L is the DR3 ligand, and Apo2L (TRAIL) is the ligand for DR4 and DR5 (reviewed in Ashkenazi and Dixit 1998). In addition, there is a set of decoy receptors, non-signalling members of the family, which function by attenuating death receptor function.

5.5.2 *Signal Transduction Through Fas*

At the physiological level, Fas and FasL play a very important role in regulating homeostasis of the immune system, the deletion of activated T cells, and elimination of cancer- or virus-infected cells. They are also involved in eliminating pro-inflammatory cells in immune privileged places (Nagata 1997). Furthermore, mutations of the genes coding for either Fas or FasL are involved in autoimmune diseases characterised by the accumulation of peripheral blood cells (PBLs) both in mice and in humans (Nagata 1997).

FasL is a homotrimer, and its binding to Fas provokes trimerisation of the latter, and the clustering of its cytoplasmic death domains (DDs; Ashkenazi and Dixit 1998). This clustering of the DDs makes it possible for Fas to recruit FADD, an adapter protein that binds to the receptor via homotypic interaction through its DDs (Chinnaiyan et al. 1995). In addition FADD contains a death effector domain (DED) that can recruit pro-caspase 8 (Muzio et al. 1996), facilitate self-processing, and thus trigger activation of the proteolytic caspase pathway.

5.5.3 *Signalling by TNF-R1*

Tumor necrosis factor was first described as an anticancer activity in 1894, when Colley realised that cell-free extracts from bacteria could cause tumour shrinkage (Ashkenazi 2002). It took almost 100 years to begin to understand the genetic and molecular basis of this process. TNF-α is a soluble factor whose binding to TNFR-1 triggers its oligomerisation and induces its activation (reviewed in Chen and Goeddel 2002). TNF-α binding finishes in the activation of the two main transcription factors, NFκB and c-Jun. Ultimately, these two pathways are responsible for inducing the genes necessary for multiple biological processes. After TNF-α binding to TNFR-1, the inhibitory silencer of death domain (SODD) proteins are released from the DDs of TNFR-1, and can then aggregate. Upon oligomerisation, TRADD, an adaptor protein, binds the TNFR-1 DD complex (Hsu et al. 1995) and allows recruitment of three additional adaptor proteins, RIP, TRAF2 (Rothe at al. 1995) and FADD (Hsu et al. 1996). These three proteins are directly involved in recruiting enzymes that initiate the key signalling events. As is the case during Fas signalling, FADD recruits pro-caspase 8, thus triggering the apoptotic cascade. TRAF2, on the other hand, recruits both cIAP1 and cIAP2, two apoptosis inhibitory proteins, which act by balancing apoptosis signalling. In addition, TRAF2 is involved in the recruiting of a not yet fully identified complex, containing a mitogen-activated protein kinase kinase kinase (MAPKKK) which results in the activation of JNK, a kinase that phosphorylates c-Jun and increases its transcriptional activity. Finally, RIP is essential for activation of NFκB (Liu et al. 1996). Through an unidentified intermediate factor, RIP recruits and activates the IKK complex, which, as we have described above, triggers NFκB activity.

5.5.4
Signalling by Other Members of the Family and Its Regulation by Decoy Receptors

Although significant progress has been made, the signalling pathways used by other death receptors are not as well known as those of Fas or TNFR-1 (Ashkenazi 2002). In summary, they all use similar adaptor molecules, although some signal apoptosis through FADD-dependent and others through FADD-independent pathways (Yeh et al. 1998). Regulation of DR signalling is complicated by the existence of a subset of decoy receptors, with the capacity to attenuate signalling of specific receptors by binding to ligands without transmitting any intracellular signal (Ashkenazi and Dixit 1999).

5.6
Control of Apoptosis by NFκB: Pro- or Anti-apoptotic Role

One of the key aspects of cellular biology where NFκB is involved is the regulation of cell death (Barkett and Gilmore 1999; Foo and Nolan 1999). Interestingly, it has a dual role. In most circumstances, NFκB plays an anti-apoptotic role, hence protecting from apoptosis induced by multiple stimuli such as TNF-α treatment, chemotherapeutic drugs or ionising radiation (Van Antwerp 1996; Wang et al. 1996). However, with other stimuli such as infection with certain viruses, treatment with hydrogen peroxide or PKR activation, NFκB activity is necessary to induce apoptosis (Dumont et al. 1999; Gil et al. 1999; Connolly et al. 2000).

5.6.1
Role of NFκB in Apoptosis Protection

The first hint of an active survival role for NFκB stemmed from the fact that RelA$^{-/-}$ mice died before birth as a consequence of hepatic failure caused by massive hepatocyte apoptosis (Beg et al. 1995). This phenotype was further associated with a defect in protection from TNF-α-induced cell death as a consequence of NFκB absence (Doi et al. 1999). This observation allows us to draw two important conclusions. First, certain NFκB target genes inhibit apoptosis. Second, and more interestingly, the activation of NFκB by TNF-α attenuates or prevents the induction of apoptosis by this cytokine, and acts as a safeguard mechanism. NFκB has been linked, as we have mentioned before, to protection from apoptosis triggered by multiple stimuli, such as protecting from TNF-α, chemotherapeutic drugs and ionising radiation. NFκB also protects B cells from apoptosis, and is upregulated in chronically HIV-infected cell lines (Bachelerie et al. 1991; De Luca et al. 1996; Kwon et al. 1998). Various NF-κB-regulated anti-apoptotic gene products have been identified as potential mediators of its protective effect. These include the zinc finger protein A20, several IAP proteins (cIAP1, cIAP2 and XIAP), c-FLIP, members of the Bcl2 family (such as A1), TRAF1 and TRAF2 or even unsuspected candidates such as gadd45, which downregulates pro-apoptotic JNK signalling (Wang et al. 1998; reviewed in Karin and Lin 2002). These proteins act at several levels. IAPs act by directly inhibiting effector caspases as well as preventing pro-caspase activation. cFLIP acts by interacting with both FADD and pro-caspase 8 by inhibiting the activation of this pathway (Irmler et al. 1997). A1 inserts itself into mitochondria and can prevent mitochondria depolarisation and the beginning of mitochondrial caspase activation. Thus, NFκB induces multiple genes acting at several levels to assure the prevention of apoptosis.

As NFκB has an important role in avoiding apoptosis, counteraction of NFκB activation is common to several pro-apoptotic stimuli. Therefore, several key components of the NFκB pathway are targeted for degradation by caspases, thus impeding NFκB signalling. For example, RIP and TRAF2 are caspase substrates (Lin et al. 1999; Leo et al. 2001). Proteolysis of RIP results in the generation of a truncated form with dominant negative effects. In the same way, cleavage of TRAF2 results in its downregulation and concomitant increased sensitivity to TNF-α apoptotic effects. Other caspase targets whose cleavage generates dominant negative mutants are IKK, IκBα, RelA, cIAP1 and XIAP (Levkau et al. 1999; Reuther and Baldwin 1999; Tang et al. 2001).

5.6.2
NFκB as a Pro-apoptotic Factor

Although NFκB is mainly considered a survival factor, it can induce apoptosis in response to several stimuli such as infection by different viruses, treatment with hydrogen peroxide, activation of PKR, glutamate-induced neurotoxicity, serum withdrawal in HEK cells or activation of double-positive thymocytes (Hettmann et al. 1999). Amongst the targets of NFκB involved in apoptosis induction we can find several death receptors or their ligands, such as DR4, DR5, DR6, Fas, and Fas L (DeLuca et al. 1999; Karin and Lin 2002). In addition, NFκB can also activate other genes, such as p53 or c-myc, which trigger cell death. Induction of apoptosis by NFκB seems to be limited to certain cell types and the nature of stimuli, since it is known that different stimuli can use NFκB to induce or protect from apoptosis in the same cells.

5.6.3
Role of NFκB in PKR-Induced Apoptosis

PKR is an IFN-induced, dsRNA-activated protein kinase with potent antiviral activity. PKR exerts its effects by controlling translation through eIF2α phosphorylation and through regulating the activation of different transcription factors such as NFκB (Meurs et al. 1990; Clemens and Elia 1997). The observation that PKR expression from recombinant vaccinia viruses induced apoptosis in HeLa cells linked PKR with programmed cell death (Lee and Esteban 1994). Since then, several reports have related PKR with apoptosis induced by multiple stimuli (reviewed in Gil and Esteban 2000b). Not only does PKR mediate apoptosis in cells infected with various viruses such as influenza (Takizawa et al. 1996), or the picornavirus EMCV (Yeung et al. 1999), but it is also involved in apoptosis in the absence of viral infections. Studies carried out with cells derived from $PKR^{0/0}$ mice have shown that the PKR-defective MEFs are more resistant to cell death triggered by LPS, TNFα, and dsRNA than wild-type MEFs (Der et al. 1997). Analysis of the role of PKR targets on mediating apoptosis induction has revealed a complex scenario (reviewed in Gil and Esteban 2000b). PKR apoptosis induction relies mainly on the FADD/caspase 8 pathway, although it can also activate the mitochondrial/caspase 9 pathway (Gil and Esteban 2000a; Gil et al. 2002; Fig. 4).

As more immediate targets, a role for eIF-2α, NFκB and p53 on PKR-induced apoptosis has been suggested by different studies (Balachandran et al. 1998, Srivastava et al. 1998; Gil et al. 1999). eIF-2α is the best-characterised PKR substrate. Upon eIF-2α phosphorylation on serine 51 by PKR, phosphorylated eIF-2α increases the affinity of eIF-2 by GDP, thus inhibiting the initiation of translation due to the lack of eIF-2-GTP-Met-$tRNA^{met}$ ternary complexes. Clear evidence linking eIF-2α phosphorylation with apoptosis came from two independent studies, showing that PKR-induced apoptosis can be inhibited by the expression of a non-phosphorylatable eIF-2α mutant (eIF-2α 51A; Srivastava et al. 1998; Gil et al. 1999). Translational control is not the only effect exerted by PKR involvement in apoptosis induction. There is increasing evidence relating transcriptional pathways to apoptosis induction triggered by PKR. Upon dsRNA treatment, PKR is responsible for activation of the transcription factor NFκB (Gil et al. 1999, 2000) and it has been demonstrated that such NFκB activity is involved in PKR-mediated apoptosis (Gil et al. 1999). The mechanism by which PKR activates NFκB is not fully understood; however, it is mediated by the IκB kinase complex (IKK) and PKR kinase activity is necessary (Gil et al. 2000, 2001). The precise PKR target in the activation of the IKK complex has not yet been identified, although TRAF family proteins seem to play an important role in the process (Gil et al. unpubl. observ.).

The identity of the precise targets upregulated by NFκB activity which are involved in cell death is unknown. Candidate genes shown to be NFκB-dependent and involved in apoptosis induction include FasL (Kasibhatla et al.

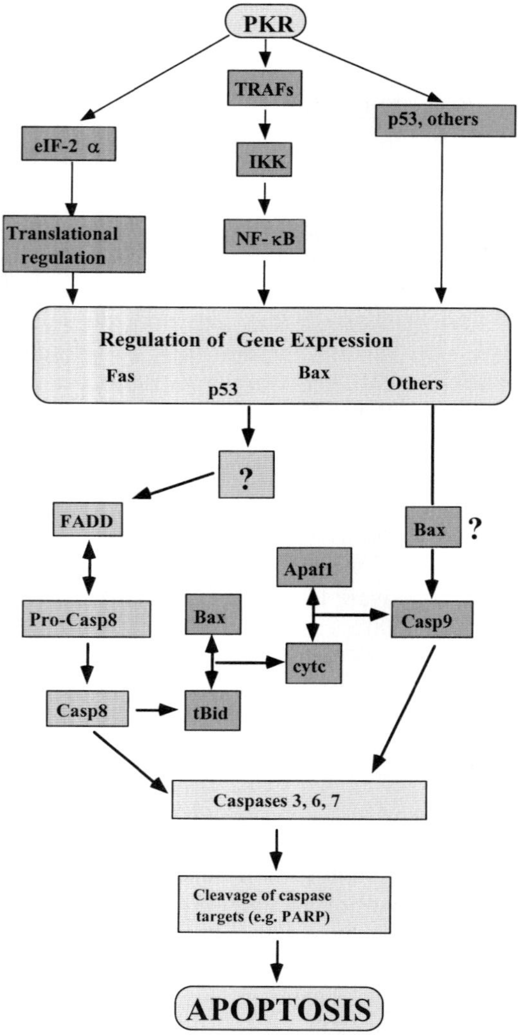

Fig. 4. Pathways of PKR-mediated apoptosis

1998), Fas (Behrman et al. 1994), IRF-1 (Kumar et al. 1997), caspase 1 (Casano et al. 1994) and p53 (Wu and Lozano 1994). p53 is one of the best-known apoptosis inducers and it has been shown that PKR upregulates transcription of p53 in an NFκB-dependent way. On the other hand, PKR interacts physically with p53 and phosphorylates it on serine 392 (Cuddihy et al. 1999b). Through this functional and physical interaction, PKR enhances p53-mediated transcriptional activation (Cuddihy et al. 1999a).

In summary, depending on the cell context and the nature and duration of the stimuli, NFκB can either protect from or induce apoptosis. To explain this apparent contradiction we cannot forget that in addition to activating NFκB, each stimulus results in activation of other pathways, and their combined effect is what determines the cell's fate.

6 Conclusion: Apoptosis, a Major Strategy in a Tale of Two Cities

The complex interactions between virus and host are the result of a long process that can be compared to a continuous war in which, much in the same way as in the Dicken's novel “A Tale of Two Cities”, changing strategies from both armies are required to survive. These strategies take place at two different levels: at a microscopic level, in which virions replicate into cells, and at a macroscopic level, in which a viral population of billions of viral particles escapes from the sophisticated machinery of the immune system. At both levels, apoptosis plays a major role and is used by the host as a potent antiviral mechanism, but is also subverted by viral infection and used to its own advantage.

At a microscopic level the challenge for the virus is to adapt to the cell machinery. Given its parasitic nature, HIV depends on the cell factory for its entire life cycle. It has to adapt to cell receptors for entry, interact with cellular factors for nuclear transport, and adapt to the cell machinery to drive reactivation from latency, mRNA transcription, protein expression and viral morphogenesis.

To fight against viral infection, cell suicide or apoptosis is a host-defence strategy to terminate viral replication and prevent propagation. In addition, viral replication itself compromises the biochemical machinery of the cell and some viral products display strong toxicity for the cellular environment thus facilitating apoptosis. Control of cell apoptosis is therefore a key mechanism in viral replication and persistence. With this aim in mind, some viral proteins, such as Vpr and Nef, counteract and delay apoptosis through different mechanisms: G2 arrest, inactivation of p53, caspase inhibition or interference with death-receptor expression. Furthermore, the induction of NFκB by HIV could also contribute to delaying apoptosis in the infected cell. Nevertheless, pro-apototic mechanisms, partly triggered by viral proteins, mainly Env, Tat and Vpr, overcome anti-apoptotic processes and result in cell death. Two major issues remain to be addressed: the relevance of the different apoptosis-induced processes described, and the mechanisms by which HIV induces apoptosis in non-infected bystander cells. With regard to the first point, many different – and sometimes contradictory – mechanisms by which viral proteins induce apoptosis have been described. As a general principle, pure biochemical experiments are not enough and pathogenic models are required to conclude the relevance of the different pathways subverted by viral proteins to induce or protect from apoptosis.

With respect to the induction of apoptosis in bystander cells, three, or maybe four, proteins could be involved in this process: Tat, Vpr, Env and Nef. We still do not know whether the activity and levels reached by these proteins in vivo can induce the toxic pro-apoptotic properties observed in vitro. Taking into account these limitations, there is nevertheless strong evidence that the viral envelope and the regulatory proteins Tat and Vpr display apoptotic activities from inside and outside the cell. The pathogenic mechanism of Nef on bystander cells is more probably related to its capacity to trigger death in activated CTLs.

From a macroscopic perspective viruses have evolved different mechanisms to evade the host immune response. Among these mechanisms, HIV escapes immune surveillance through latency and fast reactivation, high variability in immunodominant epitopes, protein masking and mimicry, interference with antigen recognition and avoiding killing by immune-mediated apoptosis. Some viral proteins contribute to escape of immune surveillance. It has been reported that Tat and the envelope can interfere with activation signalling in lymphocytes and induce a state of anergy and/or apoptosis in immunocompetent cells. It has also been suggested that Nef plays a prominent role in this process through downregulation of MHC I, thus avoiding antigen recognition by CTLs. Moreover, by means of the induction of CD95L, Nef could trigger apoptosis of specific anti-HIV-armed CTLs. Due to the low number of infected lymphocytes in peripheral blood it is difficult to study the relevance of these phenomena in vivo, but experimental evidence is consistent enough to consider that some of the viral escape from immune surveillance could be related to subversion of apoptosis regulation in the immune system, particularly in HIV-specific lymphocytes.

In addition, the NFκB system, given its central role in regulating cell death and immune responses, is an indispensable part of this machinery and HIV has developed different strategies for using it: relying on NFκB for its replication, inhibiting NFκB for downregulating immune responses, or sometimes merely taking advantage of its role in apoptosis regulation to achieve its life cycle.

Viruses are the ultimate cell hijackers, and can turn the cell machinery to their own advantage including the control of key cell regulatory elements involved in apoptotic pathways. The final goal of this process is to allow viral replication despite the strong toxicity of viral products and the immune response elicited by viral infection.

Acknowledgements. Our work is supported by grants from Ministerio de Ciencia y Tecnología, Fundación Caja Madrid, FIPSE and Red de Investigación en SIDA (RIS). We acknowledge the expert secretarial assistance provided by Olga Palao.

References

Aceituno E, Castanon S, Jimenez C, Subira D, De Gorgolas M, Fernandez-Guerrero M, Ortiz F, Garcia R (1997) Circulating immune complexes from HIV-1+ patients induces apoptosis on normal lymphocytes. Immunology 92:317–320

Akari H, Bour S, Kao S, Adachi A, Strebel K (2001) The human immunodeficiency virus type 1 accessory protein Vpu induces apoptosis by suppressing the nuclear factor kappaB-dependent expression of antiapoptotic factors J Exp Med 194:1299–1311

Alcami J, Lain de Lera T, Folgueira L, Pedraza MA, Jacque JM, Bachelerie F, Noriega AR, Hay RT, Harrich D, Gaynor RB, Virelizier JL, Arenzana F (1995) Absolute dependence on kB responsive elements for initiation and Tat-mediated amplification of HIV transcription in blood CD4 T lymphocytes. EMBO J 14:1552–1560

Ameisen JC (2001) Apoptosis subversion: HIV-Nef provides both armor and sword. Nat Med 7:1181–1182

Ameisen JC, Capron A (1991) Cell dysfunction and depletion in AIDS: the programmed cell death hypothesis. Immunol Today 12:102–105

Arenzana-Seisdedos F, Turpin P, Rodriguez M, Thomas D, Hay RT, Virelizier JL, Dargemont C (1997) Nuclear localization of IkBa promotes active transport of NFκB from the nucleus to the cytoplasm. J Cell Sci 110:369–378

Ashkenazi A (2002) Targeting death and decoy receptors of the tumour-necrosis factor superfamily. Nat Rev Cancer 2:420–430

Ashkenazi A, Dixit VM (1998) Death receptors: signaling and modulation. Science 281:1305–1308

Ashkenazi A, Dixit VM (1999) Apoptosis control by death and decoy receptors. Curr Opin Cell Biol 11:255–260

Ayavoo V, Mahboubi A, Mahalingam S, Ramalingam R, Kudchodkar S, Williams WV, Green DR, Weiner DB (1997) HIV-1 Vpr suppresses immune activation and apoptosis through regulation of nuclear factor kappa B. Nat Med 3:1117–1123

Badley AD, Pilon AA, Landay A, Lynch DH (2000) Mechanisms of HIV associated lymphocyte apoptosis. Blood 96:2951–2964

Bachelerie F, Alcami J, Arenzana F, Virelizier JL (1991). HIV enhancer activity perpetuated by NF-kappa B induction on infection of monocytes. Nature 350 709–712

Baeuerle PA, Baltimore D (1988) Activation of DNA-binding activity in an apparently cytoplasmic precursor of the NF-kappa B transcription factor. Cell 53:211–217

Balachandran S, Kim CN, Yeh WC, Mak TW, Bhalla K, Barber GN (1998) Activation of the dsRNA-dependent protein kinase, PKR, induces apoptosis through FADD-mediated death signaling. EMBO J 17:6888–6902

Banda NK, Bernier J, Kurahara DK, Kurrle R, Haigwood N, Sekaly RP, Finkel TH (1992) Crosslinking CD4 by human immunodeficiency virus gp120 primes T cells for activation-induced apoptosis. J Exp Med 176:1099–1106

Barkett M, Gilmore TD (1999) Control of apoptosis by Rel/NF-kappaB transcription factors. Oncogene 18:6910–6924

Bartz SR, Emerman M (1999) Human immunodeficiency virus type 1 Tat induces apoptosis and increases sensitivity to apoptotic signals by up-regulating FLICE/caspase-8. J Virol 73:1956–1963

Basanez G, Zimmerberg J (2001) HIV and apoptosis death and the mitochondrion. J Exp Med 193:509–519

Beg AA, Sha WC, Bronson RT, Ghosh S, Baltimore D (1995) Embryonic lethality and liver degeneration in mice lacking the RelA component of NF-kappa B. Nature 376:167–170

Behrmann I, Walczak H, Krammer PH (1994) Structure of the human APO-1 gene. Eur J Immunol 24:3057–3062

Ben Neriah Y (2002) Regulatory functions of ubiquitination in the immune system. Nat Immunol 3:20–26

Benedict CA, Norris PS, Ware CF (2002) To kill or be killed: viral evasion of apoptosis. Nat Immunol 3:1013–1018

Berger EA, Murphy PM, Farber JM (1999) Chemokine receptors as HIV-1 coreceptors: roles in viral entry, tropism, and disease. Annu Rev Immunol 17:657–700

Bermejo M, Martin-Serrano J, Oberlin E, Pedraza MA, Serrano A, Santiago B, Caruz A, Loetscher P, Baggiolini M, Arenzana-Seisdedos F, Alcami J (1998) Activation of blood T lymphocytes down-regulates CXCR4 expression and interferes with propagation of X4 HIV strains. Eur J Immunol 28:3192–3204

Blanco J, Barretina J, Henson G, Bridger G, De Clercq E, Clotet B, Este JA (2000) The CXCR4 antagonist AMD3100 efficiently inhibits cell-surface-expressed human immunodeficiency virus type 1 envelope-induced apoptosis. Antimicrob Agents Chemother 44:51–56

Blanco J, Barretina J, Ferri KF, Jacotot E, Gutierrez A, Armand-Ugon M, Cabrera C, Kroemer G, Clotet B, Este JA (2003) Cell-surface-expressed HIV-1 envelope induces the death of CD4 T cells during gp41-mediated hemifucion-like events. Virology 305:318–329

Blackson JN, Persaud D, Siliciano RF (2002) The challenge of viral reservoirs in HIV-1 infection. Annu Rev Med 53:557–593

Boise LH, Thompson CB (1996) Hierarchical control of lymphocyte survival. Science 274:67–68

Brenner C, Kroemer G (2003) The mitochondriotoxic domain of Vpr determines HIV-1 virulence. J Clin Invest 111:1455–1457

Brodie SJ, Lewinsohn DA, Patterson BK, Jiyamapa D, Krieger J, Corey L, Greenberg PD, Riddell SR (1999) In vivo migration and function of transferred HIV-1-specific cytotoxic T cells. Nat Med 5:34–41

Casano FJ, Rolando AM, Mudgett, JS, Molineaux SM (1994) The structure and complete nucleotide sequence of the murine gene encoding interleukin-1 beta converting enzyme (ICE). Genomics 20:474–481

Casella CR, Rapaport EL, Finkel TH (1999) Vpu increases susceptibility of human immunodeficiency virus type 1-infected cells to fas killing J Virol 73:92–100

Castedo M, Roumier T, Blanco J, Ferri KF, Barretina J, Tintignac LA, Andreau K, Perfettini JL, Amendola A, Nardacci R, Leduc P, Ingber DE, Druillennec S, Roques B, Leibovitch SA, Vilella-Bach M, Chen J, Este JA, Modjtahedi N, Piacentini M, Kroemer G (2002) Sequential involvement of Cdk1, mTOR and p53 in apoptosis induced by the HIV-1 envelope. EMBO J 21:4070–4080

Chen D, Wang M, Zhou S, Zhou Q (2002) HIV-1 Tat targets microtubules to induce apoptosis, a process promoted by the pro-apoptotic Bcl-2 relative Bim. EMBO J 21:6801–6810

Chen G, Goeddel DV (2002) TNF-R1 signaling: a beautiful pathway. Science 296:1634–1635

Chinnaiyan AM, O'Rourke K, Tewari M, Dixit VM (1995) FADD, a novel death domain-containing protein, interacts with the death domain of Fas and initiates apoptosis. Cell 81:505–512

Chun TW, Stuyver L, Mizell SB, Ehler LA, Mican JA, Baseler M, Lloyd AL, Nowak MA, Fauci AS (1997) Presence of an inducible HIV-1 latent reservoir during highly active antiretroviral therapy. Proc Natl Acad Sci USA 94:13193–13197

Chu WM, Ostertag D, Li ZW, Chang L, ChenY, Hu Y, Williams B, Perrault J, Karin M (1999) JNK2 and IKKbeta are required for activating the innate response to viral infection. Immunity 11:721–731

Cicala C, Arthos J, Rubbert A, Selig S, Wildt K, Cohen OJ, Fauci AS (2000) HIV-1 envelope induces activation of caspase-3 and cleavage of focal adhesion kinase in primary human CD4(+) T cells. Proc Natl Acad Sci USA 97:1178–1183

Clemens MJ, Elia A (1997) The double-stranded RNA-dependent protein kinase PKR: structure and function. J Interferon Cytokine Res 17:503–524

Cloyd MW, Chen JJ, Adeqboyega P, Wang L (2001) How does HIV cause depletion of CD4 lymphocytes? A mechanism involving virus signaling through its cellular receptors. Curr Mol Med 1:545–550

Connolly JL, Rodgers SE, Clarke P, Ballard DW, Kerr LD, Tyler KL, Dermody TS (2000) Reovirus-induced apoptosis requires activation of transcription factor NF-kappaB. J Virol 74:2981–2989

Coultas L, Strasser A (2000) The molecular control of DNA damage-induced cell death. Apoptosis 5:491–507

Cuddihy AR, Li S, Tam NW, Wong AH, Taya Y, Abraham N, Bell JC, Koromilas AE (1999a) Double-stranded-RNA-activated protein kinase PKR enhances transcriptional activation by tumor suppressor p53. Mol Cell Biol 19:2475–2484

Cuddihy AR, Wong AH, Tam NW, Li S, Koromilas AE (1999b) The double-stranded RNA activated protein kinase PKR physically associates with the tumor suppressor p53 protein and phosphorylates human p53 on serine 392 in vitro. Oncogene 18:2690–2702

Cullen BR (1993) Does HIV-1 Tat induce a change in viral initiation rights? Cell 73:417–420

Delhase M, Hayakawa M, Chen Y, Karin M (1999) Positive and negative regulation of IkappaB kinase activity through IKKbeta subunit phosphorylation. Science 284:309–313

DeLuca C, Kwon H, Lin R, Wainberg M, Hiscott J (1999) NF-kappaB activation and HIV-1 induced apoptosis. Cytokine Growth Factor Rev 10:235–253

DeLuca C, Roulston A, Koromilas A, Wainberg MA, Hiscott J (1996) Chronic human immunodeficiency virus type 1 infection of myeloid cells disrupts the autoregulatory control of the NF-kappaB/Rel pathway via enhanced IkappaBalpha degradation. J Virol 70:5183–5193

De Martino GN, Slaughter CA (1999) The proteasome, a novel protease regulated by multiple mechanisms. J Biol Chem 274:22123–22126

Der SD, Yang YL, Weissmann C, Williams BR (1997) A double-stranded RNA-activated protein kinase-dependent pathway mediating stress-induced apoptosis. Proc Natl Acad Sci USA 94:3279–3283

DiDonato J, Mercurio F, Rosette C, Wu LJ, Suyang H, Ghosh S, Karin M (1996) Mapping of the inducible IkappaB phosphorylation sites that signal its ubiquitination and degradation. Mol Cell Biol 16:1295–1304

Doi TS, Marino MW, Takahashi T, Yoshida T, Sakakura T, Old LJ, Obata Y (1999) Absence of tumor necrosis factor rescues RelA-deficient mice from embryonic lethality. Proc Natl Acad Sci USA 96:2994–2999

Dumont A, Hehner SP, Hofmann TG, Ueffing M, Droge W, Schmitz ML (1999) Hydrogen peroxide-induced apoptosis is CD95-independent, requires the release of mitochondria-derived reactive oxygen species and the activation of NF-kappaB. Oncogene 18:747–757

Emerman M, Malim MH (1998) HIV-1 regulatory/accessory genes: keys to unraveling viral and host cell biology. Science 280:1880–1884

Estaquier J, Lelievre JD, Petit F, Brunner T, Moutouh-De Parseval L, Richman DD, Ameisen JC, Corbeil J (2002) Effects of antiretroviral drugs on human immunodeficiency virus type 1-induced CD4(+) T-cell death. J Virol 76:5966–5973

Fackler OT, Baur AS (2002) Live and let die: Nef functions beyond HIV replication. Immunity 16:493–497

Ferri KF, Jacotot E, Blanco J, Este JA, Kroemer G (2000) Mitochondrial control of cell death induced by HIV-1-encoded proteins. Ann NY Acad Sci 926:149–164

Finkel TH, Tudor-Williams G, Banda NK, Cotton MF, Curiel T, Monks C, Baba TW, Ruprecht RM, Kupfer A (1995) Apoptosis occurs predominantly in bystander cells and not in productively infected cells of HIV- and SIV-infected lymph nodes. Nat Med 1:129–134

Foo SY, Nolan GP (1999) NF-kappaB to the rescue: RELs, apoptosis and cellular transformation. Trends Genet 15:229–235

Geiss GK, Bumgarner RE, An MC, Agy MB, van 't Wout AB, Hammersmark E, Carter VS, Upchurch D, Mullins JI, Katze MG (2000) Large-scale monitoring of host cell gene expression during HIV-1 infection using cDNA microarrays. Virology 266:8–16

Geleziunas R, Xu W, Takeda K, Ichijo H, Greene WC (2001) HIV-1 Nef inhibits ASK1-dependent death signalling providing a potential mechanism for protecting the infected host cell. Nature 410:834–838

Ghosh S, May MJ, Kopp EB (1998) NF-kappa B and Rel proteins: evolutionarily conserved mediators of immune responses. Annu Rev Immunol 16:225–260

Gibellini D, Re MC, Ponti C, Maldini C, Celeghini C, Cappellini A, La Placa M, Zauli G (2001) HIV-1 Tat protects CD4+ Jurkat T lymphoblastoid cells from apoptosis mediated by TNF-related apoptosis-inducing ligand. Cell Immunol 207:89–99

Gil J, Alcamí J, Esteban M (1999) Induction of apoptosis by double-stranded RNA-dependent protein kinase (PKR) involves the alpha subunit of eukaryotic translation initiation factor 2 and NF-kappaB. Mol Cell Biol 19:4653–4663

Gil J, Alcamí J, Esteban M (2000) Activation of NF-kappaB by the dsRNA-dependent protein kinase, PKR involves the IkappaB kinase complex. Oncogene 19:1369–1378

Gil J, Esteban M (2000a) FADD-mediated activation of caspase 8 by a mechanism independent of Fas and TNFR is involved in PKR-induced apoptosis. Oncogene 19:3665–3674

Gil J, Esteban M (2000b) Role of the protein kinase PKR in apoptosis induction. Apoptosis 5:107–114

Gil J, Rullas J, García M, Alcamí J, Esteban M (2001) The catalytic activity of the dsRNA-dependent protein kinase, PKR, is required for NF-kB activation. Oncogene 20: 385–394

Gil J, Garcia M, Esteban M (2002) Caspase 9 activation by the dsRNA-dependent protein kinase, PKR: molecular mechanism and relevance, FEBS Lett 529:249–255

Glushakova S, Grivel JC, Fitzgerald W, Sylwester A, Zimmerberg J, Margolis LB (1998) Evidence for the HIV-1 phenotype switch as a causal factor in acquired immunodeficiency. Nat Med 4:346–349

Glushakova S, Yi Y, Grivel JC, Singh A, Schols D, De Clercq E, Collman RG, Margolis L (1999) Preferential coreceptor utilization and cytopathicity by dual-tropic HIV-1 in human lymphoid tissue ex vivo. J Clin Invest 104:R7-R11

Gougeon ML (2003) Apoptosis as an HIV strategy to escape immune attack. Nat Rev Immunol 3:392–404

Gougeon ML, Montagnier L (2000) Programmed cell death as a mechanism of CD4 and CD8 T cell deletion in AIDS. Molecular control and effect of highly active anti-retroviral therapy. Ann NY Acad Sci 887:199–212

Greene WC, Peterlin BM (2002). Charting HIV's remarkable voyage through the cell: basic science as a passport to future therapy. Nat Med 8 673–680

Greenway AL, McPhee DA, Allen K, Johnstone R, Holloway G, Mills J, Azad A, Sankovich S, Lambert P (2002) Human immunodeficiency virus type 1 Nef binds to tumor suppressor p53 and protects cells against p53-mediated apoptosis. J Virol 76:2692–2702

Grivel JC, Margolis LB (1999) CCR5- and CXCR4-tropic HIV-1 are equally cytopathic for their T-cell targets in human lymphoid tissue. Nat Med 5:344–346

Groux H, Torpier G, Monte D, Mouton Y, Capron A, Ameisen JC (1992) Activation-induced death by apoptosis in CD4+ T cells from human immunodeficiency virus-infected asymptomatic individuals. J Exp Med 175:331–340

Guillerm C, Coudronniere N, Robert-Hebmann V, Devaux C (1998) Delayed human immunodeficiency virus type 1-induced apoptosis in cells expressing truncated forms of CD4. Virology 72:1754–1761

Haase AT, Henry K, Zupancic M, Sedgewick G, Faust RA, Melroe H, Cavert W, Gebhard K, Staskus K, Zhang ZQ, Dailey PJ, Balfour HH Jr, Erice A, Perelson AS (1996) Quantitative image analysis of HIV-1 infection in lymphoid tissue. Science 274:985–989

Hashimoto F, Oyaizu N, Kalyanaraman VS, Pahwa S (1997) Modulation of Bcl-2 protein by CD4 cross-linking: a possible mechanism for lymphocyte apoptosis in human immunodeficiency virus infection and for rescue of apoptosis by interleukin-2. Blood 90:745–753

Haughey NJ, Mattson MP (2002) Calcium dysregulation and neuronal apoptosis by the HIV-1 proteins Tat and gp120. J Acquir Immune Defic Syndr 31(Suppl 2):S55–61

Hay S, Kannourakis G (2002) A time to kill: viral manipulation of the cell death program. J Gen Virol 83:1547–1564

Hazenberg MD, Otto SA, Cohen Stuart JW, Verschuren MC, Borleffs JC, Boucher CA, Coutinho RA, Lange JM, Rinke de Wit TF, Tsegaye A, van Dongen JJ, Hamann D, de Boer RJ, Miedema F (2000a) Increased cell division but not thymic dysfunction rapidly affects the T-cell recep-

tor excision circle content of the naive T cell population in HIV-1 infection. Nat Med 6:1036–1042
Hazenberg MD, Hamann D, Schuitemaker H, Miedema F (2000b) T cell depletion in HIV-1 infection: how CD4+ T cells go out of stock. Nat Immunol 1:285–289
Hazenberg MD, Borghans JA, de Boer RJ, Miedema F (2003) Thymic output: a bad TREC record. Nat Immunol 4:97–99
Hettmann T, DiDonato J, Karin M, Leiden JM (1999) An essential role for nuclear factor kappaB in promoting double positive thymocyte apoptosis. J Exp Med 189:145–158
Hilderman DA, Zhu Y, Mitchell TC, Kappler J, Marrack P (2002) Molecular mechanisms of activated T cell death in vivo. Curr Opin Immunol 14:354–359
Holtzman MJ, Green JM, Jayaraman S, Arch RH (2000) Regulation of T cell apoptosis. Apoptosis 5:459–471
Hsu H, Xiong J, Goeddel DV (1995) The TNF receptor 1-associated protein TRADD signals cell death and NF-kappa B activation. Cell 81:495–504
Hsu H, Shu HB, Pan MG, Goeddel DV (1996) TRADD-TRAF2 and TRADD-FADD interactions define two distinct TNF receptor 1 signal transduction pathways.Cell 84:299–308
Irmler M, Thome M, Hahne M, Schneider P, Hofmann K, Steiner V, Bodmer JL, Schroter M, Burns K, Mattmann C, Rimoldi D, French LE, Tschopp J (1997) Inhibition of death receptor signals by cellular FLIP. Nature 388:190–195
Jacotot E, Ravagnan L, Loeffler M, Ferri KF, Vieira HL, Zamzami N, Costantini P, Druillennec S, Hoebeke J, Briand JP, Irinopoulou T, Daugas E, Susin SA, Cointe D, Xie ZH, Reed JC, Roques BP, Kroemer G (2000) The HIV-1 viral protein R induces apoptosis via a direct effect on the mitochondrial permeability transition pore. J Exp Med 191:33–46
Jaworowski A, Crowe SM (1999) Does HIV cause depletion of CD4+ T cells in vivo by the induction of apoptosis? Immunol Cell Biol 77:90–98
Jekle A, Keppler OT, De Cleck E, Weinstein M, Goldsmith MA (2003) In vivo evolution of human immunodeficiencecy virus type I toward increased pathogenicity through CXR4-mediated killing of uninfected CD4 T cells. J Virol 77:5846–5854
Jimenez A, Molero L, Jimenez A, Castanon S, Subira D, De Gorgolas M, Fedz-Guerrero M, Garcia R (2002) Role of antiretroviral regimes in HIV-1 patients in reducing immune activation. Immunology 106:80–86
Kameoka M, Kimura T, Zheng YH, Suzuki S, Fujinaga K, Luftig RB, Ikuta K (1997) Protease-defective, gp120-containing human immunodeficiency virus type 1 particles induce apoptosis more efficiently than does wild-type virus or recombinant gp120 protein in healthy donor-derived peripheral blood T cells. J Clin Microbiol 35:41–47
Karin M (1999) The beginning of the end: IkappaB kinase (IKK) and NF-kappaB activation. J Biol Chem 274:27339–27342
Karin M, Lin A (2002) NF-kappaB at the crossroads of life and death. Nat Immunol 3:221–227
Kasibhatla S, Brunner T, Genestier L, Echeverri F, Mahboubi A, Green DR (1998) DNA damaging agents induce expression of Fas ligand and subsequent apoptosis in T lymphocytes via the activation of NF-kappa B and AP-1. Mol Cell 1:543–551
Khaled AR, Durum SK (2002) Lymphocide: cytokines and the control of lymphoid homeostasis. Nat Rev Immunol 2:817–830
Kim TA, Avraham HK, Koh YH, Jiang S, Park IW, Avraham S (2003) HIV-1 Tat-mediated apoptosis in human brain microvascular endothelial cells J Immunol 170:2629–37
Kinter A, Arthos J, Cicala C, Fauci AS (2000) Chemokines, cytokines and HIV: a complex network of interactions that influence HIV pathogenesis Immunol Rev 177:88–98
Krammer PH (2000) CD95's deadly mission in the immune system. Nature 407:789–795
Kumar A, Yang YL, Flati V, Der S, Kadereit S, Deb A, Haque J, Reis L, Weissmann C, Williams BR (1997) Deficient cytokine signaling in mouse embryo fibroblasts with a targeted deletion in the PKR gene: role of IRF-1 and NF-kappaB. EMBO J 16:406–416
Kwon H, Pelletier N, DeLuca C, Genin P, Cisternas S, Lin R, Wainberg MA, Hiscott J (1998) Inducible expression of IkappaBalpha repressor mutants interferes with NF-kappaB activity and HIV-1 replication in Jurkat T cells. J Biol Chem 273:7431–7440

LaBonte JA, Madani N, Sodroski J (2003) Cytolysis by CCR5 using human immunodeficiency virus type 1 envelope glycoproteins is dependent on membrane fusion and can be inhibited by high levels of CD4 expression J Virol 77:6645–6659
Laurent-Crawford AG, Krust B, Riviere Y, Desgranges C, Muller S, Kieny MP, Dauguet C, Hovanessian AG (1993) Membrane expression of HIV envelope glycoproteins triggers apoptosis in CD4 cells. AIDS Res Hum Retroviruses 9:761–773
Lee SB, Esteban M (1994) The interferon-induced double-stranded RNA-activated protein kinase induces apoptosis. Virology 199:491–496
Leo E, Deveraux QL, Buchholtz C, Welsh K, Matsuzawa S, Stennicke HR, Salvesen GS, Reed JC (2001) TRAF1 is a substrate of caspases activated during tumor necrosis factor receptor-alpha-induced apoptosis. J Biol Chem 276:8087–8093
Levkau B, Scatena M, Giachelli CM, Ross R, Raines EW (1999) Apoptosis overrides survival signals through a caspase-mediated dominant-negative NF-kappa B loop. Nat Cell Biol 1999 1:227–233
Li CJ, Friedman DJ, Wang C, Metelev V, Pardee AB (1995) Induction of apoptosis in uninfected lymphocytes by HIV-1 Tat protein. Science 268:429–431
Li Q, Verma I (2002) NF-kB regulation in the immune system. Nat Rev Immunol 2:725–734
Lin Y, Devin A, Rodriguez Y, Liu ZG (1999) Cleavage of the death domain kinase RIP by caspase-8 prompts TNF-induced apoptosis. Genes Dev 13:2514–2526
Liu ZG, Hsu H, Goeddel DV, Karin M (1996) Dissection of TNF receptor 1 effector functions: JNK activation is not linked to apoptosis while NF-kappaB activation prevents cell death. Cell 87:565–576
Lum JJ, Cohen OJ, Nie Z, Weaver JG, Gomez TS, Yao XJ, Lynch D, Pilon AA, Hawley N, Kim JE, Chen Z, Montpetit M, Sanchez-Dardon J, Cohen EA, Badley AD (2003) Vpr R77Q is associated with long-term nonprogressive HIV infection and impaired induction of apoptosis J Clin Invest 111:1547–1554
Martinou JC, Green DR (2001) Breaking the mitochondrial barrier. Nat Rev Mol Cell Biol 2:63–67
May MJ, Ghosh S (1997) Rel/NF-kappa B and I kappa B proteins: an overview. Semin Cancer Biol 8:63–73
Matarrese P, Gambardella L, Cassone A, Vella S, Cauda R, Malorni W (2003) Mitochondrial membrane hyperpolarization hijacks activated T lymphocytes toward the apoptotic-prone phenotype: homeostatic mechanisms of HIV protease inhibitors. J Immunol 170:6006–6015
McCloskey TW, Ott M, Tribble E, Khan SA, Teichberg S, Paul MO, Pahwa S, Verdin E, Chirmule N (1997) Dual role of HIV Tat in regulation of apoptosis in T cells. J Immunol 158:1014–1019
McCune JM (2001) The dynamics of CD4+ T-cell depletion in HIV disease. Nature 410:974–979
Meurs E, Chong K, Galabru J, Thomas NS, Kerr IM, Williams BR, Hovanessian AG (1990) Molecular cloning and characterization of the human double-stranded RNA-activated protein kinase induced by interferon. Cell 62:379–390
Moutouh L, Estaquier J, Richman DD, Corbeil J (1998) Molecular and cellular analysis of human immunodeficiency virus-induced apoptosis in lymphoblastoid T-cell-line-expressing wild-type and mutated CD4 receptors. J Virol 72:8061–8072
Muthumani K, Hwang DS, Desai BM, Zhang D, Dayes N, Green DR, Weiner DB (2002) HIV-1 Vpr induces apoptosis through caspase 9 in T cells and peripheral blood mononuclear cells J Biol Chem 277:37820–37831
Muthumani K, Choo AY, Hwang DS, Chattergoon MA, Dayes NN, Zhang D, Lee MD, Duvvuri U, Weiner DB (2003) Mechanism of HIV-1 viral protein R-induced apoptosis Biochem Biophys Res Commun 304:583–592
Muzio M, Chinnaiyan AM, Kischkel FC, O'Rourke K, Shevchenko A, Ni J, Scaffidi C, Bretz JD, Zhang M, Gentz R, Mann M, Krammer PH, Peter ME, Dixit VM (1996) FLICE, a novel FADD-homologous ICE/CED-3-like protease, is recruited to the CD95 (Fas/APO-1) death-inducing signaling complex. Cell 85:817–827
Nagata S (1997)Apoptosis by death factor. Cell 86:355–365
Ogg GS, Jin X, Bonhoeffer S, Dunbar PR, Nowak MA, Monard S, Segal JP, Cao Y, Rowland-Jones SL, Cerundolo V, Hurley A, Markowitz M, Ho DD, Nixon DF, McMichael AJ (1998) Quantita-

tion of HIV-1 specific cytotoxic T-lymphocytes and plasma load of viral RNA. Science 279:2103–2106

Oyaizu N, McCloskey TW, Coronesi M, Chirmule N, Kalyanaraman VS, Pahwa S (1993) Accelerated apoptosis in peripheral blood mononuclear cells (PBMCs) from human immunodeficiency virus type-1 infected patients and in CD4 cross-linked PBMCs from normal individuals. Blood 82:3392–3400

Oyaizu N, McCloskey TW, Than S, Hu R, Kalyanaraman VS, Pahwa S (1994) Cross-linking of CD4 molecules upregulates Fas antigen expression in lymphocytes by inducing interferon-gamma and tumor necrosis factor-alpha secretion. Blood 84:2622–2631

Pahl HL (1999) Activators and target genes of Rel/NF-kappaB transcription factors. Oncogene 18:6853–6866

Parker NG, Notermans DW, de Boer RJ, Roos MT, de Wolf F, Hill A, Leonard JM, Danner SA, Miedema F, Schellekens PT (1998) Biphasic kinetics of peripheral blood T cells after tripe combination therapy in HIV-1 infection: a composite of redistribution and proliferation. Nat Med 4:208–214

Penn ML, Grivel JC, Schramm B, Goldsmith MA, Margolis L (1999) CXCR4 utilization is sufficient to trigger CD4+ T cell depletion in HIV-1-infected human lymphoid tissue Proc Natl Acad Sci USA 96:663–668

Perelson AS, Neumann AU, Markowitz M, Leonard JM, Ho DD (1996) HIV1 dynamics in vivo: virion clearance rate, infected cell life-span, and viral generation time. Science 271:1582–1586

Perkins ND, Schmid RM, Ducket CS, Leung K, Rice NR, Nabel GJ (1992) Distinct combinations of NF-kappa B subunits determine the specificity of transcriptional activation. Proc Natl Acad Sci USA 89:1529–1533

Persaud D, Zhou Y, Siliciano JM, Siliciano RF (2003) Latency in human immunodeficiency virus type 1 infection: no easy answers J Virol 77:1659–65

Peterlin BM and Trono D (2003). Hide, shield and strike back: how HIV-infected cells avoid immune eradication. Nat Rev Immunol 3 97–107

Phenix BN, Cooper C, Owen C, Badley AD (2002) Modulation of apoptosis by HIV protease inhibitors. Apoptosis 7:295–312

Reuther JY, Baldwin AS Jr (1999) Apoptosis promotes a caspase-induced amino-terminal truncation of IkappaBalpha that functions as a stable inhibitor of NF-kappaB. J Biol Chem 274:20664–20670

Rothe M, Sarma V, Dixit VM, Goeddel DV (1995) TRAF2-mediated activation of NF-kappa B by TNF receptor 2 and CD40. Science 269:1424–1427

Rothwarf DM, Zandi E, Natoli G, Karin M (1998) IKK-gamma is an essential regulatory subunit of the IkappaB kinase complex. Nature 395:297–300

Roulston A, Marcellus RC, Branton PE (1999) Viruses and apoptosis. Annu Rev Microbiol 53 577–628

Sen R, Baltimore D (1986) Multiple nuclear factors interact with the immunoglobulin enhancer sequences. Cell 46:705–716

Shostak LD, Ludlow J, Fisk J, Pursell S, Rimel BJ, Nguyen D, Rosenblatt JD, Planelles V (1999) Roles of p53 and caspases in the induction of cell cycle arrest and apoptosis by HIV-1 vpr. Exp Cell Res 251:156–165

Sodroski J, Goh WC, Rosen C, Campbell K, Haseltine WA (1986) Role of the HTLV-III/LAV envelope in syncytium formation and cytopathicity. Nature 322:470–474

Somasundaran M, Sharkey M, Brichacek B, Luzuriaga K, Emerman M, Sullivan JL, Stevenson M (2002) Evidence for a cytopathogenicity determinant in HIV-1 Vpr. Proc Natl Acad Sci USA 99:9503–9508

Speth C, Dierich MP (1999) Modulation of cell surface protein expression by infection with HIV-1. Leukemia 13(Suppl 1):S99–105

Srivastava SP, Kumar KU, Kaufman RJ (1998) Phosphorylation of eukaryotic translation initiation factor 2 mediates apoptosis in response to activation of the double-stranded RNA-dependent protein kinase. J Biol Chem 273:2416–2423

Stark GR, Kerr IM, Williams BR, Silverman RH, Schreiber RD (1998) How cells respond to interferons. Annu Rev Biochem 67:227–264

Stewart SA, Poon B, Song JY, Chen IS (2000) Human immunodeficiency virus type 1 vpr induces apoptosis through caspase activation. J Virol 74:3105–3111

Sylwester AW, Grivel JC, Fitzgerald W, Rossio JL, Lifson JD, Margolis LB (1998) D4(+) T-lymphocyte depletion in human lymphoid tissue ex vivo is not induced by noninfectious human immunodeficiency virus type 1 virions. J Virol 72:9345–9347

Takahashi M, Osono E, Nakagawa Y, Wang J, Berzofsky JA, Margulies DH (2002) Rapid induction of apoptosis in CD8+ HIV-1 envelope-specific murine CTLs by short exposure to antigenic peptide. J Immunol 169:6588–6593

Takizawa T, Ohashi K, Nakanishi Y (1996) Possible involvement of double-stranded RNA-activated protein kinase in cell death by influenza virus infection. J Virol 70:8128–8132

Tang G, Yang J, Minemoto Y, Lin A (2001) Blocking caspase-3-mediated proteolysis of IKKbeta suppresses TNF-alpha-induced apoptosis. Mol Cell 8:1005–1016

Van Antwerp D, Martin SJ, Kafri T, Green DR, Verma IM (1996) Suppression of TNF-alpha-induced apoptosis by NF-kappaB. Science 274:787–789

van 't Wout AB, Lehrman GK, Mikheeva SA, O'Keeffe GC, Katze MG, Bumgarner RE, Geiss GK, Mullins JI (2003) Cellular gene expression upon human immunodeficiency virus type 1 infection of CD4(+)-T-cell lines. J Virol 77:1392–1402

Wang CY, Mayo MW, Baldwin AJ (1996) TNF- and cancer therapy-induced apoptosis: potentiation by inhibition of NF-kappaB. Science 274:784–787

Wang CY, Mayo MW, Korneluk RG, Goeddel DV, Baldwin, AJ (1998) NF-kappaB antiapoptosis: induction of TRAF1 and TRAF2 and c-IAP1 and c- IAP2 to suppress caspase-8 activation. Science 281:1680–1683

Weiss RA (1996) HIV receptors and the pathogenesis of AIDS. Science 272:1885–1886

Westendorp MO, Frank R, Ochsenbauer C, Stricker K, Dhein J, Walczak H, Debatin KM, Krammer PH (1995) Sensitization of T cells to CD95-mediated apoptosis by HIV-1 Tat and gp120. Nature 375:497–500

Wolf D, Witte V, Laffert B, Blume K, Stromer E, Trapp S, d'Aloja P, Schurmann A, Baur AS (2001) HIV-1 Nef associated PAK and PI3-kinases stimulate Akt-independent Bad-phosphorylation to induce anti-apoptotic signals Nat Med 7:1217–1224

Wolf T, Findhammer S, Nolte B, Helm EB, Brodt HR (2003) Inhibition of TNF-alpha mediated cell death by HIV-1 specific protease inhibitors. Eur J Med Res 28:17–24

Wu H, Lozano G (1994) NF-kappa B activation of p53. A potential mechanism for suppressing cell growth in response to stress. J Biol Chem 269:20067–20074

Wyatt R, Sodroski J (1998) The HIV-1 envelope glycoproteins: fusogens, antigens, and immunogens. Science 280:1949–1953

Xu XN, Screaton G (2001) HIV-1 Nef: negative effector of Fas? Nat Immunol 2:384–385

Xu XN, Screaton GR, Gotch FM, Dong T, Tan R, Almond N, Walker B, Stebbings R, Kent K, Nagata S, Stott JE, McMichael AJ (1997) Evasion of cytotoxic T lymphocyte (CTL) responses by nef-dependent induction of Fas ligand (CD95L) expression on simian immunodeficiency virus-infected cells J Exp Med 186:7–16

Yang Y, Dong B, Mittelstadt PR, Xiao H, Ashwell JD (2002) HIV Tat binds Egr proteins and enhances Egr-dependent transactivation of the Fas ligand promoter. J Biol Chem 277:19482–19487

Yang Y, Tikhonov I, Ruckwardt TJ, Djavani M, Zapata JC, Pauza CD, Salvato MS (2003) Monocytes treated with human immunodeficiency virus Tat kill uninfected CD4(+) cells by a tumor necrosis factor-related apoptosis-induced ligand-mediated mechanism. J Virol 77:6700–6708

Yeh WC, Pompa JL, McCurrach ME, Shu HB, Elia AJ, Shahinian A, Ng M, Wakeham A, Khoo W, Mitchell K, El DW, Lowe SW, Goeddel DV, Mak TW (1998) FADD: essential for embryo development and signaling from some, but not all, inducers of apoptosis. Science 279:1954–1958

Yeh WC, Hakem R, Woo M, Mak TW (1999) Gene targeting in the analysis of mammalian apoptosis and TNF receptor superfamily signaling. Immunol Rev 169:283–302

Yeung MC, Chang DL, Camantigue RE, Lau AS (1999) Inhibitory role of the host apoptogenic gene PKR in the establishment of persistent infection by encephalomyocarditis virus in U937 cells. Proc Natl Acad Sci USA 96:11860–11865

Yuan H, Xie YM, Chen IS (2003) Depletion of Wee-1 kinase is necessary for both human immunodeficiency virus type 1 Vpr- and gamma irradiation-induced apoptosis. J Virol 77:2063–2070

Zack JA, Arrigo SJ, Weitsman SR, Go AS, Haislip A, Chen IS (1990) HIV-1 entry into quiescent primary lymphocytes: molecular analysis reveals a labile, latent viral structure. Cell 61:213–222

Zauli G, Gibellini D, Secchiero P, Dutartre H, Olive D, Capitani S, Collette Y (1999) Human immunodeficiency virus type 1 Nef protein sensitizes CD4(+) T lymphoid cells to apoptosis via functional upregulation of the CD95/CD95 ligand pathway. Blood 93:1000–1010

Zhang M, Li X, Pang X, Ding L, Wood O, Clouse KA, Hewlett I, Dayton AI (2002) Bcl-2 upregulation by HIV-1 Tat during infection of primary human macrophages in culture. J Biomed Sci 9:133–139

Zhu Y, Roshal M, Li F, Blackett J, Planelles V (2003) Upregulation of survivin by HIV-1 Vpr. Apoptosis 8:71–79

Poliovirus and Apoptosis

B. Blondel[1], T. Couderc[1], Y. Simonin[1], A.-S. Gosselin[1] and F. Guivel-Benhassine[1]

1 Introduction

Poliovirus (PV) is the causal agent of paralytic poliomyelitis, an acute disease of the central nervous system (CNS). A killed and an oral live attenuated vaccine were both developed in the 1950s (Salk 1955; Sabin and Boulger 1973), and subsequent massive vaccination campaigns resulted in near total eradication of PV from most industrialized countries: currently, wild strains are endemic in only 10 countries. However, there are two current problems concerning PV: the occurrence, in rare cases, of poliomyelitis outbreaks due to oral-vaccine-derived PV strains (Nomoto and Arita 2002); and the development of a new neuro-muscular pathology, called the post-polio syndrome, in poliomyelitis survivors long after the acute disease (Dalakas 1995).

PV is an enterovirus belonging to the Picornaviridae family and it is classified into three serotypes: PV-1, PV-2, and PV-3. Because of its very simple structure, PV has been used as a model to study non-retroviral RNA viruses and, consequently, PV is now one of the best-characterized animal viruses. The development of new animal and cell models has allowed the key step of the pathogenesis of poliomyelitis to be investigated at the molecular level. Thus, the interactions between PV and cells of the CNS (Blondel et al. 1998), notably the involvement of apoptosis in CNS injury during paralytic poliomyelitis, have been studied.

Apoptosis is an active process of cell death that occurs in response to various stimuli, including viral infection (Roulston et al. 1999). It involves a number of distinct morphological and biochemical features, such as cell shrinkage, plasma membrane blebbing, chromatin condensation, and internucleosomal DNA cleavage. These changes are mediated in particular by a family of proteases called caspases (cysteine proteases with aspartate-specificity; Earnshaw et al. 1999). The apoptotic pathways leading to cell death can be generally divided into two non-exclusive signaling cascades involving death receptors (extrinsic pathway) or mitochondria (intrinsic pathway; Kaufmann and Hengartner 2001). The death receptor pathway is activated by the binding of ligand

[1]Unité de Neurovirologie et Régénération du Système Nerveux, Institut Pasteur, 75724 Paris cedex 15, France,

Progress in Molecular and Subcellular Biology
C. Alonso (Ed.): Viruses and Apoptosis
© Springer-Verlag Berlin Heidelberg 2004

to the membrane receptor leading to the formation of the death-inducing signaling complex (DISC) which allows caspase-8 and/or caspase-10 autoactivation followed by caspase-3 activation (Kischkel et al. 2001). Apoptosis via the mitochondrial pathway involves specific signals that lead to the release of proapoptotic molecules from mitochondria to the cytosol (Li et al. 1997). These molecules include cytochrome c which forms a caspase-activating complex by interaction with the apoptosis protease-activating factor 1 (Apaf-1) and pro-caspase 9. This event triggers caspase-9 activation and initiates the apoptotic cascade by processing executive caspase-3.

In this review, we will briefly review the molecular biology of PV and of the pathogenesis of poliomyelitis, and then focus on several models of PV-induced apoptosis; we will also consider the role of the cellular receptor of PV (CD155) in the modulation of apoptosis.

2 Poliovirus

2.1 Structure of the Virion

The PV is composed of a single-stranded RNA genome of positive polarity surrounded by a non-enveloped icosahedral protein capsid. The mature virion is approximately 30 nm in diameter and the three-dimensional structures of the three serotypes of PV have been determined by X-ray crystallography (Hogle et al. 1985; Filman et al. 1989; Lentz et al. 1997). The capsid consists of 60 copies of each of the four viral structural proteins VP1, VP2, VP3, and VP4 (Fig. 1A). A deep surface depression, called the "canyon", surrounds each five-

Fig. 1. A Schematic structure of the PV capsid. The two-, three- and five-fold axes of symmetry and the positions of capsid proteins *VP1*, *VP2* and *VP3* are indicated for one protomer. Five molecules of VP1 surround the five-fold axis of symmetry, whereas VP2 and VP3 alternate around the three-fold axis of symmetry; VP4 is exclusively internal. The depression surrounding the five-fold axis, called the canyon, is formed by residues of VP1, VP2 and VP3, and contains the site for cell receptor binding (Adapted from Hogle et al. 1985). B Schematic structure of CD155. The PV receptor has three extracellular domains (D1 to D3), a transmembrane domain and an intracytoplasmic domain. The binding site for PV has been located in the D1 domain of CD155. The 5′ and 3′ noncoding regions, indicated as *5′NCR* and *3′NCR*, respectively, flank the single open-reading frame, encoding the polyprotein which is shown as an *elongated rectangle*. The protein precursors, *P1*, *P2* and *P3*, are designated by *arrows* above the genome. The viral proteins are indicated in the *rectangles*. The small viral protein *VPg* is covalently linked to the 5′ end of the RNA genome. Proteolytic cleavages occur between the amino acid pairs Asn-Ser, Gln-Gly and Tyr-Gly, as indicated by *empty*, *solid*, and *cross-hatched arrowheads*, respectively. The cleavage sites of proteases *2A*, *3CD* and *3C* are shown. The mechanism of cleavage of the precursor VP0 giving VP4 and VP2 is not known (Adapted from Kitamura et al. 1981). C Genetic organization of PV-1/Mahoney

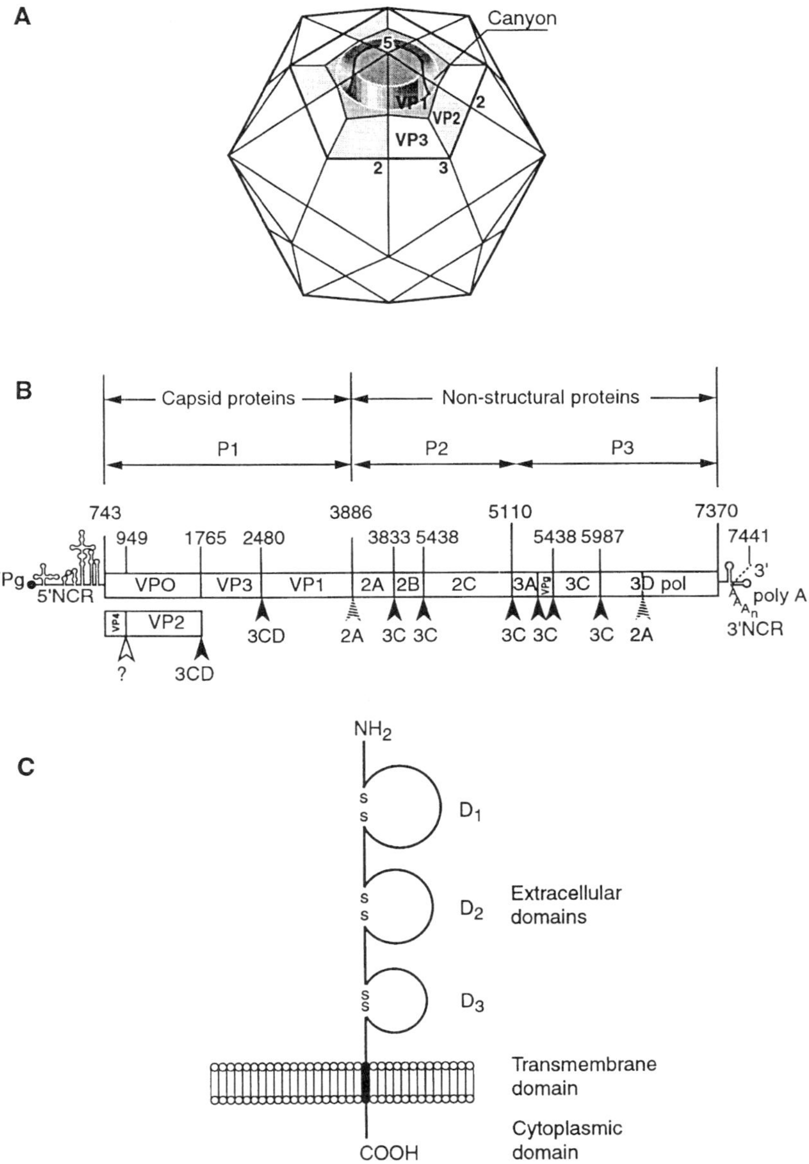
A
Canyon
5
VP1
VP2
VP3
2
2
3
B
Capsid proteins
Non-structural proteins
P1
P2
P3
743
3886
5110
7370
949
1765
2480
3833
5438
5438
5987
7441
VPg
5'NCR
VPO
VP3
VP1
2A
2B
2C
3A
VPg
3C
3D pol
3'
poly A
3'NCR
VP4
VP2
3CD
2A
3C
3C
3C
3C
3C
2A
?
3CD
C
NH2
S
S
D1
D2
Extracellular
domains
D3
Transmembrane
domain
Cytoplasmic
domain
COOH

fold axis of symmetry and contains the site for cell receptor binding (Colston and Racaniello 1994, 1995; Belnap et al. 2000; He et al. 2000; Xing et al. 2000).

The PV RNA genome is about 7500 nucleotides long (Fig. 1C). It is polyadenylated at its 3′-terminus and covalently linked to a small viral protein, VPg (3B), at its 5′-terminus (for review, see Racaniello 2001). It contains a long 5′ non-coding region (NCR) followed by a single large open reading frame (ORF) and a short 3′ NCR that includes the poly(A) tail. The ORF is translated to produce a 247-kDa polyprotein that can be considered as three distinct regions corresponding to structural (P1) and non-structural (P2 and P3) proteins.

2.2 Poliovirus Receptor

The human PV receptor, CD155, and its simian counterpart, are members of the immunoglobulin superfamily (Mendelsohn et al. 1989; Koike et al. 1990, 1992). They are related to the nectin family of adhesion molecules found at intercellular junctions (Eberle et al. 1995; Lopez et al. 1995; Takahashi et al. 1999). CD155 is predicted to contain three extracellular Ig-like domains in the order V-C2-C2, followed by a transmembrane region and a short cytoplasmic tail (Fig. 1B). The binding site for PV has been mapped to the V-like Ig domain (domain 1; Koike et al. 1991a; Aoki et al. 1994; Bernhardt et al. 1994; Morrison et al. 1994; Belnap et al. 2000; He et al. 2000). In human epithelial HeLa cells, two different bases (G and A) have been found at nucleotide position 199 in the mRNA encoding CD155. Consequently, amino-acid position 67, within domain 1 of CD155, is either Ala or Thr (Mendelsohn et al. 1989; Koike et al. 1990). Both these CD155 forms have been found in humans (Lundstöm et al. 2001) suggesting that CD155 with a Thr residue at amino-acid position 67 is an allelic form of CD155.

The ectodomain of CD155 binds specifically to vitronectin, a multifunctional adhesive glycoprotein (Lange et al. 2001), whereas its cytoplasmic tail interacts with Tctex-1, a light chain of the dyneins, microtubule-based molecular motor complexes (Ohka and Nomoto 2001; Mueller et al. 2002). Expression of CD155 is activated by the secreted morphogen sonic hedgehog protein (Solecki et al. 2002). Furthermore, both the expression of CD155 and vitronectin production are associated with regions of the CNS involved in the differentiation of motor neurons during embryonic development (Martinez-Moralez et al. 1997; Gromeier et al. 2000b). CD155 is also overexpressed in human colorectal carcinoma (Masson et al. 2001) and in malignant gliomas (Gromeier et al. 2000a). However, despite these numerous data, the cellular role of CD155 is still unclear.

2.3 Viral Cycle

In vitro, PV multiplies exclusively in primate cell lines either human or simian. The viral cycle of PV occurs entirely in the cytoplasm of the host cell (Racaniello 2001). It is among the fastest known viral cycles, lasting approximately 8 h at 37 °C in cell culture.

2.3.1 Early Steps of Cell Entry

The initial event of the viral cycle is attachment of the virion to the receptor CD155: this receptor is only found on the surface of primate cells. On binding to CD155, the PV capsid undergoes a conformational alteration. The altered particle, named the A particle, may be a cell entry intermediate necessary for RNA release (Hogle 2002; Hogle and Racaniello 2002).

2.3.2 Translation of Viral RNA

After PV RNA has been released into the cytoplasm of infected cells, translation of PV RNA is initiated by the binding of ribosomes to a specialized region in the 5′ NCR, called the internal ribosomal entry site (IRES; Ehrenfield and Teterina 2002). Efficient IRES-dependent translation of PV RNA requires IRES-specific cellular factors as well as canonical initiation factors (Jackson 2002). Translation produces the large polyprotein which is processed co-translationally by viral proteases 3C, its precursor 3CD, and 2A to yield the structural and non-structural proteins responsible for the proteolytic activities, RNA synthesis and biochemical and structural changes that occur in the infected cell.

2.3.3 Replication of Viral RNA

Viral RNA replicates on the surface of membranous vesicles that bud from various host cell organelles (Carrasco et al. 2002; Egger et al. 2002). RNA is replicated by the viral RNA-dependent RNA polymerase 3Dpol. Most of the other PV non-structural proteins and also cellular factors are involved in viral RNA synthesis. RNA replication starts with the formation of a complementary negative-stranded RNA molecule which serves as template for the synthesis of progeny positive-stranded viral RNAs (Xiang et al. 1997; Andino et al. 1999; Gromeier et al. 1999; Paul 2002).

2.3.4
Virion Assembly and Release

The formation of viral particles seems to be coupled to RNA synthesis (Hellen and Wimmer 1995; Ansardi et al. 1996; Nugent et al. 1999). VP0 (the precursor of VP2 and VP4), VP1 and VP3 aggregate with the viral RNA to form the provirion (Hogle 2002). During the last step of virus assembly, VP0 is cleaved to give VP2 and VP4 by an unknown mechanism. Once assembled, the virions accumulate in the cytoplasm of infected cells in the form of crystalline inclusions which are liberated by the bursting of vacuoles at the cell surface (Dunnebacke et al. 1969; Bienz et al. 1973). Vectorial release has been described in polarized human intestinal epithelial cells (Tucker et al. 1993). Cell lysis is accompanied by the massive release of new progeny virions.

2.3.5
Effect of Poliovirus Replication on the Host Cell

As PV infection progresses in vitro, the host cell undergoes substantial metabolic and morphological changes commonly referred to as cytopathic effects (Haller and Semler 1995; Schlegel et al. 1996; Carrasco et al. 2002). Early in the infectious cycle, PV proteases mediate the shutoff of both host cell translation and transcription. The 2A protease of PV cleaves or induces the cleavage of translation initiation factors, such as eIF4G, thus inhibiting host cap-dependent mRNA translation (Etchison et al. 1982; Novoa and Carrasco 1999; Kuechler et al. 2002). The protease 3C inhibits pol III transcription, presumably by cleavage of the transcription factor TF3C (Clark and Dasgupta 1990). In addition, non-structural PV proteins have large effects on host intracellular-membrane structure and function. In particular, protein 2C induces membrane vesiculation (Cho et al. 1994; Aldabe and Carrasco 1995; Teterina et al. 1997), whereas proteins 2B and 3A are each sufficient to inhibit protein traffic through the host secretory pathway (Doedens and Kirkegaard 1995; Doedens et al. 1997; Dodd et al. 2001; Neznanov et al. 2001). Finally, PV infection can trigger the development of apoptosis, as described below.

3
Pathogenesis of Poliomyelitis and Post-polio Syndrome

Humans are the only natural host of PV. It is transmitted mainly via the fecal-oral route or via droplets from the pharynx. Following oral ingestion, the virus infects the oropharynx and the gut, and is excreted in the oropharyngeal secretions and in the stool for several weeks (for review, see Ohka and Nomoto 2001; Pallansch and Roos 2001; Gromeier and Nomoto 2002). There is some evidence that PV enters the gut by translocation through the epithelial M cells over the Peyer's patches, and multiplies extensively in the tonsils and the Peyer's patches. However, the exact target cell where initial multiplication occurs is still unidentified. The ability of PV to replicate in mononuclear phagocytic cells from the blood suggests that resident phagocytic cells in the tonsils and the Peyer's patches may be the site of the initial rounds of PV replication. Moreover, the PV receptor CD155 has been detected at the cell surface of enterocytes of the follicle-associated epithelium as well as in follicular dendritic cells and B cells of germinal centers within the Peyer's patches. However, the susceptibility of these cells to PV infection is not yet known.

The virus is released from lymphoid tissues into the blood stream. In a minority of cases, this results in viral spread to other lymphoid tissues, amplifying viremia. The circulating virus invades the CNS probably through the blood-brain barrier and this does not require the CD155 molecule. In some cases, virus may enter the CNS by the retrograde axonal pathway (Bodian 1955; Nathanson and Langmuir 1963; Ren and Racaniello 1992; Gromeier and Wimmer 1998; Ohka et al. 1998; Ohka and Nomoto 2001; Gromeier and Nomoto 2002). In the CNS, the main target cell of PV is the motor neuron of the spinal cord and the brainstem. The destruction of motor neurons, a consequence of PV replication, results in paralysis. Studies with mice transgenic for the PV receptor molecule CD155 (Tg-CD155) have shown that the specific tropism for motor neurons is due to their expression of CD155 molecule. In Tg mice expressing CD155 under the transcriptional control of a ubiquitous promoter, glial cells as well as neurons were susceptible to PV infection (Ida-Hosonuma et al. 2002). Within the CNS, PV probably travels through a cell-to-cell spread via an axonal pathway (Ponnuraj et al. 1998).

Infections with PV result in poliomyelitis in 1–2% of cases. Following decades of clinical stability, many poliomyelitic patients develop a disease called post-polio syndrome, characterized notably by slowly progressive muscle weakness (Dalakas 1995). The presence of PV RNA sequences or PV-related RNA (Leon-Monzon and Dalakas 1995; Muir et al. 1995; Leparc-Goffart et al. 1996) and of anti-PV IgM antibodies (Sharief et al. 1991) suggests that PV persistence may be involved in this syndrome. Moreover, PV can establish persistent infections in human cell cultures of neuronal origin (Colbère-Garapin et al. 1989, 2002; Pavio et al. 1996).

Experimentally, poliomyelitis can be transmitted to monkeys and in some cases to mice by intracerebral inoculation of PV. Monkeys and Tg-CD155 mice are susceptible to wild strains of all three PV serotypes (Ren et al. 1990; Koike et al. 1991b). In these mice, infection of the CNS is usually extensive and most often results in fatal poliomyelitis. In non-Tg-CD155 mice, only a small number of mouse-adapted PV strains induce poliomyelitis. We have isolated and characterized a PV mutant pathogenic for mice, PV-1/Mah-T1022I (Couderc et al. 1993, 1996). Interestingly, this mutant induced paralytic poliomyelitis in Swiss mice that is, as in humans, not always lethal. With this model, we have shown that PV persists in the CNS throughout the life of animals and that PV RNA replication in the CNS of persistently infected mice is restricted

(Destombes et al. 1997), mainly as a consequence of inhibition of plus-strand RNA synthesis (Girard et al. 2002). The mouse models described above allowed the study principally of the neurologic phase of poliomyelitis but not the digestive phase. Recently, a new Tg-CD155 mouse model has been reported for a mucosal route of infection with PV (Crotty et al. 2002).

4
Poliovirus and Apoptosis

Cell damage in the CNS in response to virus infection can involve apoptosis (Shen and Shenk 1995). This has been illustrated in vivo with human and murine neurotropic RNA viruses including HIV (Petito and Roberts 1995), HTLV-1 (Umehara et al. 1994), reovirus (Oberhaus et al. 1997), La Crosse virus (Pekosz et al. 1996), rabies virus (Jackson and Rossiter 1997a), mouse hepatitis virus (Lin et al. 1997), dengue virus (Desprès et al. 1998), Sindbis virus (Levine et al. 1993; Lewis et al. 1996), Venezuelan equine encephalitis virus (Jackson and Rossiter 1997b), and Theiler murine encephalomyelitis virus (Tsunoda et al. 1997), a member of the Picornaviridae family. Coxsackie virus B3 (Carthy et al. 1998) and HAV (Brack et al. 1998) both other picorna viruses, induce apoptosis in cell cultures.

4.1
Poliovirus-Induced Apoptosis in Nerve Cells In Vivo and Ex Vivo

PV-induced paralytic poliomyelitis results from the destruction of motor neurons. However, the process leading to the death of motor neurons was until recently unknown.

Work with mouse models (Tg-CD155 and non-Tg mice) shows that PV-infected motor neurons in the spinal cord died by apoptosis (Fig. 2): the extent of apoptosis correlates with viral load and its onset coincides with that of paralysis (Girard et al. 1999). Moreover, CNS injury may be enhanced by apoptosis in uninfected nonneuronal cells, probably glial or inflammatory cells, contiguous to the PV-infected neurons (Girard et al. 1999). PV-induced apoptosis is therefore an important component of tissue injury in the CNS of infected mice that leads to paralysis.

To investigate PV-induced apoptosis in nerve cells, a culture of mixed mouse primary nerve cells from the cerebral cortex of Tg-CD155 mice has been developed (Couderc et al. 2002). These cultures contain all three main cell types of the CNS, i.e. neurons, astrocytes and oligodendrocytes. All these cell types are susceptible to PV infection and viral replication leads to DNA fragmentation characteristic of apoptosis (Couderc et al. 2002). It is interesting to note that, in contrast to the observations in vivo, mixed primary nerve cell cultures harbor PV-infected glial cells in addition to PV-infected neurons. This

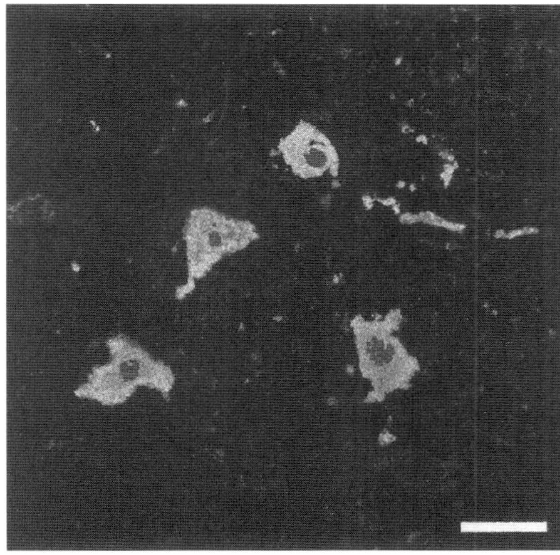

Fig. 2. PV-infected motor neurons (immunofluorescence labeling, *white*) with an apoptotic nucleus (terminal deoxynucleotidyltransferase-mediated dUTP nick end labeling, *gray*) in the mouse spinal cord. *Bar* 20 µm

is probably due to a difference in the expression of CD155 ex vivo and in vivo. In these nerve cell cultures, we have shown that PV-induced apoptosis involved caspases. This new cellular model could allow analysis of the molecular mechanisms of PV-induced apoptosis in nerve cells and lead to the identification of specific neuronal pathways by comparing PV-induced apoptosis in neurons with that in glial cells.

4.2
Poliovirus-Induced Apoptosis In Vitro

PV can trigger apoptosis in human tissue cultures of enterocyte-like cells (CaCo-2) and promonocytic cells (U937; Ammendolia et al. 1999; Lopez-Guerrero et al. 2000). If enterocytes and monocytic cells are indeed infected by PV during viral infection in humans, PV-induced apoptosis in these two target cells could allow the virus to reach the CNS with a limited activation of immune inflammatory responses.

Tolskaya et al. (1995) showed that no apoptotic reaction occurred in a subline of human epithelial cells (HeLa) following productive infection. In contrast, apoptosis can develop upon non-permissive infection with various PV mutants (guanidine-sensitive, guanidine-dependent, or temperature-sensitive mutants). In permissive conditions, apoptosis induced by apoptotic inducers, such as metabolic inhibitors, is suppressed by PV infection (Tolskaya et al. 1995; Koyama et al. 2001). Therefore, PV appears to encode two separate functions which can have opposite effects: one triggering and the other suppressing apoptosis. According these data, Agol et al. (2000) propose a model in which, during the PV productive infection, the commitment of cells switches in the middle of the viral cycle from apoptosis to an anti-apoptotic state and cytopathic effect. Analysis of specific apoptotic pathways in this model suggested that early induced PV apoptosis involves translocation of cytochrome c from mitochondria to the cytosol and thus activates caspase-9 and caspase-3 (Belov et al. 2003). The subsequent suppression of the apoptotic program may be due, at least in part, to aberrant processing and degradation of procaspase-9 (Belov et al. 2003). Apoptosis triggered by PV in HeLa cells may require the activity of RNase L and involve a Bcl-X_L-sensitive pathway and caspases (Castelli et al. 1997; Liu et al. 1999; Agol et al. 2000).

Expression of 2A or 3C PV proteases is sufficient to trigger apoptosis (Barco et al. 2000; Goldstaub et al. 2000), suggesting that PV proteases are able to activate an endogenous cell suicide program. Apoptosis induced by 3C seems to depend on the caspase pathway (Barco et al. 2000), whereas 2A-induced apoptosis may be caspase-independent (Goldstaub et al. 2000). When elucidating the exact molecular mechanism used by 3C to provoke apoptosis it should be remembered that this protease cleaves a variety of host proteins, including transcription factors (Clark et al. 1991, 1993) and the cytoskeletal protein MAP4 (Joachims et al. 1995). Cleavage of these proteins could thus trigger apoptosis via a transcriptional inhibition mechanism or cytoskeleton changes, respectively. Protease 2A is involved in the cleavage of eIF4GI and eIF4GII, and the poly(A)-binding protein (PABP), which are factors involved in cellular protein synthesis activity (Etchison et al. 1982; Joachims et al. 1999; Goldstaub et al. 2000). Thus, the PV 2A protease may induce apoptosis either by arresting cap-dependent translation of some cellular mRNAs encoding proteins that are required for maintaining cellular viability. Alternatively, the protease 2A may induce apoptosis by allowing preferential cap-independent translation of cellular mRNAs encoding proteins that induce apoptosis. Proteases 2A and 3C may also trigger the apoptotic process via the cleavage of other unidentified cellular substrates.

PV-mediated suppression of apoptosis may involve mechanisms in addition to caspase-9 cleavage. Expression of 3A specifically suppresses the host secretory pathway, resulting notably in the loss of tumor necrosis factor (TNF) receptor from the cell surface (Neznanov et al. 2001). This causes a decrease in cell sensitivity to TNF and thus may be another mechanism preventing apoptosis. Furthermore, although expression of PV 3C protease induces apoptosis (Barco et al. 2000), 3C may also be able to delay or prevent apoptosis as it mediates the degradation of the transcriptional activator and tumor suppressor p53 (Weidman et al. 2001).

4.3 CD155 and Apoptosis

PV can establish persistent infections in cells of neural origin (neuroblastoma cells) and in human fetal brain cell cultures (Colbère-Garapin et al. 1989; Pavio et al. 1996; Colbère-Garapin et al. 2002). During persistent infection in human neuroblastoma IMR-32 cells, specific mutations were selected in CD155 that affected domain 1, the binding site for PV. These mutations included the Ala 67→Thr substitution, corresponding to a switch from one allelic form of the PV receptor to the other form previously identified. This mutated form of CD155 was not expressed in IMR-32 cells. The two forms of PV receptors, mutated ($CD155_{Thr67}$) and non-mutated ($CD155_{IMR}$), were expressed independently in murine LM cells lacking the CD155 gene. Interestingly, although virus adsorption and viral growth were identical in the two cell lines, PV-induced cytopathic effect in cells expressing the mutated form, $CD155_{Thr67}$ (LM-$CD155_{Thr67}$), appeared later than those in cells expressing the non-mutated form $CD155_{IMR}$ (LM-$CD155_{IMR}$; Pavio et al. 2000). A role for CD155 in cell cytopathic effect was previously suggested with mutated forms of CD155 generated by site-directed mutagenesis affecting residues in domain 1 (Morrison et al. 1994).

Analysis of the PV-induced death of LM-$CD155_{Thr67}$ or LM-$CD155_{IMR}$ cells showed that PV infection triggers DNA fragmentation characteristic of apoptosis in both LM cell lines, but at a lower level in LM-$CD155_{Thr67}$ cells than in LM-$CD155_{IMR}$ cells (Gosselin et al. 2003). Intrinsic (mitochondrial disfunction) and extrinsic (death receptor mediated) apoptotic pathways were investigated in both cell lines. The mitochondrial dysfunction has been analyzed by detection of the translocation of cytochrome c from mitochondria and by measuring the activation of caspases-3 and -9. Levels of cytochrome c release and caspase-9 and caspase-3 activation were lower in PV-infected LM-$CD155_{Thr67}$ cells than in LM-$CD155_{IMR}$ cells (Gosselin et al. 2003). The death receptor-mediated pathway was checked by measuring the activation of caspase-8 and caspase-10. In both cell lines caspase-8 and caspase-10 activation paralleled that of caspase-9 and caspase-3. Thus, it appeared that the two main apoptotic signaling pathways are simultaneously initiated in response to PV infection. However, this observation does not exclude a possible cross-talk between the intrinsic and the extrinsic apoptotic cascades. Indeed, caspase-8 can generate a truncated form of the Bid protein which then translocates to mitochondria and initiates the mitochondrial pathway (Kaufmann and Hengartner 2001). Moreover, some of the caspases activated by the mitochondrial pathway may also activate caspase-8 in a feedback loop (Viswanath et al. 2001).

Altogether these data indicated that the level of PV-induced apoptosis is lower in cells expressing mutated $CD155_{Thr67}$ selected during persistent PV infection of IMR-32 cells than in cells expressing $CD155_{IMR}$. As stated above, expression of 2A and 3C PV proteases is sufficient to induce cell apoptosis (Barco et al. 2000; Goldstaub et al. 2000). Thus, CD155 may be an additional cellular factor involved in the modulation of PV-induced apoptosis. One attractive possibility is the modulation of an apoptotic signal triggered by the PV-CD155 interaction. The triggering of apoptosis by the binding of a virus to a receptor, and also by a post-binding entry step, has been shown for other viruses, including reovirus (Tyler et al. 2001), Sindbis virus (Jan and Griffin 1999) and HIV (Banda et al. 1992). However, no data are currently available supporting the notion that an apoptotic signal is directly transduced via CD155. Alternatively, a molecule interacting with CD155 may transduce this signal. One candidate is CD44, the major receptor for hyaluronic acid, which is physically associated with CD155 and is able to induce apoptotic signals (Freistadt et al. 1997; Foger et al. 2000; Okamoto et al. 2001; Shepley et al. 1994). Another possible mechanism involves interaction between CD155 and cellular factors known to be involved in apoptotic pathways.

5 Conclusions

During recent years, the amount of effort put into the study of PV-induced apoptosis has increased. This work has revealed the complexity of the apoptotic process in PV-infected cells. PV-induced apoptosis in the CNS seems to be a significant factor in poliomyelitis pathogenesis and it would be interesting to identify the molecular mechanisms of PV-induced apoptosis in neurons. New ex vivo cellular models will allow the analysis of apoptotic pathways triggered by PV infection in nerve cells. In vitro studies using various cellular models have shown that PV can either trigger, or in contrast, suppress the development of apoptosis according to the conditions of infection and to the host cell type. Several viral proteins are good candidates for involvement in the apoptotic process in PV-infected cells. Proteases 2A and 3C are able to trigger apoptosis whereas protein 3A can inhibit it. Protease 3C can also prevent apoptosis. However, most of these data were obtained with individual protein expression systems and the situation might be more complex in the context of cellular infection. PV interaction with CD155 could also trigger apoptosis either via CD155 itself or more probably via another molecule interacting with CD155. However the molecular apoptotic pathways leading to cell death following PV–CD155 interaction remain to be determined. PV-induced apoptosis is less extensive in cells expressing mutated $CD155_{Thr67}$ selected during persistent PV infection of IMR-32 cells than in cells expressing non-mutated CD155IMR. Thus, CD155 may also be a cellular factor involved in the modulation of PV-induced apoptosis. This modulation could be profitable to the virus. If apoptosis is not induced correctly or not induced at all in a context where it should be, it may allow the establishment of a persistent infection by PV. Thus, it would be interesting to determine which CD155 allelic forms are expressed in patients developing poliomyelitis and post-polio syndrome.

Acknowledgements. The authors are very grateful to Laurent Blondel for his precious help with the figures. This work was supported by grants from the Institut Pasteur and the Association Française contre les Myopathies (contract N° 6932 and 7290). The Fondation pour la Recherche Médicale is acknowledged for the fellowship awarded to A.S.G.

References

Agol VI, Belov GA, Bienz K, Egger D, Kolesnikova MS, Romanova LI, Sladkova LV, Tolskaya EA (2000) Competing death programs in poliovirus-infected cells: commitment switch in the middle of the infectious cycle. J Virol 74:5534–5541

Aldabe R, Carrasco L (1995) Induction of membrane proliferation by poliovirus proteins 2C and 2BC. Biochem Biophys Res Commun 206:64–76

Ammendolia MG, Tinari A, Calcabrini A, Superti F (1999) Poliovirus infection induces apoptosis in CaCo-2 cells. J Med Virol 59:122–129

Andino R, Böddeker N, Silvera D, Gamarnik A (1999) Intracellular determinants of picornavirus replication. Trends Microbiol 7:76–82

Ansardi D, Porter D, Anderson M, Morrow C (1996) Poliovirus assembly and encapsidation of genomic RNA. Adv Virus Res 46:1–68

Aoki J, Koike S, Ise I, Satoyoshida Y, Nomoto A (1994) Amino acid residues on human poliovirus receptor involved in interaction with poliovirus. J Biol Chem 269:8431–8438

Banda NK, Bernier J, Kurahara DK, Kurrle R, Haigwood N, Sekaly RP, Finkel T (1992) Crosslinking CD4 by human immunodeficiency virus gp120 primes T cells for activation-induced apoptosis. J Exp Med 176:1099–1106

Barco A, Feduchi E, Carrasco L (2000) Poliovirus protease 3Cpro kills cells by apoptosis. Virology 266:352–360

Belnap DM, McDermott BM, Filman DJ, Cheng NQ, Trus BL, Zuccola HJ, Racaniello VR, Hogle JM, Steven AC (2000) Three-dimensional structure of poliovirus receptor bound to poliovirus. Proc Natl Acad Sci USA 97:73–78

Belov GA, Romanova LI, Tolskaya EA, Kolesnikova MS, Lazebnik YA, Agol VI (2003) The major apoptotic pathway activated and suppressed by poliovirus. J Virol 77:45–56

Bernhardt G, Harber J, Zibert A, de Crombrugghe M, Wimmer E (1994) The poliovirus receptor: identification of domains and amino acid residues critical for virus binding. Virology 203:344–356

Bienz K, Egger D, Wolff DA (1973) Virus replication, cytopathology, and lysosomal enzyme response of mitotic and interphase Hep-2 cells infected with poliovirus. J Virol 11:565–574

Blondel B, Duncan G, Couderc T, Delpeyroux F, Pavio N, Colbère-Garapin F (1998) Molecular aspects of poliovirus biology with a special focus on the interactions with nerve cells. J Neurovirol 4:1–26

Bodian D (1955) Viremia, invasiveness, and the influence of injections. Ann NY Acad Sci 61:877–882

Brack K, Frings W, Dotzauer A, Vallbracht A (1998) A cytopathogenic, apopotosis-inducing variant of hepatitis A virus. J Virol 72:3370–3376

Carrasco L, Guinea R, Irurzun A, Barco A (2002) Effects of viral replication on cellular membrane metabolism and function. In: Semler BL, Wimmer E (eds) Poliovirus receptors and cell entry. ASM Press, Washington, DC, pp 337–354

Carthy CM, Granville DJ, Watson KA, Anderson DR, Wilson JE, Yang D, Hunt DWC, McManus BM (1998) Caspase activation and specific cleavage of substrates after coxsackievirus B3-induced cytopathic effect in HeLa cells. J Virol 72:7669–7675

Castelli JC, Hassel BA, Wood KA, Li XL, Amemiya K, Dalakas MC, Torrence PF, Youle RJ (1997) A study of the interferon antiviral mechanism—apoptosis activation by the 2–5a system. J Exp Med 186:967–972

Cho MW, Teterina N, Egger D, Bienz K, Ehrenfeld E (1994) Membrane rearrangement and vesicle induction by recombinant poliovirus 2C and 2BC in human cells. Virology 202:129–145

Clark ME, Dasgupta A (1990) A transcriptionally active form of TFIIIC is modified in poliovirus-infected HeLa cells. Mol Cell Biol 10:5106–5113

Clark ME, Hammerle T, Wimmer E, Dasgupta A (1991) Poliovirus proteinase-3C converts an active form of transcription factor-IIIC to an inactive form: a mechanism for inhibition of host cell polymerase-III transcription by poliovirus. EMBO J 10:2941–2947

Clark ME, Lieberman PM, Berk AJ, Dasgupta A (1993) Direct cleavage of human TATA-binding protein by poliovirus protease 3C in vivo and in vitro. Mol Cell Biol 13:1232–1237

Colbère-Garapin F, Christodoulou C, Crainic R, Pelletier I (1989) Persistent poliovirus infection of human neuroblastoma cells. Proc Natl Acad Sci USA 86:7590–7594

Colbère-Garapin F, Pelletier I, Ouzilou L (2002) Persistent infection by poliovirus. In: Semler BL, Wimmer E (eds) Molecular biology of picornaviruses. ASM Press, Washington, DC, pp 437–448

Colston E, Racaniello VR (1994) Soluble receptor-resistant poliovirus mutants identify surface and internal capsid residues that control interaction with the cell receptor. EMBO J 13:5855–5862

Colston EM, Racaniello VR (1995) Poliovirus variants selected on mutant receptor-expressing cells identify capsid residues that expand receptor recognition. J Virol 69:4823–4829

Couderc T, Hogle J, Le Blay H, Horaud F, Blondel B (1993) Molecular characterization of mouse-virulent poliovirus type 1 Mahoney mutants: involvement of residues of polypeptides VP1 and VP2 located on the inner surface of the capsid protein shell. J Virol 67:3808–3817

Couderc T, Delpeyroux J, Le Blay H, Blondel B (1996) Mouse adaptation determinants of poliovirus type 1 enhance viral uncoating. J Virol 70:305–312

Couderc T, Guivel-Benhassine F, Calaora V, Gosselin AS, Blondel B (2002) An ex vivo murine model to study poliovirus-induced apoptosis in nerve cells. J Gen Virol 83:1925–30

Crotty S, Hix L, Sigal LJ, Andino R (2002) Poliovirus pathogenesis in a new poliovirus receptor transgenic mouse model: age-dependent paralysis and a mucosal route of infection. J Gen Virol 83:1707–1720

Dalakas MC (1995) The post-polio syndrome as an evolved clinical entity. Definition and clinical description. In: Dalakas MC, Bartfeld H, Kurland LT (eds) The post-polio syndrome. The New York Academy of Sciences, New York, pp 68–80

Desprès P, Frenkiel MP, Ceccaldi PE, Duarte dos Santos C, Deubel V (1998) Apoptosis in the mouse central nervous system in response to infection with mouse-neurovirulent dengue viruses. J Virol 72:823–829

Destombes J, Couderc T, Thiesson D, Girard S, Wilt SG, Blondel B (1997) Persistent poliovirus infection in mouse motoneurons. J Virol 71:1621–1628

Dodd DA, Giddings TH, Kirkegaard K (2001) Poliovirus 3A protein limits interleukin-6 (IL-6), IL-8, and beta interferon secretion during viral infection. J Virol 75:8158–8165

Doedens JR, Kirkegaard K (1995) Inhibition of cellular protein secretion by poliovirus proteins 2B and 3A. EMBO J 14:894–907

Doedens JR, Giddings TH, Kirkegaard K (1997) Inhibition of endoplasmic reticulum-to-golgi traffic by poliovirus protein 3a—genetic and ultrastructural analysis. J Virol 71:9054–9064

Dunnebacke TH, Levinthal JD, Williams RC (1969) Entry and release of poliovirus as observed by electron microscopy of cultured cells. J Virol 4:504–513

Earnshaw W, Martins L, Kaufmann S (1999) Mammalian caspases: structure, activation, substrates, and functions during apoptosis. Annu Rev Biochem 68:383–424

Eberle F, Dubreuil P, Mattei MG, Devilard E, Lopez M (1995) The human PRR2 gene, related to the human poliovirus receptor gene (PVR), is the true homolog of the murine MPH gene. Gene 159:267–272

Egger D, Gosert R, Bienz K (2002) Role of cellular structures in viral RNA replication. In: Semler BL, Wimmer E (eds) Molecular biology of picornaviruses. ASM Press, Washington, DC, pp 247–253

Ehrenfield E, Teterina N (2002) Initiation of translation of picornavirus RNAs: structure and function of internal ribosome entry site. In: Semler BL, Wimmer E (eds) Molecular biology of picornaviruses. ASM Press, Washington, DC, pp 159–169
Etchison D, Milburn SC, Edery I, Sonenberg N, Hershey JW (1982) Inhibition of HeLa cell protein synthesis following poliovirus infection correlates with the proteolysis of a 220,000-dalton polypeptide associated with eucaryotic initiation factor 3 and a cap binding protein complex. J Biol Chem 257:14806–14810
Filman DJ, Syed R, Chow M, Macadam AJ, Minor PD, Hogle JM (1989) Structural factors that control conformational transitions and serotype specificity in type 3 poliovirus. EMBO J 8:1567–1579
Foger N, Marhaba R, Zoller M (2000) CD44 supports T cell proliferation and apoptosis by apposition of protein kinases. Eur J Immunol 30:2888–2899
Freistadt MS, Eberle KE (1997) Physical association between CD155 and CD44 in human monocytes. Mol Immunol 34:1247–1257
Girard S, Couderc T, Destombes J, Thiesson D, Delpeyroux F, Blondel B (1999) Poliovirus induces apoptosis in the mouse central nervous system. J Virol 73:6066–6072
Girard S, Gosselin AS, Pelletier I, Colbere-Garapin F, Couderc T, Blondel B (2002) Restriction of poliovirus RNA replication in persistently infected nerve cells. J Gen Virol 83:1087–1093
Goldstaub D, Gradi A, Bercovitch Z, Grosmann Z, Nophar Y, Luria S, Sonenberg N, Kahana C (2000) Poliovirus 2A protease induces apoptotic cell death. Mol Cell Biol 20:1271–1277
Gosselin AS, Simonin Y, Guivel-Benhassine F, Rincheval V, Vayssiere JL, Mignotte B, Colbere-Garapin F, Couderc T, Blondel B (2003) Poliovirus-induced apoptosis is reduced in cells expressing a mutant CD155 selected during persistent poliovirus infection in neuroblastoma cells. J Virol 77:790–798
Gromeier M, Nomoto A (2002) Determinants of poliovirus pathogenesis. In: Semler BL, Wimmer E (eds) Molecular biology of picornaviruses. ASM Press, Washington, DC, pp 367–379
Gromeier M, Wimmer E (1998) Mechanism of injury-provoked poliomyelitis. J Virol 72:5056–5060
Gromeier M, Wimmer E, Gorbalenya AE (1999) Genetics, pathogenesis and evolution of picornaviruses. In: Domingo E, Webster RG, Holland JJ (eds) Origin and evolution of viruses. Academic Press, New York, pp 287–343
Gromeier M, Lachmann S, Rosenfeld MR, Gutin PH, Wimmer E (2000a) Intergeneric poliovirus recombinants for the treatment of malignant glioma. Proc Natl Acad Sci USA 97:6803–6808
Gromeier M, Solecki D, Patel DD, Wimmer E (2000b) Expression of the human poliovirus receptor/CD155 gene during development of the central nervous system: implications for the pathogenesis of poliomyelitis. Virology 273:248–257
Haller A, Semler B (1995) Translation and host cell shutoff. In: Rotbart HA (eds) Human enterovirus infection. ASM Press, Washington, DC, pp 113–133
He YN, Bowman VD, Mueller S, Bator CM, Bella J, Peng XH, Baker TS, Wimmer E, Kuhn RJ, Rossmann MG (2000) Interaction of the poliovirus receptor with poliovirus. Proc Natl Acad Sci USA 97:79–84
Hellen C, Wimmer E (1995) Enterovirus structure and assembly. In: Rotbart HA (ed) Human enterovirus infection. ASM Press, Washington, DC, pp. 155–174
Hogle JM (2002) Poliovirus cell entry: common structural themes in viral cell entry pathways. Annu Rev Microbiol 56:677–702
Hogle JM, Racaniello VR (2002) Poliovirus receptors and cell entry. In: Semler BL, Wimmer E (eds) Molecular biology of picornaviruses. ASM Press, Washington, DC, pp 71–83
Hogle JM, Chow M, Filman DJ (1985) Three dimensional structure of poliovirus at 2.9 Å resolution. Science 229:1358–1365
Ida-Hosonuma M, Iwasaki T, Taya C, Sato Y, Li J, Nagata N, Yonekawa H, Koike S (2002) Comparison of neuropathogenicity of poliovirus in two transgenic mouse strains expressing human poliovirus receptor with different distribution patterns. J Gen Virol 83:1095–1105

Jackson R (2002) Proteins involved in the function of picornavirus internal ribosomal entry sites. In: Semler BL, Wimmer E (eds) Molecular biology of picornaviruses. ASM Press, Washington, DC, pp 171–183

Jackson AC, Rossiter JP (1997a) Apoptosis plays an important role in experimental rabies virus infection. J Virol 71:5603–5607

Jackson AC, Rossiter JP (1997b) Apoptotic cell death is an important cause of neuronal injury in experimental Venezuelan equine encephalitis virus infection of mice. Acta Neuropathol 93:349–353

Jan J-T, Griffin DE (1999) Induction of apoptosis by Sindbis virus occurs at cell entry and does not require virus replication. J Virol 73:10296–10302

Joachims M, Harris KS, Etchison D (1995) Poliovirus protease 3C mediates cleavage of microtubule-associated protein 4. Virology 211:451–461

Joachims M, Van Breugel PC, Lloyd RE (1999) Cleavage of poly(A)-binding protein by enterovirus proteases concurrent with inhibition of translation in vitro. J Virol 73:718–727

Kaufmann SH, Hengartner MO (2001) Programmed cell death: alive and well in the new millennium. Trends Cell Biol 11:526–534

Kischkel FC, Lawrence DA, Tinel A, LeBlanc H, Virmani A, Schow P, Gazdar A, Blenis J, Arnott D, Ashkenazi A (2001) Death receptor recruitment of endogenous caspase-10 and apoptosis initiation in the absence of caspase-8. J Biol Chem 276:46639–46646

Kitamura N, Semler BL, Rothberg PG, Larsen GR, Adler CJ, Dorner AJ, Emini EA, Hanecak R, Lee JJ, Van der Werf S, Anderson CW, Wimmer E (1981) Primary structure, gene organization and polypeptide expression of poliovirus RNA. Nature 291:547–553

Koike S, Horie H, Ise I, Okitsu A, Yoshida M, Iizuka N, Takeuchi K, Takegami T, Nomoto A (1990) The poliovirus receptor protein is produced both as membrane-bound and secreted forms. EMBO J 9:3217–3224

Koike S, Ise I, Nomoto A (1991a) Functional domains of the poliovirus receptor. Proc Natl Acad Sci USA 88:4104–4108

Koike S, Taya C, Kurata T, Abe S, Ise I, Yonekawa H, Nomoto A (1991b) Transgenic mice susceptible to poliovirus. Proc Natl Acad Sci USA 88:951–955

Koike S, Ise I, Sato Y, Yonekawa H, Gotoh O, Nomoto A (1992) A 2nd gene for the African green monkey poliovirus receptor that has no putative N-glycosylation site in the functional N-terminal immunoglobulin-like domain. J Virol 66:7059–7066

Koyama AH, Irie H, Ueno F, Ogawa M, Nomoto A, Adachi A (2001) Suppression of apoptotic and necrotic cell death by poliovirus. J Gen Virol 82:2965–2972

Kuechler E, Seipelt J, Liebig H-D, Sommergruber W (2002) Picornavirus proteinase-mediated shutoff of host cell translation: direct cleavage of a cellular initiation factor. In: Semler BL, Wimmer E (eds) Molecular biology of picornaviruses. ASM Press, Washington, DC, pp 301–311

Lange R, Peng X, Wimmer E, Lipp M, Bernhard G (2001) The poliovirus receptor CD155 mediates cell-to-matrix contacts by specifically binding to vitronectin. Virology 285:218–227

Lentz KN, Smith AD, Geisler SC, Cox S, Buontempo P, Skelton A, Demartino J, Rozhon E, Schwartz J, Girijavallabhan V, Oconnell J, Arnold E (1997) Structure of poliovirus type 2 lansing complexed with antiviral agent sch48973 – comparison of the structural and biological properties of the three poliovirus serotypes. Structure 5:961–978

Leon-Monzon ME, Dalakas MC (1995) Detection of poliovirus antibodies and poliovirus genome in patients with post-polio syndrome (PPS). In: Dalakas MC, Bartfeld H, Kurland LT (eds) The post-polio syndrome. New York Academy of Sciences, New York, pp 208–218

Leparc-Goffart I, Julien J, Fuchs F, Janatova I, Aymard M, Kopecka H (1996) Evidence of presence of poliovirus genomic sequences in cerebrospinal fluid from patients with postpolio syndrome. J Clin Microbiol 34:2023–2026

Levine B, Huang Q, Isaacs JT, Reed JC, Griffin DE, Hardwick JM (1993) Conversion of lytic to persistent alphavirus infection by the bcl-2 cellular oncogene. Nature 361:739–742

Lewis J, Wesselingh SL, Griffin DE, Hardwick JM (1996) Alphavirus-induced apoptosis in mouse brains correlates with neurovirulence. J Virol 70:1828–1835

Li P, Nijhawan D, Budihardjo I, Srinivasula SM, Ahmad M, Alnemri ES, Wang X (1997) Cytochrome c and dATP-dependent formation of Apaf-1/caspase-9 complex initiates an apoptotic protease cascade. Cell 91:479–489
Lin MT, Stohlman SA, Hinton DR (1997) Mouse hepatitis virus is cleared from the central nervous system of mice lacking perforin-mediated cytolysis. J Virol 71:383–391
Liu XH, Castelli JC, Youle RJ (1999) Receptor-mediated uptake of an extracellular Bcl-x(L) fusion protein inhibits apoptosis. Proc Natl Acad Sci USA 96:9563–9567
Lopez M, Eberle F, Mattei MG, Gabert J, Birg F, Bardin F, Maroc C, Dubreuil P (1995) Complementary DNA characterization and chromosomal localization of a human gene related to the poliovirus receptor-encoding gene. Gene 155:261–265
Lopez-Guerrero JA, Alonso M, Martin-Belmonte F, Carrasco L (2000) Poliovirus induces apoptosis in the human U937 promonocytic cell line. Virology 272:250–256
Lundstöm A, Pöyry T, Vaarala O, Ilonen O, Hovi T, Roivanen M, Hyypiä T (2001) Abstr 6th Int Symp Positive-Strand RNA Viruses, abstr. P1–44
Martinez-Moralez JR, Barbas JA, Marti E, Bovolenta P, Edgar D, Rodriguez-Tébar A (1997) Vitronectin is expressed in the ventral region of the neural tube and promotes the differentiation of motor neurons. Development 124:5139–5147
Masson D, Jarry A, Baury B, Blanchardie P, Laboisse C, Lustenberger P, Denis MG (2001) Overexpression of the CD155 gene in human colorectal carcinoma. Gut 49:236–240
Mendelsohn CL, Wimmer E, Racaniello VR (1989) Cellular receptor for poliovirus: molecular cloning, nucleotide sequence and expression of a new member of the immunoglobulin superfamily. Cell 56:855–865
Morrison ME, He YJ, Wien MW, Hogle JM, Racaniello VR (1994) Homolog-scanning mutagenesis reveals poliovirus receptor residues important for virus binding and replication. J Virol 68:2578–2588
Mueller S, Cao X, Welker R, Wimmer E (2002) Interaction of the poliovirus receptor CD155 with the dynein light chain Tctex-1 and its implication for poliovirus pathogenesis. J Biol Chem 277:7897–904
Muir P, Nicholson F, Sharief MK, Thompson EJ, Cairns NJ, Lantos P, Spencer GT, Kaminski HJ, Banatvala JE (1995) Evidence for persistent enterovirus infection of the central nervous system in patients with previous paralytic poliomyelitis. In: Dalakas MC, Bartfeld H, Kurland LT (eds) The post-polio syndrome. New York Academy of Sciences, New York, pp 219–232
Nathanson N, Langmuir AD (1963) The Cutter incident. Am J Hyg 78:16–81
Neznanov N, Kondratova A, Chumakov KM, Angres B, Zhumabayeva B, Agol VI, Gudkov AV (2001) Poliovirus protein 3A inhibits tumor necrosis factor (TNF)-induced apoptosis by eliminating the TNF receptor from the cell surface. J Virol 75:10409–10420
Nomoto A, Arita I (2002) Eradication of poliomyelitis. Nat Immunol 3:205–208
Novoa I, Carrasco L (1999) Cleavage of eukaryotic translation initiation factor 4G by exogenously added hybrid proteins containing poliovirus 2A(pro) in HeLa cells: effects on gene expression. Mol Cell Biol 19:2445–2454
Nugent CI, Johnson KL, Sarnow P, Kirkegaard K (1999) Functional coupling between replication and packaging of poliovirus replicon RNA. J Virol 73:427–435
Oberhaus SM, Smith RL, Clayton GH, Dermody TS, Tyler KL (1997) Reovirus infection and tissue injury in the mouse central nervous system are associated with apoptosis. J Virol 71:2100–2106
Ohka S, Nomoto A (2001) Recent insights into poliovirus pathogenesis. Trends Microbiol 9:501–6
Ohka S, Yang W-X, Terada E, Iwasaki K, Nomoto A (1998) Retrograde transport of intact poliovirus through the axon via the fast transport system. Virology 250:67–75
Okamoto I, Kawano Y, Murakami D, Sasayam T, Araki N, Miki T, Wong AJ, Saya H (2001) Proteolytic release of CD44 intracellular domain and its role in the CD44 signaling pathway. J Cell Biol 155:755–762
Pallansch M, Roos R (2001) Enteroviruses: polioviruses, coxsackieviruses, echoviruses, and newer enteroviruses. In: Knipe DM, Howley PM (eds) Fields virology. Lippincott Williams and Wilkins, Philadelphia, pp 723–775

Paul AV (2002) Possible unifying mechanisms of picornavirus genome replication. In: Semler BL, Wimmer E (eds) Molecular biology of picornaviruses. ASM Press, Washington, DC, pp 227–246

Pavio N, Buc-Caron M-H, Colbère-Garapin F (1996) Persistent poliovirus infection of human fetal brain cells. J Virol 70:6395–6401

Pavio N, Couderc T, Girard S, Sgro JY, Blondel B, Colbère-Garapin F (2000) Expression of mutated receptors in human neuroblastoma cells persistently infected with poliovirus. Virology 274:331–342

Pekosz A, Phillips J, Pleasure D, Merry D, Gonzalez-Sacarano F (1996) Induction of apoptosis by La Crosse virus infection and role of neuronal differentiation and human bcl-2 expression in its prevention. J Virol 70:5329–5335

Petito CK, Roberts B (1995) Evidence of apoptotic cell death in HIV encephalitis. Am J Pathol 146:1121–1130

Ponnuraj EM, John TJ, Levin MJ, Simoes EAF (1998) Cell-to-cell spread of poliovirus in the spinal cord of bonnet monkeys (*Macaca radiata*). J Gen Virol 79:2393–2403

Racaniello VR (2001) Picornaviridae: the viruses and their replication. In: Knipe DM, Howley PM (eds) Fields virology. Lippincott Williams and Wilkins, Philadelphia, pp 685–722

Ren R, Racaniello VR (1992) Poliovirus spreads from muscle to the central nervous system by neural pathways. J Infect Dis 166:747–752

Ren RB, Costantini F, Gorgacz EJ, Lee JJ, Racaniello VR (1990) Transgenic mice expressing a human poliovirus receptor: a new model for poliomyelitis. Cell 63:353–362

Roulston A, Marcellus R, Branton PE (1999) Virus and apoptosis. Annu Rev Microbiol 53:577–628

Sabin AB, Boulger LR (1973) History of Sabin attenuated poliovirus oral live vaccine strains. J Biol Stand 1:115–118

Salk JE (1955) Consideration in the preparation and use of poliomyelitis virus vaccine. J Am Med Ass 1548:1239–1248

Schlegel A, Giddings TH, Ladinsky MS, Kirkegaard K (1996) Cellular origin and ultrastructure of membranes induced during poliovirus infection. J Virol 70:6576–6588

Sharief MK, Hentges MR, Ciardi M (1991) Intrathecal immune response in patients with the post-polio syndrome. N Engl J Med 325:749–755

Shen Y, Shenk TE (1995) Viruses and apoptosis. Curr Opin Genet Develop 5:105–111

Shepley MP, Racaniello VR (1994) A monoclonal antibody that blocks poliovirus attachment recognizes the lymphocyte homing receptor CD44. J Virol 68:1301–1308

Solecki DJ, Gromeier M, Mueller S, Bernhardt G, Wimmer E (2002) Expression of the human poliovirus receptor/CD155 gene is activated by sonic-hedgehog. J Biol Chem 277:25697–25702

Takahashi K, Nakanishi H, Miyahara M, Mandai K, Satoh K, Satoh A, Nishioka H, Aoki J, Nomoto A, Mizoguchi A, Takai Y (1999) Nectin/PRR: an immunoglobulin-like cell adhesion molecule recruited to cadherin-based adherens junctions through interaction with afadin, a PDZ domain-containing protein. J Cell Biol 145:539–549

Teterina NL, Gorbalenya AE, Egger D, Bienz K, Ehrenfeld E (1997) Poliovirus 2C protein determinants of membrane binding and rearrangements in mammalian cells. J Virol 71:8962–8972

Tolskaya EA, Romanova L, Kolesnikova MS, Ivannikova TA, Smirnova EA, Raikhlin NT, Agol VI (1995) Apoptosis-inducing and apoptosis-preventing functions of poliovirus. J Virol 69:1181–1189

Tsunoda I, Kurtz CIB, Fujinami RS (1997) Apoptosis in acute and chronic central nervous system disease induced by Theiler's murine encephalomyelitis virus. Virology 228:388–393

Tucker SP, Thornton CL, Wimmer E, Compans RW (1993) Vectorial release of poliovirus from polarized human intestinal epithelial cells. J Virol 67:4274–4282

Tyler KL, Clarke P, DeBiasi RL, Kominsky D, Poggioli GJ (2001) Reoviruses and the host cell. Trends Microbiol 9:560–564

Umehara F, Nakamura A, Izumo S, Kubota R, Ijchi S, Kashio N, Hashimoto K-I, Usuku K, Sato E, Osame M (1994) Apoptosis of T lymphocytes in the spinal cord lesions in HTLV-I-associated myelopathy: a possible mechanism to control viral infection in the central nervous system. J Neuropathol Exp Neurol 53:617–624

Viswanath V, Wu Y, Boonplueang R, Chen S, Stevenson FF, Yantiri F, Yang L, Beal MF, Andersen JK (2001) Caspase-9 activation results in downstream caspase-8 activation and bid cleavage in 1-methyl-4-phenyl-1,2,3,6-tetrahydropyridine-induced Parkinson's disease. J Neurosci 21:9519–9528
Weidman MK, Yalamanchili P, Ng B, Tsai W, Dasgupta A (2001) Poliovirus 3C protease-mediated degradation of transcriptional activator p53 requires a cellular activity. Virology 291:260–271
Xiang W, Paul AV, Wimmer E (1997) RNA signals in entero- and rhinovirus genome replication. Semin Virol 8:256–273
Xing L, Tjarnlund K, Lindqvist B, Kaplan GG, Feigelstock D, Cheng RH, Casasnovas JM (2000) Distinct cellular receptor interactions in poliovirus and rhinoviruses. EMBO J 19:1207–1216

Flaviviruses and Apoptosis Regulation

A. Catteau[1], M.-P. Courageot[1] and P. Desprès[1]

1 Introduction

The flaviviruses comprise a large genus of medically important arthropod-transmitted, enveloped viruses. Flaviviruses cause a variety of human diseases ranging from mild febrile illnesses to severe hemorrhagic manifestations (yellow fever [YF], dengue [DEN]; Monath 2001; Guzman and Kouri 2002) or meningo-encephalitic syndromes (Japanese encephalitis [JE], Saint Louis encephalitis [SLE], West Nile [WN] fever and tick-borne encephalitis [TBE]; Gould 1999; Igarashi 1999; Porterfield 1999; Brinton 2002) (Table 1). DEN, JE and YF viruses are the most significant human viral pathogens transmitted by mosquitoes. WN fever has been an emerging problem in Europe, in the Middle East, and more recently in the United States (Brinton 2002). The natural life cycles of vector-borne flaviviruses involve complex relationships among arthropod vectors, reservoirs, and mammalians. Knowledge of the molecular interactions between virus and its host cells are of particular importance for the understanding of flavivirus pathogenicity. Flaviviruses can replicate lytically in a variety of mammalian cells and cytopathic effects appear to be due to the induction of apoptosis. The purpose of this review is to summarize what is presently known about the molecular signaling mechanisms and viral components that contribute to flavivirus-induced apoptosis.

2 Flavivirus

Flavivirus is a genus of the Flaviviridae family which consists of more than 70 members. *Flavivirus* genus includes mostly arthropod-borne pathogens (Table 1). Flaviviruses are capable of infecting their vertebrate hosts through persistently infected mosquito or tick vectors. Mosquito-borne diseases DEN, JE and YF cause millions of cases of disease annually in the tropics and subtropics.

[1]U.P. Flavivirus-Host Molecular Interactions, Virology Department, Pasteur Institute, 25 rue du Dr Roux, 75724 Paris cedex 15, France, e-mail: pdespres@pasteur.fr

Progress in Molecular and Subcellular Biology
C. Alonso (Ed.): Viruses and Apoptosis
© Springer-Verlag Berlin Heidelberg 2004

Table 1. Apoptosis-inducing flaviviruses

Complex	Disease	Virus	Apoptotic induction seen in	Reference
Dengue	Fever, hemorragic fever	Dengue	Neurons	Desprès et al. (1996, 1998); Marianneau et al. (1998b); Duarte Dos Santos et al. (2000); Jan et al. (2000); Su et al. (2001)
			Epithelial cells	Shafee and AbuBakar (2002)
			Endothelial cells	Avirutnan et al. (1998)
			Hepatocytes	Marianneau et al. (1997); Couvelard et al. 1999); Duarte Dos Santos et al. (2000)
			Kupffer cells	Marianneau et al. (1999b)
JE encephalitis	Encephalitis	Japanese encephalitis	Neuroblastoma cells	Liao et al. (2001); Su et al. (2002)
		West Nile	Neuroblastoma cells	Parquet et al. (2001) Parquet et al. (2001)
		St. Louis encephalitis	Mononuclear cells	Parquet et al. (2002) Parquet et al. (2002)
Yellow fever	Fever, hepatonephrite	Yellow fever	Hepatocytes	Marianneau et al. (1999a); Xiao et al. (2001)
Tick-borne encephalitis	Encephalitis	Langat	Neuroblastoma cells	Pridhod'ko et al. (2001)

2.1 Taxonomy

The members of the *Flavivirus* genus are positive sense, single-stranded RNA viruses that replicate in the cytoplasm of infected cells (Fig. 1, 1). The virion is composed of three structural proteins, designated C (core protein), M (membrane protein) and E (envelope protein) (Chambers et al. 1990; Rice 1996). The three-dimensional image reconstruction has shown that the flavivirion has a well-organized outer protein shell and a lipid bilayer membrane (Kuhn et al. 2002). Protein E, which is exposed on the surface of the virus particle, is responsible for the main biological functions of the virion including virus

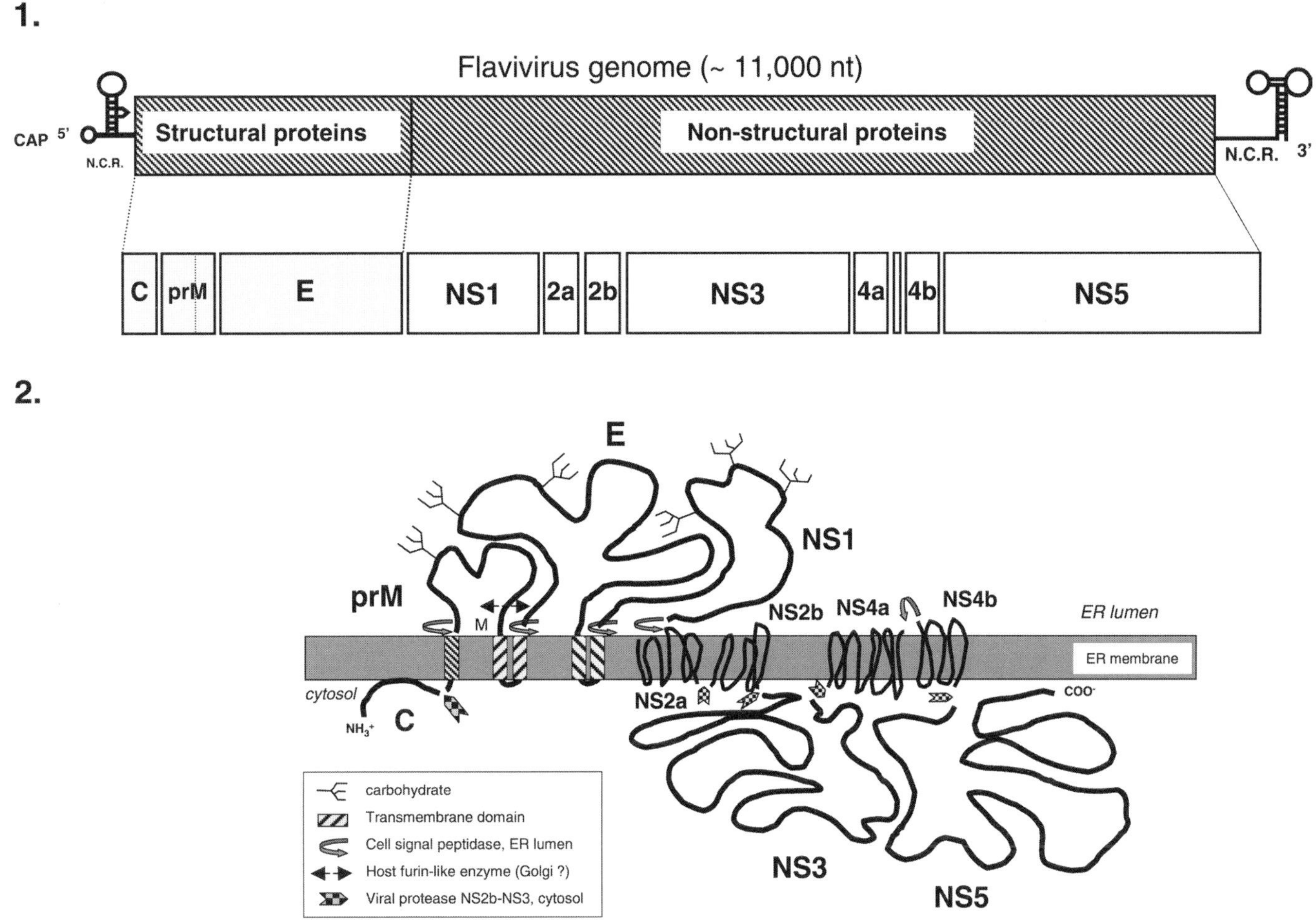

Fig. 1. Schematic representation of **1** the flavivirus genome organization and **2** the polyprotein precursor. *NCR* Non-coding region; *ER* endoplasmic reticulum

attachment and virus-specific membrane fusion in acid pH endosomes (Chambers et al. 1990; Rice 1996). The E protein is also the principal antigen that elicits neutralizing antibodies. The genomic RNA is translated to give rise to a large polyprotein precursor, which is cotranslationally processed by host-cell and virus-specified proteases to yield the individual viral proteins (Fig. 1, 2; Chambers et al. 1990; Rice 1996). The structural proteins are C, prM (the intracellular precursor of the M protein), and E. The nonstructural (NS) proteins are NS1, NS2a, NS2b, NS3, NS4a, NS4b and NS5. NS proteins are assumed to be involved primarily in the replication of viral RNA as a part of the replication complex (RC) (Chambers et al. 1990; Rice 1996; Brinton 2002). Flavivirus NS1 appears to be essential for virus viability although no precise function has yet been ascribed to it. In mammalian cells, NS1 may be released into the extracellular fluid as a unique hexameric species (Flamand et al. 1999). NS3 protein exhibits protease, nucleoside triphosphatase (NTPase), helicase, and RNA triphosphatase activities. Flavivirus NS5 proteins have been shown to possess RNA-dependent RNA polymerase activity (Chambers et al. 1990; Rice 1996; Brinton 2002). The functions of small membrane-associated NS2a, 2b, 4a and 4b remain largely unknown.

2.2 Virus Replication

Infection of host cells by flaviviruses proceeds by initial attachment to cellular receptors that remain unidentified. Entry to cells is believed to occur by receptor-mediated endocytosis, in which conformational changes in the E protein lead to fusion of viral and endosomal membranes and then release of the RNA genome into the cytosol (Chambers et al. 1990; Rice 1996; Brinton 2002). Replication of the flavivirus is associated with virus-induced membrane structures within the cytoplasm of infected cells. The active flavivirus RC is composed of NS1, NS2a, NS3, NS4a, NS5 proteins and viral RNA template (Chambers et al. 1990; Rice 1996; Brinton 2002). Specific cell proteins which interact with the stem-loops of the untranslated regions of RNA flavivirus may also be a part of the RC (Brinton 2002).

The first steps of flavivirus assembly take place in association with the membranes of the endoplasmic reticulum (ER). The virion is first assembled as an immature particle that contains prM noncovalently associated with E in a heterodimeric complex. Virions are transported to the plasma membrane in vesicles and are released by exocytosis (Chambers et al. 1990; Rice 1996). Late in virus morphogenesis, glycoprotein prM is proteolytically processed by Furin-like proteases in the exocytotic pathway of the *trans*-Golgi network, resulting in the small membrane M protein (Chambers et al. 1990; Rice 1996). Proteolysis of prM leads to the formation of homodimeric forms of E in virus particles before their release from the cell.

2.3 Flaviviruses Can Trigger Apoptosis in Host Cells

As shown in Table 1, DEN, JE, SLE, WN, YF and tick-borne Langat (LGT) viruses can trigger apoptosis in various cell types of different origin. Apoptosis is an active process of cell death involving a number of distinct morphological changes including cell shrinkage, fragmentation of the cell nucleus, and chromatin condensation, (Kimura et al. 2000). The morphological and biochemical changes associated with apoptosis are orchestrated by the activity of a family of cysteine proteases called caspases (Earnshaw et al. 1999; Hengartner 2000).

Apoptosis is induced via the activation of intracellular signaling systems, a number of which converge on mitochondrial membranes (Adrain and Martin 2001; Ravagnan et al. 2002; Fig. 2). Mitochondrial membrane permeabilization plays a major role in apoptosis triggered by many stimuli, releasing caspase-activating proteins that are normally confined to the mitochondrial intermembrane space (Adrain and Martin 2001; Ravagnan et al. 2002). One important route to caspase activation involves the translocation of cytochrome (cyt.) *c* from the mitochondrial intermembrane space into the cytosol. Cytosolic cyt.*c* interacts with adapter protein Apaf-1, and the complex forms an oligomer with initiator pro-caspase-9 (Adrain and Martin 2001; Desagher and Martinou 2000; Fig. 2). Upon activation, caspase-9 instigates a caspase cascade involving the downstream caspase-3, which culminates in the cleavage of the proteome (Kumar 1999). The apoptotic demise of a cell results in the formation of apoptotic bodies which are consumed by the phagocytic action of neighboring cells or macrophages.

The Bcl-2 protein family is able to modulate the cyt.*c*/Apaf-1/caspase-9 pathway essentially by regulating the liberation of cyt.*c* (Desagher and Martinou 2000; Adams and Corey2001; Fig. 2). Members of the Bcl-2 family are classified into two groups, pro-survival and pro-apoptotic groups (Adams and Corey 2001). Increased levels of Bcl-2 and Bcl-X_L lead to cell survival whereas excess of Bax is associated with apoptosis. The pro-survival group acts to mitigate the release of cyt.*c* from the intermembrane space of the mitochondria, while the pro-apoptotic group encourages the release of the Apoptotic factors into the cytosol (Desagher and Martinou 2000; Adams and Corey 2001).

Cellular perturbations caused by virus replication can inadvertently trigger signaling pathways or specific sensors that initiate the apoptotic pathway (Everett and McFadden 1999; Barber 2001). It is proposed that virus-induced apoptosis may serve to limit infection and spread of progeny (Shen and Shenk 1995; Teodoro and Branton 1997; Hardwick 1998; O'Brien 1998; Hay and Kannourakis 2002), implying that the cell suicide program represents a defense mechanism that acts by rapidly eliminating the infected cell (Everett and McFadden 1999; Barber 2001). Another idea postulated that RNA viruses induce apoptosis to facilitate release of virions (Shen and Shenk 1995; Hay and Kannourakis 2002). At present, it is unclear if flaviviruses induce apoptosis by assisting in virus dissemination.

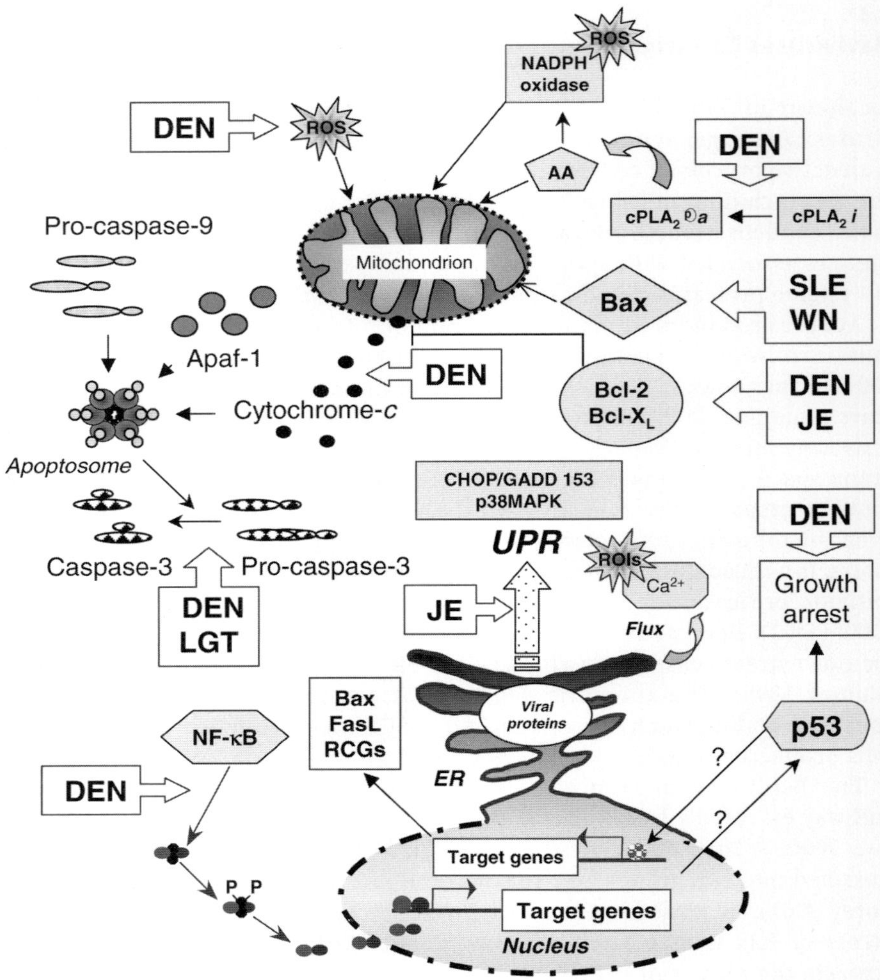

Fig. 2. Intracellular pathways involved in flavivirus-induced apoptosis. *i* inactive; *a* active; *AA* arachidonic acid; *DEN* dengue; *JE* Japanese encephalitis; *LGT* Langat; *cPLA*$_2$ cytosolic phospholipase A$_2$; *RCG* redox control genes; *ROIs* radical oxygen intermediates; *ROS* reactive oxygen species; *SLE* Saint Louis encephalitis; *UPR* unfolded protein response; *WN* West Nile

3 Apoptosis-Inducing Flaviviruses

3.1 Dengue Virus

DEN fever has emerged as an important mosquito-borne viral disease in tropical areas, second only to malaria (McBride and Bielefeldt-Ohmann 2000). It is estimated that 100 million cases of dengue fever occur annually. DEN disease is considered as a major public health problem in southeast Asia and South America (Guzman and Kouri 2002). DEN virus is classified into four serotypes designated DEN-1, DEN-2, DEN-3 and DEN-4 (McBride and Bielefeldt-Ohmann 2000). All four serotypes of DEN virus are capable of causing severe human diseases.

3.1.1
Pathogenesis

DEN virus causes human disease with different degrees of severity. In some cases, DEN virus infection can cause dengue hemorrhagic fever (DHF) or dengue shock syndrome (DSS), a potentially fatal plasma leakage syndrome (Rothman and Ennis 1999; McBride and Bielefeldt-Ohmann 2000; Guzman and Kouri 2002). The average mortality rate for fulminant DHF/DSS can exceed 5%. The underlying mechanisms of the pathogenesis of DHF/DSS remain poorly understood (Rothman and Ennis 1999). Hemorrhage and shock caused by DEN infections are probably due to the occurrence of abnormal immune responses, involving production of cytokines or chemokines and activation of T-lymphocytes and disturbed homeostasis of the blood clotting system (Rothman and Ennis 1999; McBride and Bielefeldt-Ohmann 2000). The intrinsic virulence among different DEN isolates may also contribute to disease severity (Desprès et al. 1998).

3.1.2
DEN-Virus-Induced Apoptosis in Vivo

The clinical features of severe DEN disease include hemorrhagic diathesis, liver involvement, and encephalopathy (Guzman and Kouri 2002). Apoptosis has been shown to be the mechanism by which DEN virus infection causes liver and neuronal cell death in vivo (Desprès et al. 1998; Marianneau et al. 1998b; Couvelard et al. 1999).

3.1.2.1
Liver Cells

The presence of viral antigens inside liver cells from biopsies of DEN patients indicates that these cells may be susceptible to infection by DEN virus (Couvelard et al. 1999). In DHF, liver involvement is a characteristic sign that the disease may be fatal (Couvelard et al. 1999; McBride and Bielefeldt-Ohmann 2000). Hepatic injury is characterized by centrolobular and midzonal necrosis with hyperplasia of Kupffer cells, the macrophage-like cells residing in the liver. Using the in situ TUNEL method, apoptotic DNA degradation was detected in liver tissue sections from patients with diagnosis of DEN at autopsy (Couvelard et al. 1999). Ex vivo, DEN virus can trigger Kupffer cells to undergo apoptosis (Marianneau et al. 1999b). As heathly Kupffer cells were able to ingest infected apoptotic bodies, these macrophage-like cells may exert antiviral activities by eliminating liver cells in the apoptotic state.

3.1.2.2 Neuronal Cells

Virus infection of the mouse central nervous system (CNS) has been used as a model system for the characterization of genetic determinants and viral factors involved in the pathogenicity of DEN virus (Desprès et al. 1996, 1998; Marianneau et al. 1998a; Duarte Dos Santos et al. 2000). Newborn mice inoculated intracerebrally with lethal doses of neurovirulent DEN virus developed fatal encephalitis (Desprès et al. 1998). Cortical and pyramidal neurons in the hippocampus are the major target cells of DEN virus in the CNS and their destruction should be responsible for mouse death (Desprès et al. 1998). Apoptotic cells were essentially detected in the cortical and hippocampal regions. There was a correlation between the distribution of apoptotic neurons and neurons positive for viral antigens (Desprès et al. 1998). The neurovirulence of DEN virus was essentially restricted to newborn mice (Desprès et al. 1998). It is unclear if inhibition of DEN-virus-induced neuronal apoptosis in the mature nervous system may be implicated in the protection of older mice against fatal DEN virus infection.

3.1.3 DEN-Virus-Induced Apoptosis in Vitro

DEN virus infection potentially induces NF-κB in cells of various origins (Marianneau et al. 1997; Avirutnan et al. 1998; Jan et al. 2000) (Fig. 2). NF-κB is a heterodimer present in the cytoplasm in an inactive complex with IκB, the inhibitor of NF-κB (Mercurio and Manning 1999). Exposure of cells to stress inducers such as viral infections induces the phosphorylation of IκB. Phosphorylated IκB is targeted for degradation, releasing the NF-κB heterodimer to translocate to the nucleus and activate target genes (Mercurio and Manning 1999; Fig. 2).

3.1.3.1 Liver Cells

It has been reported that DEN virus infection of human hepatoma cells activates the transcription factor NF-κB, which, in turn, promotes the induction of apoptosis (Marianneau et al. 1997). The transcriptional regulation of target genes such as the tumor suppressor p53 is believed to be the mechanism by which NF-κB promotes apoptosis (Barkett and Gimore 1999). Transcriptional factor p53, normally a very short-lived protein, can induce growth arrest and apoptosis in response to intracellular disruptions (Gottlieb and Oren 1998; Sheikh and Fornace 2000) (Fig. 2). Genes such as Bax, FasL and genes involved in the control of redox states are transactivated during p53-induced apoptosis (Gottlieb and Oren 1998; Sheikh and Fornace 2000). It was reported that endocytotic trafficking mutant Trf1 isolated from HuH-7 cells was resistant to DEN-virus-mediated cell death (Hilgard et al. 2000). Protection was linked to a low level of casein kinase (CK) 2 expression in Trf1 cells. The C-terminal regulatory domain of p53 is a target for phosphorylation by CK2 (Hilgard et al. 2000). Whether CK2-mediated phosphorylation of p53 is required for DEN-virus-induced apoptosis remains to be investigated. At present, it is difficult to judge the relative contribution of NF-κB to DEN-virus-induced apoptosis (Liao et al. 2001). NF-κB induction closely parallels the time course of viral protein synthesis in DEN-virus-infected hepatoma cells (Marianneau et al. 1997). Flaviviruses utilize the ER as the primary site of polyprotein processing, envelope glycoprotein biogenesis, and virion formation (Chambers et al. 1990; Rice 1996; Su et al. 2000; Brinton 2002). Analysis of DEN protein processing showed that most viral envelope glycoproteins were degraded or may aggregate, leading to a nonproductive pathway in liver cells (Marianneau et al. 1997; Duarte Dos Santos et al. 2000). As a protein-folding compartment the ER is exquisitely sensitive to alterations in homeostasis (Ferri and Kroemer 2002). Most evidence supports the hypothesis that accumulation of misfolded proteins can lead to ER stress (Ferri and Kroemer 2002). The ER generates two second messengers in response to stress: calcium and reactive oxygen intermediates (Kaufman 1999; Fig. 2). Changes in the intracellular redox status elicited by the generation of oxidants may result in the activation of NF-κB. It is of interest to note that DEN virus infection can lead to oxidative stress in infected liver cells (Lin et al. 2000).

3.1.3.2 Endothelial Cells

DEN virus can trigger apoptosis in infected endothelial cells (Avirutnan et al. 1998). Apoptotic breakdown of the endothelial barrier could contribute to local vascular leakage (McBride and Bielefeldt-Ohmann 2000). Early in infection, NF-κB activation occurred in infected endothelial cells (Avirutnan et al. 1998; Fig. 2). It remains to be determined whether NF-κB plays a critical event in DEN-virus-induced apoptosis in vascular endothelium.

It has recently been observed that mouse anti-NS1 antibodies induced endothelial cell apoptosis in a caspase-dependent manner (Lin et al. 2002). Inducible nitric oxide (NO) synthase expression and the production of NO are involved in endothelial cell apoptosis in response to anti-NS1 antibodies. The role of NO production is the downregulation of Bcl-2 and Bcl-X_L and the upregulation of p53 and Bax expression (Lin et al. 2002).

3.1.3.3
Neuronal Cells

Cell growth arrest in the G1 phase has been shown to be associated with the occurrence of apoptosis in neuroblastoma cells infected with DEN-2 virus (Su et al. 2002; Fig. 2). The causal relationship between G1-growth arrest and apoptosis during DEN virus infection remains elusive. DEN virus infection of neuroblastoma cells has been shown to trigger activation of cytosolic phospholipase A2 ($cPLA_2$) (Jan et al. 2000) (Fig. 2). $cPLA_2$ has been implicated in the receptor-mediated release of arachidonic acid (AA) from membrane phospholipids, playing a key role in many signal transduction reactions. AA can activate membrane-associated NADPH oxidase to generate reactive oxygen species which, in turn, activate NF-κB (Jan et al. 2000). AA may also cause dysfunctional changes in the mitochondrial membrane potential, resulting in the activation of the apoptotic cascade (Fig. 2). AA, superoxide anion and NF-κB were involved in DEN-virus-mediated neuronal cell death (Jan et al. 2000; Liao et al. 2001; Chen et al. 2002). Mitochondrial release of cyt.*c* and caspase-3 activation are critical for induction of apoptosis by DEN virus (Jan et al. 2000; Fig. 2). Overexpression of Bcl-x_L but not Bcl-2 protein has been shown to protect neuronal cells from DEN-virus-induced apoptosis (Su et al. 2001) (Fig. 2).

3.1.3.4
Epithelial Cells

DEN virus can trigger apoptosis in infected epithelial cells (Shafee and AbuBakar 2002). The induction of apoptosis was more pronounced in the presence of high extracellular Zn^{2+} concentrations, presumably by enhancing the virus-induced activation of NF-κB (Shafee and AbuBakar 2002) (Fig. 2).

3.2
Yellow Fever Virus

3.2.1
Pathogenesis

Yellow fever (YF) virus is the prototype of the flavivirus genus. The YF disease is an important cause of morbidity and mortality in tropical regions of Africa and South America (Monath 2001). The clinical disease varies from nonspecific illness to fatal hemorrhagic fever (Monath 2001).

3.2.2
Mechanisms of Cell Death

Pathologic changes in YF virus-infected liver include swelling and necrosis of hepatocytes (Councilman bodies; Monath 2001). The cell death in the YF-infected liver may start with apoptosis that is triggered by the infecting virus (Xiao et al. 2001). Hepatocytic apoptosis has been demonstrated in vitro in human hepatoma cells infected with the wild-type strain of YF virus (Marianneau et al. 1999a; Xiao et al. 2001).

3.3
Japanese Encephalitis Virus

Japanese encephalitis (JE) is an zoonotic infection that can cause acute encephalitis in humans, frequently resulting in a high mortality rate. JE is prevalent in some east Asian countries (Igarashi 1999).

3.3.1
Pathogenesis

Humans are a dead-end host for the JE virus. The primary sites for JE virus multiplication are most likely in myeloid and lymphoid cells or in vascular endothelial cells (Igarashi 1999). The principal target cells for JE virus in the CNS are neurons (Igarashi 1999). Massive neuronal dysfunction and/or destruction are proposed to be responsible for the manifestations of encephalitis. A wide variety of continuous cell lines can support the productive growth of JE virus (Igarashi 1999).

3.3.2
Mechanisms of Cell Death

Recent work has shown that JE virus can trigger the unfolded protein response (UPR), a signaling cascade from the ER to the nucleus (Kaufman 1999; Su et al. 2002). The JE-virus-mediated UPR appears to induce *CHOP/GADD153* expression and p38 mitogen-activated protein kinase (p38 MAPK) activation (Zinszner et al. 1998; Su et al. 2002) (Fig. 2). *CHOP/GADD153* is a transcription factor with a leucine zipper motif, whose induction has been closely linked to the perturbation of homeostasis in the ER (Zinszner et al. 1998). The stress-inducible p38MAPK is a post-translational activator of *CHOP/GADD153*.

Overexpression of *CHOP/GADD153* leads to transcriptional downregulation of the anti-apoptosis *bcl-2* gene and promotes apoptosis (Zinszner et al. 1998). Bcl-2 overexpression and treatment with a general inhibitor of caspases inhibited *CHOP/GADD153* induction and diminished JE-virus-induced apoptosis (Liao et al. 1997; Su et al. 2002). Thus, JE virus infection could activate ER stress signaling pathways that contribute to apoptosis of infected cells.

Ectopic expression of *bcl-2* can restrain the apoptotic process in infected cells, subsequently facilitating the establishment of persistent JE virus infection in surviving cells (Fig. 2). Generation of truncated NS1 proteins occurred in *bcl-2*-expressing cells persistently infected by JE virus (Liao et al. 1997, 1998).

3.4 Saint-Louis Encephalitis Virus

Saint-Louis encephalitis (SLE) virus is endemic to the Americas where it causes outbreaks of encephalitic disease, predominantly in the United States. The primary transmission cycle involves birds and mosquitoes (Porterfield 1999).

3.4.1 Pathogenesis

Humans are a dead-end host for SLE virus. The severity of the disease caused by SLE virus ranges from a mild influenza-like syndrome to an acute CNS disease (Porterfield 1999). Death during the clinical course is mainly due to the direct CNS damage caused by the virus. High levels of CPEs are observed when some vertebrate continuous cell lines are infected with SLE virus (Porterfield 1999).

3.4.2 Mechanisms of Cell Death

Apoptosis is the mechanism by which SLE virus causes cell death in infected human mononuclear and mouse neuroblastoma cell lines (Parquet et al. 2002). Active SLE virus infection appeared to be necessary in order to trigger apoptosis in these cells. The levels of Bax were upregulated in SLE-virus-infected cells (Fig. 2). The disturbance of the balance between Bax and Bcl-x_L might be critical for the induction of apoptosis by SLE virus [47].

3.5 West Nile Virus

West Nile (WN) Fever is a zoonotic infection. Birds are natural hosts for WN virus and serve as a reservoir from which vector mosquitoes may infect humans and other mammals (Brinton 2002). Encephalitis is a common mammals complication of WN fever. In the last 5 years, WN fever has been an emerging concern for public health in Europe, in the Middle East, and more recently in North America (Brinton 2002).

3.5.1 *Pathogenesis*

WN virus causes occasional febrile illness in humans, and patients infected may exhibit severe encephalitic syndromes. Fatal neurological disease involves damage to neurons (Porterfield 1999; Brinton 2002).

3.5.2 *Mechanisms of Cell Death*

WN virus has been shown to trigger apoptosis in infected human mononuclear and mouse neuroblastoma cells (Parquet et al. 2001). Viral replication may play a critical role for the induction of apoptosis. The levels of Bax were upregulated in WN-virus-infected cells (Fig. 2). It is believed that increased expression of Bax is responsible for the WN-virus-induced apoptosis (Parquet et al. 2001).

3.6 Langat Virus

Langat (LGT) virus, a member of the tick-borne encephalitis (TBE) complex, is mainly found in the northern hemisphere (Gould 1999; Porterfield 1999). LGT virus does not naturally infect humans.

3.6.1 *Pathogenesis*

LGT virus has been studied as a potential live virus vaccine to protect against encephalitis caused by viruses of the TBE complex (Gould 1999).

3.6.2 *Mechanisms of Cell Death*

LGT virus has been shown to induce apoptosis in infected epithelial and neuroblastoma cells (Pridhod'ko et al. 2001). Caspase-3 activation is critical for induction of apoptosis by LGT virus (Pridhod'ko et al. 2001, 2002). It is hypothesized that NF-κB, p53, and Bax are necessary for the activation of apoptosis by LGT virus (Fig. 2).

4 Viral Determinants That May Influence Virus-Induced Apoptosis

Neurovirulent variant FGA/NA d1d was generated during the adaptation of human isolate of DEN-1 virus strain FGA/89 in mouse brain (Desprès et al. 1998). Apoptosis has been shown to be the mechanism by which FGA/NA d1d infection causes cell death in the infected mouse CNS (Desprès et al. 1998). FGA/89 and its mouse-passaged, neurovirulent variant FGA/NA d1d differed in their efficiency to induce apoptosis in mouse neuroblastoma and human hepatoma cells (Desprès et al. 1998; Duarte Dos Santos et al. 2000). Apoptosis in hepatoma cells was much less pronounced after infection with FGA/NA d1d than with the parental strain (Duarte Dos Santos et al. 2000). In contrast, apoptosis in neuroblastoma cells progressed more rapidly during FGA/NA d1d virus infection. Changes in virus life cycle may account for the differences of DEN-virus-induced apoptosis in a cell-specific manner. FGA/NA d1d virus infection was characterized by the late accumulation of viral antigens (Duarte Dos Santos et al. 2000). The differential expression of FGA/NA d1d proteins was a result of different patterns of viral RNA synthesis in infected cells (Duarte Dos Santos et al. 2000). The productive assembly pathways of FGA/NA d1d envelope glycoproteins were more efficient than those for the parental strain (Duarte Dos Santos et al. 2000).

Complete sequencing of FGA/NA d1d genes identified only three amino acid substitutions in the envelope E protein (E-196, E-365, and E-405) and one in the viral helicase NS3 (NS3-435; Desprès et al. 1998; Duarte Dos Santos et al. 2000). All four amino acid substitutions in FGA/NA d1d lie at regions that are likely to be important for conformational changes during the virus life cycle. The identified substitutions in the E and NS3 proteins may account for the differences in virus replicative functions and virus morphogenesis (Duarte Dos Santos et al. 2000). Thus, the envelope E protein and viral RNA helicase NS3 may modulate the induction of DEN-virus-induced apoptosis by altering viral growth.

5 The M, E and NS3 Proteins Have Pro-apoptotic Properties

The LGT E and NS3 proteins have pro-apoptotic activities (Pridhod'ko et al. 2001, 2002; Fig. 1). Transient expression of LGT E protein induces apoptosis in transfected mouse neuroblastoma and primate epithelial cells (Pridhod'ko et al. 2001). Expression of the E protein induced activation of caspase-3-like proteases in neuroblastoma cells. Caspase inhibitor baculovirus p35 has the ability to block E-protein-mediated cytotoxicity (Pridhod'ko et al. 2001). The helicase domain of NS3 has a weak homology with the caspase recruitment domain of adapter proteins such as Apaf-1 (Pridhod'ko et al. 2002). Transient expression of LGT NS3 induces apoptosis in transfected neuroblastoma and epithelial cells (Pridhod'ko et al. 2002). Caspase-8 binds to the protease and helicase domain of NS3, while only the intact protease domain is required for induction of apoptosis by the protein. Caspase-8-specific inhibitor blocked NS3-mediated apoptosis (Pridhod'ko et al. 2002). Further experiments will be necessary to better delineate the precise mechanisms by which LGT NS3 protein induce apoptosis via caspase-8 activation.

More recently, it has been reported that the 40-residue ectodomain of the M protein has the ability to induce apoptosis in cells of various origins (Catteau et al., 2003a; Fig. 1). The M ectodomains of all four serotypes of DEN virus and the wild-type strains of JE, YF, and WN viruses have pro-apoptotic properties. The transport of the M ectodomain through the secretory pathway appears to be essential for the induction of apoptosis. The M ectodomain may exert its cytotoxicity by activating a mitochondrial apoptotic pathway (Catteau et al., 2003b). The death-promoting activity of the M ectodomain reflects the pro-apoptotic properties of the nine carboxy-terminal amino acids referred to as ApoptoM. The identification of an intrinsic sequence responsible for the pro-apoptotic activity of the M ectodomain may provide novel insights into the role of M in flavivirus pathogenicity.

6
Concluding Remarks

Recent advances in cell biology have resulted in advances in our understanding of the mechanisms of flavivirus-induced cell death, which determine the outcome of virus infection. Cytotoxicity results from apoptotic cell death which may contribute to the clinical manifestations associated with flavivirus infection. It is tempting to speculate that flaviviruses may perform apoptosis serving to spread virus progeny to neighboring cells through the phagocytosis of the resulting apoptotic bodies, in the process minimizing an immune response.

The mechanisms by which flaviviruses induce the apoptosis of infected cells remain elusive. Flavivirus infection may activate biochemically different apoptotic pathways converging in the alteration of mitochondrial function. Oxidative stress, transcription factors NF-κB and *CHOP/GADD153*, p38MAPK, and pro-apoptotic factor Bax may have an important role in induction of apoptosis by flaviviruses (Fig. 2). Thus, detailed elucidation of virus-triggered apoptotic pathways may be critical to our understanding of flavivirus pathogenicity. The intracellular production of viral proteins is thought to be essential for the induction of apoptosis by flaviviruses. It is of great interest to consider the role of M, E and NS3 in flavivirus-induced apoptosis. What it remains to be determined is whether initiation of apoptosis results from the interaction of particular viral components and cellular apoptosis-signaling proteins.

Flaviviruses can cause persistent infection in invertebrate and vertebrate cells, and this process is often associated with diminished virus production and reduced cytopathic effects. Establishment of virus persistence is a process that results from the interactions between the virus and the host. Flavivirus persistence has been generally viewed as a sign of an impaired or defective immune system of the host. Anti-apoptosis proteins of the Bcl-2 family were shown to delay the process of flavivirus-induced apoptosis in a cell-dependent manner causing surviving cells to become persistently infected (Fig. 2). Thus, the persistence of flavivirus may be interpreted as a consequence of the resistance of host cells to virus-induced apoptosis. Identification of the cellular and viral proteins that regulate the apoptosis in host cells will undoubtedly enrich our understanding of viral replication strategies and help to elucidate the mechanisms involved in flavivirus pathogenicity.

Acknowledgements. The research programs were supported by grants from the Direction Générale de l'Armement (DSP/STTC, program 99.34.031), the CAPES/COFECUB (program 254/98), and by the Pasteur Institute (Direction de la Valorisation et des Partenariats Industriels). A.C. is funded by scolarship funds from the French Ministère de l'Education Nationale, de la Recherche et de la Technologie.

References

Adams JM, Cory S (2001) Life-or-death decisions by the Bcl-2 protein family. Trends Biochem Sci 26:61–66

Adrain C, Martin SJ (2001) The mitochondrial apoptosome: a killer unleashed by the cytochrome seas. Trends Biochem Sci 26:390–397

Avirutnan P, Malasit P, Seliger B, Bhakdi S, Husmann M (1998) Dengue virus infection of human endothelial cells leads to chemokine production, complement activation, and apoptosis. J Immunol 161:6338–6346

Barber GN (2001) Host defense, viruses and apoptosis. Cell Death Differ 8:113–126

Barkett M, Gimore TD (1999) Control of apoptosis by Rel/NF-κB transcription factors. Oncogene 18:6910–6924

Brinton MA (2002) The molecular biology of West Nile virus: a new invader of the western hemisphere. Annu Rev Microbiol 56:371–402

Catteau A, Kalinina O, Wagner MC, Deubel V, Courageot MP, Desprès, P (2003a) Dengue M ectodomain contains a proapoptotic sequence referred to as ApoptoM. J Gen Virol 84:2781–2793

Catteau A, Roué G, Yuste VJ, Susin SA, Desprès P (2003b) Expression of dengue ApoptoM sequence results in disruption of mitochondrial potential and caspase activation. Biochimie 85:789–793

Chambers TJ, Hahn CS, Galler R, Rice CM (1990) Flavivirus genome organization, expression, and replication. Annu Rev Microbiol 44:649–688
Chen CJ, Raung SL, Kuo MD, Wang YM (2002) Suppression of Japanese encephalitis virus infection by non-steroidal anti-inflammatory drugs. J Gen Virol 83:1897–1905
Couvelard A, Marianneau P, Bedel C, Drouet MT, Vachon F, Hénin D, Deubel V (1999) Report of a fatal case of dengue infection with hepatitis: demonstration of dengue antigens in hepatocytes and liver apoptosis. Hum Pathol 30:1106–1110
Desagher S, Martinou JC (2000) Mitochondria as the central control point of apoptosis. Trends Cell Biol 10:369–377
Desprès P, Flamand M, Ceccaldi PE, Deubel V (1996) Human isolates of dengue virus type-1 induce apoptosis in mouse neuroblastoma cells. J Virol 70:4090–4096
Desprès P, Frenkiel MP, Ceccaldi PE, Duarte dos Santos C, Deubel V (1998) Apoptosis in the mouse central nervous system in response to infection with mouse-neurovirulent dengue viruses. J Virol 72:823–829
Duarte Dos Santos CN, Frenkiel MP, Courageot MP, Rocha CS, Vazeille-Falcoz MC, Wien M, Rey F, Deubel V, Desprès P (2000) Determinants in the envelope E protein and viral RNA helicase NS3 that influence the induction of apoptosis in response to infection with dengue type-1 virus. Virology 74:564–572
Earnshaw WC, Martins LM, Kaufmann SH (1999) Mammalian caspases: structure, activation, substrates, and functions during apoptosis. Annu Rev Biochem 68:383–424
Everett H, McFadden G (1999) Apoptosis: an innate immune response to virus infection. Trends Microbiol 7:160–165
Ferri KF, Kroemer G (2002) Organelle-specific initiation of cell death pathways. Nat Cell Biol 3:255–263
Flamand M, Mégret F, Mathieu M, Lepault J, Rey FA, Deubel V (1999) Dengue virus type 1 nonstructural glycoprotein NS1 is secreted from mammalian cells as a soluble hexamer in a glycosylation-dependent fashion. J Virol 73:6104–6110
Gottlieb TM, Oren M (1998) p53 and apoptosis. Semin Cancer Biol 8:359–368
Gould EA (1999) Tick-borne encephalitis and Wesselsbron viruses. In: Granoff A, Webster RG (eds) Encyclopedia of virology, 2nd edn. Academic Press, New York, pp 430–437
Guzman MG, Kouri G (2002) Dengue: an update. Lancet 2:33–42
Hardwick JM (1998) Viral interference with apoptosis. Semin Cell Dev Biol 9:339–349
Hay S, Kannourakis G (2002) A time to kill: viral manipulation of the cell death program. J Gen Virol 83:1547–1564
Hengartner MO (2000) The biochemistry of apoptosis. Nature 407:770–776
Hilgard P, Czaja MJ, Stockert RJ (2000) Resistance of the casein kinase deficient cell line Trf1 to dengue virus-induced cell death. Hepatology Meeting, October 2000, 215A
Igarashi A (1999) Japanese encephalitis virus (*Flaviviridae*). In: Granoff A, Webster RG (eds) Encyclopedia of virology, 2nd edn. Academic Press, New York, pp 871–876
Jan JT, Chen BH, Ma SH, Liu CI, Tsai HP, Wu HC, Jiang SY, Yang KD, Shiao MF (2000) Potential dengue virus-triggered apoptotic pathway in human neuroblastoma cells: arachidonic acid, superoxide anion, and NF-κB are sequentially involved. J Virol 74:8680–8691
Kaufman RJ (1999) Stress signaling from the lumen of the endoplasmic reticulum: coordination of gene transcriptional and translational controls. Genes Dev 13:1211–1233
Kimura K, Sasano H, Shimosegawa T, Mochizuki S, Nagura H, Toyota T (2000) Ultrastructure of cells undergoing apoptosis. Vitam Horm 58:257–266
Kuhn RJ, Zhang W, Rossman MG, Pletnev SV, Corver J, Lenches E, Jones CT, Mukhopadhyay S, Chipman PR, Strauss EG, Baker TS, Strauss JH (2002) Structure of dengue virus: implications for flavivirus organization, maturation, and fusion. Cell 108:717–725
Kumar S (1999) Mechanisms mediating caspase activation in cell death. Cell Death Differ 6:1060–1066

Liao CL, Lin YJ, Wang JJ, Huang YL, Yeh CT, Ma SH, Chen LK (1997) Effect of enforced expression of human *bcl-2* on Japanese encephalitis virus-induced apoptosis. J Virol 71:5963–5971
Liao CJ, Lin YL, Shen SC, Shen JY, Su HL, Huang YL, Ma SH, Sun YC, Chen KP, Chen LK (1998) Antiapoptotic but not antiviral function of human *bcl-2* assists establishment of Japanese encephalitis virus persistence in cultured cells. J Virol 72:9844–9854
Liao CL, Lin YL, Wu BC, Tsao CH, Wang MC, Liu CI, Huang YL, Chen JH, Wang JP, Chen LK (2001) Salicylates inhibit flavivirus replication independently of blocking nuclear factor kappa B activation. J Virol 75:7828–7839
Lin YL, Liu CC, Chuang JI, Lei HT, Yeh TM, Lin YS, Huan YH, Liu HS (2000) Involvement of oxidative stress, NF-IL-6, and RANTES expression in dengue-2-virus-infected human liver cells. Virology 276:114–126
Lin CF, Lei HY, Shiau AY, Liu HH, Yeh TM, Chen SH, Liu CC, Lin YS (2002) Endothelial cell apoptosis induced by antibodies against dengue virus nonstructural protein 1 via production of nitric oxide. J Immunol 169:657–664
Marianneau P, Cardona A, Edelman L, Deubel V, Desprès P (1997) Dengue virus replication in human hepatoma cells activates NF-kappa B which in turn induces apoptotic cell death. J Virol 71:3244–3249
Marianneau P, Flamand M, Deubel V, Desprès P (1998a) Apoptotic cell death in response to dengue virus infection: the pathogenesis of dengue hemorrhagic fever revisited. Clin Diagn Virol 10:113–119
Marianneau P, Flamand M, Deubel V, Desprès P (1998b) Induction of programmed cell death (apoptosis) by dengue virus in vitro and in vivo. Acta Cient Venez 49:13–17
Marianneau P, Steffan AM, Royer C, Drouet MT, Kirn A, Deubel V (1999a) Differing infection of dengue and yellow fever viruses in a human hepatoma cells. J Infect Dis 178:1270–1278
Marianneau P, Steffan AM, Royer C, Drouet MT, Jaeck D, Kirn A, Deubel V (1999b) Infection of primary cultures of human Kupffer cells by dengue virus: no viral progeny synthesis but cytokine production is evident. J Virol 73:5201–5206
McBride WJH, Bielefeldt-Ohmann H (2000) Dengue viral infections; pathogenesis and epidemiology. Microb Infect 2:1041–1050
Mercurio F, Manning AM (1999) Multiple signals converging on NF-κB. Curr Opin Cell Biol 11:226–232
Monath T (2001) Yellow fever: an update. Lancet 1:11–20
O'Brien V (1998) Viruses and apoptosis J Gen Virol 79:1833–1845
Parquet MDC, Kumatori A, Hasebe F, Morita K, Igarashi A (2001) West Nile virus-induced bax dependent apoptosis. FEBS Lett 500:17–24
Parquet MC, Kumatori A, Hasebe F, Mathenge EG, Morita K (2002) St. Louis encephalitis virus induced pathology in cultured cells. Arch Virol 147:1105–1119
Porterfield JS (1999) Encephalitis viruses and related viruses causing hemorrhagic disease. In: Granoff A, Webster RG (eds) Encyclopedia of virology, 2nd edn. Academic Press, New York, pp 424–430
Pridhod'ko G, Pridhod'ko E, Cohen JI, Pletnev AG (2001) Infection with Langat flavivirus or expression of the envelope protein incudes apoptotic cell death. Virology 286:328–335
Pridhod'ko G, Pridhod'ko E, Pletnev AG, Cohen JI (2002) Langat flavivirus NS3 binds caspase-8 and induces apoptosis. J Virol 76:5701–5710
Ravagnan L, Roumier T, Kroemer G (2002) Mitochondria, the killer organelles and their weapons. J Cell Physiol 192:131–137
Rice CM (1996) *Flaviviridae*: the viruses and their replication. In: Fields BN, Knipe DM, Howley PM, et al. (eds) Fields virology, 3rd edn. Lippincott-Raven, Philadelphia, pp 931–959
Rothman AL, Ennis FA (1999) Immunopathogenesis of dengue hemorrhagic fever. Virology 257:1–6
Shafee N, AbuBakar S (2002) Zinc accelerates dengue virus type-2-induced apoptosis in Vero cells. FEBS Lett 524:20–24

Sheikh MS, Fornace AJ Jr (2000) Role of p53 family members in apoptosis. J Cell Physiol 182:171–181
Shen Y, Shenk TE (1995) Viruses and apoptosis. Curr Opin Genet Dev 5:105–111
Su HL, Lin YL, Tsao CH, Chen LK, Liu YT, Liao CL (2001) The effect of human *bcl-2* and *bcl-X* genes on dengue virus-induced apoptosis in cultured cells. Virology 282:141–153
Su HL, Liao CL, Lin YL (2002) Japanese encephalitis virus infection initiates endoplasmic reticulum stress and an unfolded protein response. J Virol 76:4162–4171
Teodoro JG, Branton PE (1997) Regulation of apoptosis by viral gene products. J Virol 71:1739–1746
Xiao SY, Zhang H, Guzman H, Tesh RB (2001) Experimental yellow fever virus infection in the golden hamster (*Mesiocricetus auratus*). II. Pathology. J Infect Dis 183:1437–1444
Zinszner H, Kuroda M, Wang XZ, Batchvarova N, Lightfoot RT, Remotti H, Stevens JL, Ron D (1998) CHOP is implicated in programmed cell death in response to impaired function of the endoplasmic reticulum. Genes Dev 12:982–995

Manipulation of Apoptosis by Herpes Viruses (Kaposi's Sarcoma Pathogenesis)

P. Feng[1], C. Scott[1], S.-H. Lee[1], N.-H. Cho[1] and J.U. Jung[1]

1 Introduction

The initial description of Kaposi's sarcoma can be traced back to 1872, when a Hungarian dermatologist, Moritz Kaposi, published a case of skin cancer with clinical presentation of so-called pigmented sarcomas. This disease was designated Kaposi's sarcoma (KS) in 1891. There are four distinct epidemiological forms of KS: classic KS occurring predominantly in Europe and the Mediterranean; endemic KS of the human immunodeficiency virus (HIV-1) negative patients in Africa; post-transplant or iatrogenic KS; and HIV-associated KS. The speculation of a viral etiology for KS was confirmed by the discovery of KSHV from HIV-associated KS lesions in 1994 (Chang et al. 1994). Subsequent epidemiological studies have demonstrated that KSHV DNA is present in all forms of KS tumors, suggesting that KSHV is the etiology agent for the development of KS (Sarid et al. 1999). In addition to KS, KSHV also associates with primary effusion lymphoma (PEL) and an immunoblast variant of Castleman's disease (CD), which are of B cell origin (Cesarman et al. 1995; Soulier et al. 1995).

Genomic sequencing classifies KSHV as a gamma-2 herpesvirus, closely related to herpesvirus saimiri (HVS; Jung et al. 1999), rhesus monkey rhadinovirus (RRV; Alexander et al. 2000), and murine gamma herpesvirus 68 (γHV-68 or MHV-68; Virgin et al. 1997). KSHV contains a double-stranded DNA genome of ~165 kilobases (kb) and encodes more than 80 open reading frames (orfs), with 15 orfs unique to KSHV. As with other gamma-2 herpesviruses, KSHV encodes numerous cellular homologues, some of which have been demonstrated to hijack cellular pathways that directly contribute to abnormal cell growth when expressed in the context of the viral genome (Moore et al. 1996; Lee et al. 1998). While KSHV infects a variety of cell types in culture, no efficient permissive cell system or animal model is currently available. This poses a difficulty for studying the contribution of the individual gene product in the context of viral lytic replication to viral lifecycle and viral pathogenesis. Therefore, RRV, HVS, and MHV-68 that have permissive cell culture systems

[1]Department of Microbiology and Molecular Genetics, Division of Tumor Virology, New England Regional Primate Research Center, Harvard Medical School, 1 Pine Hill Drive, Southborough, Massachusetts 01772, USA, e-mail: pinghui_feng@hms.harvard.edu

Progress in Molecular and Subcellular Biology
C. Alonso (Ed.): Viruses and Apoptosis
© Springer-Verlag Berlin Heidelberg 2004

in vitro and animal models in vivo have been used to study viral lytic replication and pathogenesis.

Apoptosis is a mechanism by which the immune system has enlisted every cell of the body to participate in its own death. Whether triggered by the oligomerization of cell surface molecules such as Fas or through signaling from intracellular molecules, apoptosis proceeds through a programmed cascade consisting of kinases, caspases, phosphatases, and ribonucleases (Leverrier and Ridley 2001). This results in the discharge of mitochondrial potential, DNA fragmentation, and plasma membrane blebbing. Upon viral infection, the host immune system can utilize apoptosis to limit the ability of virus to replicate. To circumvent this, viruses have evolved a variety of elaborate mechanisms to subvert host apoptosis pathways, ensuring continuous viral production and persistent viral infection. Therefore, subversion of apoptosis by viruses has been an ancient and forever novel theme for the virologist. Since the signaling pathways of apoptosis have been extensively reviewed elsewhere (Ferri and Kroemer 2001; Martinou and Green 2001), we briefly summarize cellular apoptosis pathways and direct our attention to the regulation imposed by herpesvirus, using KSHV as an example.

2 Cellular Apoptosis Pathways

2.1 Death-Receptor-Mediated Apoptosis

Apoptosis can be initiated by extrinsic stimuli if triggered by death receptor engagement at the cell surface or intrinsic stimuli where disturbed intracellular homeostasis is sensed (Kroemer and Reed 2000). Virus-infected cells can be targeted for cytotoxic T cell lysis by virtue of ligation of death receptors displayed on the surface, for instance, Fas and tumor necrosis factor receptor (TNFR). Upon cognate ligand binding, the cytoplasmic tails of death receptors recruit multiple adaptor molecules of the death domain (DD) family, leading to the assembly of a death-inducing signaling complex (DISC) or apoptosome (Fig. 1; Kitson et al. 1996; Irmler et al. 1997; Thome et al. 1997). In addition to the DD, adaptor molecules also contain various death effector domains (DEDs), both of which mediate protein interactions between components of the DISC. The cell surface associated apoptosome or the DISC cleaves pro-caspase 8, a typical initiation caspase, whose signal is then either amplified by the mitochondrial cascade or relayed directly to effector caspases such as caspase 3 and 7 depending on cell type (Muzio et al. 1996; Li et al. 1998).

Not surprisingly, multiple regulatory proteins have been found to modulate death signaling triggered through death receptor ligation. One of these proteins is the DED-containing molecule that inhibits Fas-associated interleukin converting enzyme (FLICE or pro-caspase 8) activation, therefore referred to

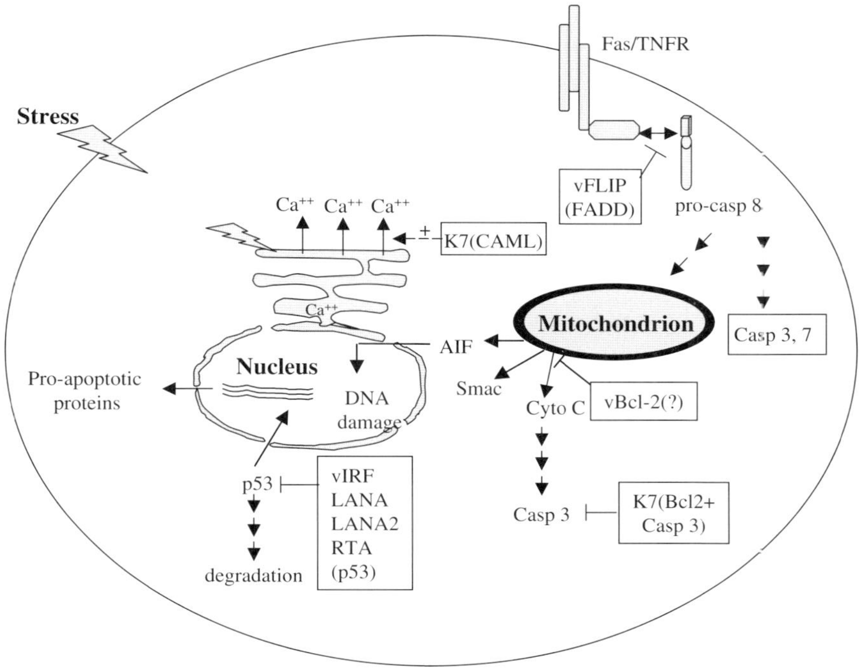

Fig. 1. Manipulation of apoptosis by KSHV gene products. The cellular apoptosis pathways are outlined here and the KSHV proteins are boxed

as FLICE-inhibitory protein (FLIP). FLIP has two isoforms due to alternative splicing. One isoform contains amino terminal DED motifs that interact with DED-containing proteins, and a carboxyl terminal region that shares significant homology with pro-caspase 8 though lacking key residues at the corresponding caspase active site. The second, shorter isoform contains only the amino terminal DED motifs (Irmler et al. 1997). As expected, FLIPs interact with Fas-associated death domain protein (FADD) or pro-caspase 8 through DEDs and interfere with the recruitment and activation of caspase 8, resulting in the inhibition of apoptosis. Interestingly, there are a number of pathogens, including the herpesviruses, which encode cellular FLIP homologues whose expression drastically prevents death-receptor-mediated apoptosis, highlighting the biological significance of FLIP functional activity in apoptosis (Bertin et al. 1997; Thome et al. 1997).

Receptor-interacting protein (RIP) and TNFR-associated factor 2 (TRAF2) exert anti-apoptotic function by activating nuclear factor κB (NFκB) activity. Upon ligand engagement, TRAF2, RIP, and IκB kinase kinase (IKK) gamma subunit are sequentially recruited to the DISC. RIP phosphorylates the IKK gamma leading to its activation and dissociation, which in turn phosphorylates IκB to induce its degradation (Liu et al. 1996; Devin et al. 2000). Subsequently,

NFκB translocates into the nucleus and activates expression of various anti-apoptotic genes, for instance, Bcl-xL, FLIP, and A20 (Krikos et al. 1992; Khoshnan et al. 2000; Micheau et al. 2001). Thus, RIP and TRAF2, in conjunction with FLIP, deploy an orchestrated anti-apoptotic signaling network that alleviates death-receptor-mediated apoptosis.

2.2 Mitochondrion-Mediated Apoptosis

In parallel with ligation of the death receptor, intracellular stress signals, such as DNA damage and alteration of ER homeostasis, can trigger apoptosis through the cellular homeostasis surveillance network. Mitochondria play a crucial role by either actively amplifying the death signal or passively allowing cells to undergo necrosis (Kroemer et al. 1998). Mitochondria contribute to apoptosis through two independent, but closely linked processes (Fig. 1). First, disruption of the mitochondrial inner membrane integrity dissipates mitochondrial potential across the membrane, leading to the depletion of cellular energy in the form of ATP and the production of reactive oxygen species (ROS). Second, mitochondria serve as a reservoir for a variety of proteins, including cytochrome c and apoptosis-inducing factor (AIF), whose release from mitochondria promotes cell death (Li et al. 1997; Susin et al. 1999; Du et al. 2000). While AIF translocates into the nucleus and triggers caspase-independent apoptotic events, cytochrome c forms a complex with Apaf-1 and drives the autoprocessing and activation of pro-caspase 9. Caspase 9 then cleaves effector caspases such as caspase 3 and 7, initiating caspase signal transduction. Crosstalk between the mitochondrial pathway and the surface death-receptor-mediated pathway has been supported by the discovery that Bid is cleaved by caspase 8 and translocates into the mitochondrial membrane where truncated Bid forms a hydrophilic pore and activates the mitochondrion-mediated signal cascade (Li et al. 1998). Accumulating evidence suggests that signals originating from mitochondrial compartments are crucial to synchronize the execution of the life-or-death decision within an individual cell (Duchen 2000).

Not surprisingly, various cellular proteins have been discovered to regulate apoptosis at the mitochondrial checkpoint. Cellular Bcl-2 protein represents a family of proteins that share four conserved Bcl-2 homology (BH) domains with a carboxyl terminal hydrophobic sequence capable of anchoring into the mitochondrial membrane (Cory and Adams 2002). Bcl-2 protein inhibits apoptosis at the mitochondria by heterodimerizing with pro-apoptotic Bcl-2 family members such as Bax and blocking their ability to induce the release of cytochrome c and other apoptogenic factors from mitochondria (Oltvai et al. 1993; Kuwana et al. 2002). Exactly how cytochrome c is released from mitochondria has not yet been elucidated, but it has been proposed that pro-apoptotic Bcl-2 family proteins either modulate pre-existing permeability transition pores (PTP) or form hydrophilic pores, resulting in the promotion of mitochondrial potential loss and apoptogenic factor efflux (Suzuki et al. 2000). To protect cells from apoptosis induced by chemotherapeutical drugs that interfere with endoplasmic reticulum (ER) calcium homeostasis, Bcl-2 expression has also been shown to facilitate calcium release from the ER compartment (Foyouzi-Youssefi et al. 2000; Pinton et al. 2001). Thus, these studies highlight the diverse functions of Bcl-2 family proteins to regulate cellular apoptosis, suggesting a possible pathway for viral piracy. Indeed, to control the cellular apoptotic pathway, all members of gamma herpesviruses encode a Bcl-2 homologue, including orf16 of KSHV, RRV and HVS, BHRF1 and BALF1 of EBV, and M11 of MHV-68 (Means et al. 2002).

Besides these multiple regulators, the complexity of the mitochondrial pathway is also reflected by the integration of diverse stress signals, initiated by DNA damage in the nucleus, calcium release from the ER, or ceramide production from the plasma membrane. p53 tumor suppressor is an apoptosis master protein that controls expression of a broad spectrum of genes that participate in apoptosis triggered by DNA damage (Ferrier 2002; Sharpless and DePinho 2002). These include cell-cycle agonist p21, mitochondrial pro-apoptotic factor Bid, PUMA, Bax, and death-domain family members Fas/Apo-1, Killer/DR5, and PIDD (Miyashita and Reed 1995; Lin et al. 2000; Yu et al. 2001; Sax et al. 2002). In addition, the transcription-independent apoptosis induced by p53 overexpression has been reported to function at mitochondria from various cell lines with p53 lacking DNA binding and transactivation activity (Chen et al. 1996; Haupt et al. 1997). To counteract p53 activity, several viral proteins have been demonstrated to influence p53 degradation and, thereby, p53-mediated apoptosis (Scheffner et al. 1990; Nakamura et al. 2001). For example, p53 has been shown to be a substrate of the E6-associated protein (E6AP), an E3 ligase, which targets p53 for degradation upon binding to human papillomavirus E6 (Scheffner et al. 1990). In contrast, Mdm2 is the E3 ubiquitin ligase in normal cells that governs ubiquitination and degradation of p53. Upon DNA damage, p53 is substantially stabilized due to its phosphorylation at multiple sites, preventing its association with Mdm2 and, therefore, dampening Mdm2-mediated protein degradation (Appella and Anderson 2001). Moreover, p53 half-life can be extended by the tumor suppressor protein p14 (ARF) that inhibits Mdm2 activity upon oncogenic insults (Sherr and Weber 2000). Finally, herpesvirus-associated ubiquitin-specific protease (HAUSP, also known as USP7) has been shown to represent a novel mechanism to control p53 protein by enhancing its de-ubiquitination (Li et al. 2002). Remarkably, HAUSP is able to inhibit Mdm2-mediated downregulation of p53, suggesting that HAUSP functions as a novel tumor suppressor (Li et al. 2002). These experiments indicate that p53 stability can be regulated by equilibrium between ubiquitination by Mdm2 and de-ubiquitination by HAUSP.

Accumulating evidence indicates that other organelles, including the ER, lysosomes, and Golgi, are also major points of integration of pro-apoptotic signaling (Ferri and Kroemer 2001). Notably, these organelles harbor numerous proteins that sense various stress signals and, therefore, coordinate cellular programs to cope with changing environment. In particular, the ER senses local stress through chaperones, Ca^{2+}-binding proteins, and Ca^{2+} channels, which transmit the ER Ca^{2+} response to other organelles such as mitochondria and the Golgi apparatus (Jackson et al. 1995; Mattson et al. 2001). This signaling pathway has been described as an unfolded protein response (Ma and Hendershot 2001). Interestingly, Bcl-2 has been reported to modulate the ER calcium status and inhibit apoptosis induced by calcium-releasing drugs, suggesting that Bcl-2 is a messenger linking different pathways and various organelles (Pinton et al. 2001; Vanden Abeele et al. 2002).

3 Herpesviral Proteins Involved in Apoptosis

3.1 Viral Proteins Interfering with Death-Receptor-Mediated Apoptosis

3.1.1 vFLIP

KSHV, RRV, and HVS encode a cellular FLIP homologue, designated vFLIP. vFLIP shares significant homology with cellular FLIP and contains two signature death-effector domains (DEDs; Irmler et al. 1997). The expression of vFLIP strongly blocks Fas-mediated apoptosis, correlating with decreased activity of caspases 3, 8 and 9 (Djerbi et al. 1999; Fig. 1). Further, KSHV vFLIP has also been shown to function as a tumor progression factor in a mouse model (Djerbi et al. 1999). These results are consistent with the fact that cellular FLIP and proteins harboring similar activity during apoptosis are normally upregulated within various malignant cell lines such as lung, colon, prostate, pancreas, and lymphocyte (Ambrosini et al. 1997). More recently, expression of KSHV FLIP has conferred leukemia cells with resistance to serum withdrawal-induced apoptosis, whereas the equine herpesvirus 2 homologue failed to do so (Sun et al. 2002). This activity is attributable to the activation of NF-κB activity, since inhibition of IκB degradation reverses the protective activity against growth factor deprivation-induced apoptosis (Sun et al. 2002). Consistent with these findings, Liu et al. (2002) have reported that KSHV vFLIP physically and functionally associates with the IκB complex, presumably to activate NF-κB transcription factor activity. However, HVS vFLIP, which also has a strong anti-apoptotic activity in culture, is not required for viral replication, transformation, and pathogenicity in cottontop tamarin, suggesting that vFLIP-mediated anti-apoptosis may play a role in viral persistent infection (Glykofrydes et al. 2000). Expressed from a bicistronic mRNA that also encodes v-Cyclin as an upstream open reading frame, vFLIP could be translated from a functional internal ribosome entry site (Bieleski and Talbot 2001). This raises the possibility that KSHV may have a novel mechanism to coordinately regulate the expression of vFLIP and v-Cyclin. In contrast to KSHV and HVS, equine herpesvirus-2, bovine herpesvirus-4, and molluscum contagiosum virus express vFLIP homologues as a monocistronic messenger RNA from their respective genome (Bertin et al. 1997; Hu et al. 1997). Nevertheless, it is apparent that an inhibition of the extrinsic apoptosis pathway is ubiquitously explored by various pathogens such as viruses to evade host defense mechanisms.

3.1.2
Other Herpesviral Proteins Implicated in Death-Receptor-Mediated Apoptosis

In a similar manner to vFLIP, expression of two mitochondrial proteins encoded by human cytomeglovirus (HCMV), vMIA and KSHV K7, efficiently inhibits death receptor ligation mediated apoptosis (Goldmacher et al. 1999; Feng et al. 2002; Wang et al. 2002). Both proteins function at the mitochondrial checkpoint, with the exception that K7 may have diverse cellular targets (Feng and Jung, unpubl. data). Numerous studies have demonstrated that mitochondrial amplification of the apoptotic signal initiated from death receptor ligation or stress stimuli is a key step to execute apoptosis. A detailed mechanism of how vMIA and K7 play their roles will be discussed in the next section.

3.2
Viral Proteins Interfering with Mitochondrion-Mediated Apoptosis

3.2.1
vBCL-2

Based on genomic sequence, KSHV encodes an apparent homologue of Bcl-2, called vBcl-2. vBcl-2 consists of 175 amino acids, which possesses approximately 15–20% homology to its cellular counterpart (Bellows et al. 2000; Fig. 1). This homology is largely restricted to the BH1 and BH2 heterodimerization and death repressor domains (Bellows et al. 2000). Similar to its cellular homologue, KSHV vBcl-2 has been shown to inhibit Bax-induced apoptosis (Sarid et al. 1997). In addition to this finding, vBcl-2 of MHV-68 and HVS also demonstrates ability to block TNF-α- and/or Fas-induced apoptosis in culture (Nava et al. 1997; Wang et al. 1999). Additionally, the HVS vBcl-2 appears to confer resistance against apoptosis induced by a broad spectrum of stimuli including dexamethasone, irradiation and ROS (Fig. 1; Nava et al. 1997). Moreover, HVS vBcl-2 and EBV BHRF1 (EBV vBcl-2) have been shown to augment the transformation ability of adenovirus E1A, c-Myc, and v-Cyclin (Fanidi et al. 1998; Ojala et al. 2000). While proto-oncogene products, E1A, c-Myc, and v-Cyclin, promote cell cycle and cell proliferation, proliferating cells are normally more susceptible to apoptosis (Tapon et al. 2002). Therefore, expression of vBcl-2 in conjunction with these oncogenes enables cells to proliferate with normal or higher resistance to apoptotic stimuli, presumably leading to malignant tumorigenesis. This suggests that vBcl-2 may contribute directly to the pathogenicity of these gamma herpesviruses. Consistent with this notion, MHV68 M11 was shown to be expressed during latency in infected mice (Virgin et al. 1999; Roy et al. 2000) and to be necessary for in vivo viral reactivation (Clambey et al. 2000).

Despite functional similarity, vBcl-2 differs from its cellular counterpart in three distinct features: first, vBcl-2 does not appear to form a heterodimer with cellular Bax or Bak, suggesting a somewhat different mechanism employed by vBcl-2 to inhibit apoptosis (Cheng et al. 2001); second, vBcl-2 escapes caspase-mediated conversion to a pro-apoptotic factor either through failure to be cleaved by active caspase 3, or the lack of pro-apoptotic activity of the cleaved carboxyl fragment (Bellows et al. 2000); third, vBcl-2 is not phosphorylated by the cellular cyclin-dependent kinase 6 (CDK6) that associates with the v-Cyclin of KSHV, whereas phosphorylation of cellular Bcl-2 protein by CDK6/vCyclin complex occurs and prevents its cleavage within the loop region (Ojala et al. 2000). Taken together, these studies indicate that gamma herpesviruses have evolved delicate mechanisms to selectively retain only anti-apoptosis function with absolute guarantee.

Additional evidence of the significance of the mitochondrial checkpoint during apoptosis comes from viral proteins that directly associate with and modulate subunits of the mitochondrial PTP complex, including the HBx protein of hepatitis B virus (Diao et al. 2001), the vMIA protein of HCMV (Goldmacher et al. 1999), the E1 19 K protein of Adenovirus (Han et al. 1998), the M11L protein of myxoma virus (Everett et al. 2002), and the Vpr of HIV (Jacotot et al. 2000). This indicates that the mitochondrial PTP complex is a common target for viral anti-apoptosis. However, none of the gamma herpesviral proteins discovered so far has been found to physically associate with mitochondrial PTP complex. Thus, it is intriguing to speculate that vBcl-2 may modulate mitochondrial PTP complex directly or indirectly to control release of apoptotic factors that are normally caged within the mitochondrial intermembrane space. Alternatively, yet-to-be-defined gamma herpesviral proteins may play such a role.

3.2.2
vMIA and K7

There are two known herpesvirus-encoded mitochondrial proteins with anti-apoptotic ability, the viral mitochondrion-localized inhibitor of apoptosis (vMIA) of HCMV and K7 of KSHV (Goldmacher et al. 1999; Feng et al. 2002; Wang et al. 2002). vMIA is a protein of 163 amino acids (a.a.) and the translation product of the first exon of UL37. Its expression renders HeLa cells resistant to Fas- and TNFR-mediated apoptosis, suggesting vMIA function at the extrinsic apoptotic pathway (Goldmacher et al. 1999). Biochemical characterization has indicated that vMIA functions at a step downstream of pro-caspase 8 activation and upstream of cytochrome c release, implying that vMIA plays an important role at the mitochondrial checkpoint (Goldmacher et al. 1999). Indeed, vMIA localizes to the mitochondrial compartment and interacts with the adenosine nucleotide translocator subunit of the PTP complex, corroborating its function in regulating cytochrome c release during Fas-mediated apoptosis (Goldmacher et al. 1999). While vMIA functionally resembles pro-survival Bcl-2 family members, it shares no sequence homology with Bcl-2 proteins. Furthermore, vMIA does not form a complex with Bax or the voltage-dependent anion channel of the PTP complex, suggesting that vMIA represents a novel class of anti-apoptotic proteins (Goldmacher et al. 1999). Studies from recombinant HCMV indicated that vMIA is necessary for HCMV viral replication, suggesting a direct contribution of viral anti-apoptotic proteins to viral lytic replication.

Two independent studies have reported that expression of KSHV K7 protects cells from a variety of apoptogenic agents, including αFas antibody, TNF-α, TRAIL, thapsigargin (TG), and staurosporine (Feng et al. 2002; Wang et al. 2002). K7 is a small protein of 126 a.a. and its internal hydrophobic sequence is sufficient for its mitochondrial localization (Feng et al. 2002). Interestingly, K7 expression in B lymphocytes confers strong protection against apoptosis induced by TG, a sarcoplasmic/endoplasmic reticulum calcium ATPase inhibitor that imposes stress signal on cells (Feng et al. 2002). This activity is attributed to its interaction with a cellular calcium-modulating cyclophilin ligand (CAML), an ER resident protein that modulates calcium release into the cytoplasm (Bram and Crabtree 1994). This suggests that KSHV K7 protein targets cellular CAML to modulate intracellular calcium effluxes which ultimately avoids stress-induced apoptosis (Fig. 1). In addition, Wang et al. (2002) have reported that K7 structurally and functionally resembles cellular survivin, a member of inhibitors of apoptosis family. In fact, anti-apoptotic activity of K7 has been shown to corroborate with its ability to mediate an interaction between Bcl-2 and activated caspase 3 through its carboxyl terminal potential BH2-like domain, therefore quenching caspase activity (Fig. 1; Wang et al. 2002). Finally, we have recently found that K7 also interacts with cellular UBL/UBA-containing regulatory factor of the ubiquitin/proteasome machinery, and that this interation leads to the rapid degradation of IkB and p53 and thereby, anti-apoptosis (unpublished results). These data suggest that KSHV K7 likely targets multiple cellular proteins to achieve its anti-apoptotic function. Further characterization of K7 will provide information concerning its function during KSHV lytic replication and persistent infection.

3.2.3
Cytokines and Chemokines

Cytokines and chemokines play essential roles during host defense, in particular, leukocyte activation, proliferation, migration, and infiltration (Adams and Lloyd 1997; Locati and Murphy 1999; Murdoch and Finn 2000). Specifically, various cytokines and chemokines are crucial to determine lymphocyte development, proliferation and activation. For example, macrophage inflammatory protein (MIP) 1 beta induces migration and proliferation of human thymocytes (Dairaghi et al. 1998). KSHV K6, K4, and K4.1 encode three vMIPs sharing up to 60% identity with cellular MIP at the amino acid level. These three chemokines are often referred to as vMIP-I, vMIP-II, and vMIP-III, respectively. vMIP-II, the most well-characterized vMIP, has promiscuous binding activity to a variety of CC chemokine receptors (CCR) as well as CXC chemokine receptors including CCR1, CCR2, CCR4, CCR5, and CXCR4 (Boshoff et al. 1997; Kledal et al. 1997; Stine et al. 2000). In addition, both vMIP-I and vMIP-II are highly angiogenic when examined by chorioallantoic assay, suggesting that vMIPs may function to counteract cytokine-deprivation induced apoptosis (Fig. 1; Boshoff et al. 1997).

In addition to vMIPs, KSHV K2 encodes a homologue of IL-6, initially discovered as a B cell differentiating factor. Cellular IL-6 secreted by CD4 helper T cells stimulates B cells to develop into plasma cells and promote antibody production. Despite its ability to replace cellular IL-6 to support proliferation of mouse myeloma cell lines, viral IL-6 (vIL-6) shows a distinct receptor usage compared to cellular IL-6 (Molden et al. 1997; Hoischen et al. 2000). vIL-6 is able to bind its cognate receptor gp130 independent of the IL-6 receptor, both of which are required for cellular IL-6 binding. This indicates that vIL-6 may overcome some of the cellular regulatory points within KSHV-infected cells. Upon binding, gp130 transduces activation signals, represented by the kinase cascade of Janus kinase/Stat1, 3/MAPK (Molden et al. 1997; Hoischen et al. 2000). This suggests that the vIL-6 expression can provide a survival signal and potentially contributes to lymphoproliferative disease associated with KSHV.

3.2.4
vIRF, RTA, LANA, and LANA2

DNA damage represents an important host stress signal under the circumstance of viral infection and replication. Specifically, herpesviruses replicate and assemble within the nucleus of host cells, which alters cellular chromosome structure and causes DNA damage. Ultimately, apoptotic pathways are activated, mainly through p53-mediated surveillance mechanisms. Thus, subversion of p53-mediated apoptosis is another necessary mechanism for herpesvirus to escape apoptosis, bypass cell cycle arrest, and promote proliferation, leading to tumorigenesis. Remarkably, four KSHV-encoded viral proteins have been reported to inhibit p53-mediated apoptosis, i.e. viral interferon regulatory factor (vIRF or K9), latency-associated nuclear antigen (LANA), LANA2 (K10.5), and replication and transcription activator (RTA or orf50; Fig. 1). Nakamura et al. (2001) found that vIRF interacts with multiple regions of p53, including the growth suppression SH3B domain, the core DNA-binding domain, and the carboxyl acidic domain. vIRF appears to inactivate p53 by preventing stress-initiated modifications which include phosphorylation and acetylation (Nakamura et al. 2001). However, the detailed mechanism by which vIRF achieves this effect is still not understood. Consistent with its interaction with p53, vIRF expression renders cells resistant to apoptosis induced by p53 overexpression and serum deprivation (Nakamura et al. 2001). Similar to vIRF, LANA has been shown to physically associate with p53 and inhibits its transactivation activity and p53-mediated apoptosis (Friborg et al. 1999). In addition, RTA and LANA2 (K10.5) have also been shown to downregulate p53 transactivation activity (Gwack et al. 2001; Rivas et al. 2001). While the mechanism of how LANA2 inhibits p53-mediated apoptosis is not clear, RTA appears to inhibit p53 transactivation through the CREB binding protein (Gwack et al. 2001; Rivas et al. 2001). In summary, KSHV encodes four different proteins that are expressed at different time points during viral life cycle to inhibit p53-mediated apoptosis, which also highlights the significance of the p53-mediated apoptosis pathway in host surveillance mechanisms.

4
Future Directions

The hallmark of herpesvirus is a persistent infection. Thus, herpesviruses exploit numerous cellular signaling pathways to escape host destruction and favor viral production to establish and maintain persistent infection. Since the discovery of KSHV in 1994, the study of this human tumor virus has yielded a body of valuable information not only to understand KSHV viral replication and KS pathogenesis, but also to understand the function of key components of the immune system. However, the progress of the study of KSHV has been handicapped by the lack of a permissive system of KSHV replication. As a consequence, most of the findings have been derived from the study of individual viral gene products outside of the viral genome context. Fortunately, a few animal models, like monkey as a non-human primate model and mouse as a small animal model, are available to investigate KSHV closely related viruses, i.e. RRV, HVS, and MHV-68. Based on the molecular mechanism of each individual gene product of KSHV that has been characterized so far, we envision that future studies will be focused on the contribution of each viral protein to pathogenesis using various animal herpesviruses. With the available genome sequence of several species, the pathogen–host interaction will be another highlight for the future study of virology.

Acknowledgements. We would like to thank other members of Jae Jung's laboratory. We are grateful to Joong Choi and Robert Means for their critical reading and discussion.

References

Adams DH, Lloyd AR (1997) Chemokines: leucocyte recruitment and activation cytokines. Lancet 349:490–495

Alexander L, Denekamp L, Knapp A, et al. (2000) The primary sequence of rhesus monkey rhadinovirus isolate 26–95: sequence similarities to Kaposi's sarcoma-associated herpesvirus and rhesus monkey rhadinovirus isolate 17577. J Virol 74:3388–3398

Ambrosini G, Adida C, Altieri DC (1997) A novel anti-apoptosis gene, survivin, expressed in cancer and lymphoma. Nat Med 3:917–921

Appella E, Anderson CW (2001) Post-translational modifications and activation of p53 by genotoxic stresses. Eur J Biochem 268:2764–2772

Bellows DS, Chau BN, Lee P, et al. (2000) Antiapoptotic herpesvirus Bcl-2 homologs escape caspase-mediated conversion to proapoptotic proteins. J Virol 74:5024–5031

Bertin J, Armstrong RC, Ottilie S, et al. (1997) Death effector domain-containing herpesvirus and poxvirus proteins inhibit both Fas- and TNFR1-induced apoptosis. Proc Natl Acad Sci USA 94:1172–1176

Bieleski L, Talbot SJ (2001) Kaposi's sarcoma-associated herpesvirus vCyclin open reading frame contains an internal ribosome entry site. J Virol 75:1864–1869

Boshoff C, Endo Y, Collins PD, et al. (1997) Angiogenic and HIV-inhibitory functions of KSHV-encoded chemokines. Science 278:290–294

Bram RJ, Crabtree GR (1994) Calcium signalling in T cells stimulated by a cyclophilin B-binding protein. Nature 371:355–358

Cesarman E, Chang Y, Moore PS, et al. (1995) Kaposi's sarcoma-associated herpesvirus-like DNA sequences in AIDS-related body-cavity-based lymphomas. N Engl J Med 332:1186–1191

Chang Y, Cesarman E, Pessin MS, et al. (1994) Identification of herpesvirus-like DNA sequences in AIDS-associated Kaposi's sarcoma. Science 266:1865–1869

Chen X, Ko LJ, Jayaraman L, et al. (1996) p53 levels, functional domains, and DNA damage determine the extent of the apoptotic response of tumor cells. Genes Dev 10:2438–2451

Cheng EH, Wei MC, Weiler S, et al. (2001) BCL-2, BCL-X(L) sequester BH3 domain-only molecules preventing Bax- and Bak-mediated mitochondrial apoptosis. Mol Cell 8:705–711

Clambey ET, VirginHWt, Speck SH (2000) Disruption of the murine gammaherpesvirus 68 M1 open reading frame leads to enhanced reactivation from latency. J Virol 74:1973–1984

Cory S, Adams JM (2002) The Bcl2 family: regulators of the cellular life-or-death switch. Nat Rev Cancer 2:647–656

Dairaghi DJ, Franz-Bacon K, Callas E, et al. (1998) Macrophage inflammatory protein-1beta induces migration and activation of human thymocytes. Blood 91:2905–2913

Devin A, Cook A, Lin Y, et al. (2000) The distinct roles of TRAF2 and RIP in IKK activation by TNF-R1: TRAF2 recruits IKK to TNF-R1 while RIP mediates IKK activation. Immunity 12:419–429

Diao J, Khine AA, Sarangi F, et al. (2001) X protein of hepatitis B virus inhibits Fas-mediated apoptosis and is associated with up-regulation of the SAPK/JNK pathway. J Biol Chem 276:8328–8340

Djerbi M, Screpanti V, Catrina A, et al. (1999) The inhibitor of death receptor signaling, FLICE-inhibitory protein, defines a new class of tumor progression factors. J Exp Med 190:1025–1032

Du C, Fang M, Li Y, et al. (2000) Smac, a mitochondrial protein that promotes cytochrome c-dependent caspase activation by eliminating IAP inhibition. Cell 102:33–42

Duchen MR (2000) Mitochondria and calcium: from cell signalling to cell death. J Physiol 529(Pt 1):57–68

Everett, H, Barry M, Sun X, et al. (2002) The myxoma poxvirus protein, M11L, prevents apoptosis by direct interaction with the mitochondrial permeability transition pore. J Exp Med 196:1127–1139

Fanidi A, Hancock DC, Littlewood TD (1998) Suppression of c-Myc-induced apoptosis by the Epstein-Barr virus gene product BHRF1. J Virol 72:8392–8395

Feng P, Park J, Lee BS, et al. (2002) Kaposi's sarcoma-associated herpesvirus mitochondrial K7 protein targets a cellular calcium-modulating cyclophilin ligand to modulate intracellular calcium concentration and inhibit apoptosis. J Virol 76:11491–11504

Ferri KF, Kroemer G (2001) Organelle-specific initiation of cell death pathways. Nat Cell Biol 3:E255–E263

Ferrier V (2002) Mini p53. Nat Cell Biol 4:E156

Foyouzi-Youssefi R, Arnaudeau S, Borner C, et al. (2000) Bcl-2 decreases the free Ca^{2+} concentration within the endoplasmic reticulum. Proc Natl Acad Sci 97:5723–5728

Friborg J Jr, Kong W, Hottiger MO, et al. (1999) p53 inhibition by the LANA protein of KSHV protects against cell death. Nature 402:889–894

Glykofrydes D, Niphuis H, Kuhn EM, et al. (2000) Herpesvirus saimiri vFLIP provides an anti-apoptotic function but is not essential for viral replication, transformation, or pathogenicity. J Virol 74:11919–11927

Goldmacher VS, Bartle LM, Skaletskaya A, et al. (1999) A cytomegalovirus-encoded mitochondria-localized inhibitor of apoptosis structurally unrelated to Bcl-2. Proc Natl Acad Sci USA 96:12536–12541

Gwack Y, Hwang S, Byun H, et al. (2001) Kaposi's sarcoma-associated herpesvirus open reading frame 50 represses p53-induced transcriptional activity and apoptosis. J Virol 75:6245–6248

Han J, Wallen HD, Nunez G, et al. (1998) E1B 19,000-molecular-weight protein interacts with and inhibits CED-4-dependent, FLICE-mediated apoptosis. Mol Cell Biol 18:6052–6062

Haupt Y, Rowan S, Shaulian E, et al. (1997) p53 mediated apoptosis in HeLa cells: transcription dependent and independent mechanisms. Leukemia 11(Suppl 3):337–339

Hoischen SH, Vollmer P, Marz P, et al. (2000) Human herpes virus 8 interleukin-6 homologue triggers gp130 on neuronal and hematopoietic cells. Eur J Biochem 267:3604–3612

Hu S, Vincenz C, Buller M, et al. (1997) A novel family of viral death effector domain-containing molecules that inhibit both CD-95- and tumor necrosis factor receptor-1-induced apoptosis. J Biol Chem 272:9621–9624

Irmler M, Thome M, Hahne M, et al. (1997) Inhibition of death receptor signals by cellular FLIP. Nature 388:190–195

Jackson SP, Schoenwaelder SM, Matzaris M, et al. (1995) Phosphatidylinositol 3,4,5-trisphosphate is a substrate for the 75 kDa inositol polyphosphate 5-phosphatase and a novel 5-phosphatase which forms a complex with the p85/p110 form of phosphoinositide 3-kinase. Embo J 14:4490–4500

Jacotot E, Ravagnan L, Loeffler M, et al. (2000) The HIV-1 viral protein R induces apoptosis via a direct effect on the mitochondrial permeability transition pore. J Exp Med 191:33–46

Jung JU, Choi JK, Ensser A, et al. (1999) Herpesvirus saimiri as a model for gammaherpesvirus oncogenesis. Semin Cancer Biol 9:231–239

Khoshnan A, Tindell C, Laux I, et al. (2000) The NF-kappa B cascade is important in Bcl-xL expression and for the anti-apoptotic effects of the CD28 receptor in primary human CD4+ lymphocytes. J Immunol 165:1743–1754

Kitson J, Raven T, Jiang YP, et al. (1996) A death-domain-containing receptor that mediates apoptosis. Nature 384:372–375

Kledal TN, Rosenkilde MM, Coulin F, et al. (1997) A broad-spectrum chemokine antagonist encoded by Kaposi's sarcoma-associated herpesvirus. Science 277:1656–1659

Krikos A, Laherty CD, Dixit VM (1992) Transcriptional activation of the tumor necrosis factor alpha-inducible zinc finger protein, A20, is mediated by kappa B elements. J Biol Chem 267:17971–17976

Kroemer G, Dallaporta B, Resche-Rigon M (1998) The mitochondrial death/life regulator in apoptosis and necrosis. Annu Rev Physiol 60:619–642

Kroemer G, Reed JC (2000) Mitochondrial control of cell death. Nat Med 6:513–519

Kuwana T, Mackey MR, Perkins G, et al. (2002) Bid, Bax, and lipids cooperate to form supramolecular openings in the outer mitochondrial membrane. Cell 111:331–342

Lee H, Veazey R, Williams K, et al. (1998) Deregulation of cell growth by the K1 gene of Kaposi's sarcoma-associated herpesvirus. Nat Med 4:435–440

Leverrier Y, Ridley AJ (2001) Apoptosis: caspases orchestrate the ROCK 'n' bleb. Nat Cell Biol 3:E91–E93
Li H, Zhu H, Xu CJ, et al. (1998) Cleavage of BID by caspase 8 mediates the mitochondrial damage in the Fas pathway of apoptosis. Cell 94:491–501
Li M, Chen D, Shiloh A, et al. (2002) Deubiquitination of p53 by HAUSP is an important pathway for p53 stabilization. Nature 416:648–653
Li P, Nijhawan D, Budihardjo I, et al. (1997) Cytochrome c and dATP-dependent formation of Apaf-1/caspase-9 complex initiates an apoptotic protease cascade. Cell 91:479–489
Lin Y, Ma W, Benchimol S (2000) Pidd, a new death-domain-containing protein, is induced by p53 and promotes apoptosis. Nat Genet 26:122–127
Liu L, Eby MT, Rathore N, et al. (2002) The human herpes virus 8-encoded viral FLICE inhibitory protein physically associates with and persistently activates the Ikappa B kinase complex. J Biol Chem 277:13745–13751
Liu ZG, Hsu H, Goeddel DV, et al. (1996) Dissection of TNF receptor 1 effector functions: JNK activation is not linked to apoptosis while NF-kappaB activation prevents cell death. Cell 87:565–576
Locati M, Murphy PM (1999) Chemokines and chemokine receptors: biology and clinical relevance in inflammation and AIDS. Annu Rev Med 50:425–440
Ma Y, Hendershot LM (2001) The unfolding tale of the unfolded protein response. Cell 107:827–830
Martinou JC, Green DR (2001) Breaking the mitochondrial barrier. Nat Rev Mol Cell Biol 2:63–67
Mattson MP, Chan SL, Camandola S (2001) Presenilin mutations and calcium signaling defects in the nervous and immune systems. Bioessays 23:733–744
Means RE, Lang SM, Chung YH, et al. (2002) Kaposi's sarcoma associated herpesvirus immune evasion strategies. Front Biosci 7:e185–e203
Micheau O, Lens S, Gaide O, et al. (2001) NF-kappaB signals induce the expression of c-FLIP. Mol Cell Biol 21:5299–5305
Miyashita T, Reed JC (1995) Tumor suppressor p53 is a direct transcriptional activator of the human bax gene. Cell 80:293–299
Molden J, Chang Y, You Y, et al. (1997) A Kaposi's sarcoma-associated herpesvirus-encoded cytokine homolog (vIL-6) activates signaling through the shared gp130 receptor subunit. J Biol Chem 272:19625–19631
Moore PS, Boshoff C, Weiss RA, et al. (1996) Molecular mimicry of human cytokine and cytokine response pathway genes by KSHV. Science 274:1739–1744
Murdoch C, Finn A (2000) Chemokine receptors and their role in inflammation and infectious diseases. Blood 95:3032–3043
Muzio M, Chinnaiyan AM, Kischkel FC, et al. (1996) FLICE, a novel FADD-homologous ICE/CED-3-like protease, is recruited to the CD95 (Fas/APO-1) death-inducing signaling complex. Cell 85:817–827
Nakamura H, Li M, Zarycki J, et al. (2001) Inhibition of p53 tumor suppressor by viral interferon regulatory factor. J Virol 75:7572–7582
Nava VE, Cheng EH, Veliuona M, et al. (1997) Herpesvirus saimiri encodes a functional homolog of the human bcl-2 oncogene. J Virol 71:4118–4122
Ojala PM, Yamamoto K, Castanos-Velez E, et al. (2000) The apoptotic v-cyclin-CDK6 complex phosphorylates and inactivates Bcl-2. Nat Cell Biol 2:819–825
Oltvai ZN, Milliman CL, Korsmeyer SJ (1993) Bcl-2 heterodimerizes in vivo with a conserved homolog, Bax, that accelerates programmed cell death. Cell 74:609–619
Pinton P, Ferrari D, Rapizzi E, et al. (2001) The Ca2+ concentration of the endoplasmic reticulum is a key determinant of ceramide-induced apoptosis: significance for the molecular mechanism of Bcl-2 action. EMBO J 20:2690–2701
Rivas C, Thlick AE, Parravicini C, et al. (2001) Kaposi's sarcoma-associated herpesvirus LANA2 is a B-cell-specific latent viral protein that inhibits p53. J Virol 75:429–438
Roy DJ, Ebrahimi BC, Dutia BM, et al. (2000) Murine gammaherpesvirus M11 gene product inhibits apoptosis and is expressed during virus persistence. Arch Virol 145:2411–2420

Sarid R, Sato T, Bohenzky RA, et al. (1997) Kaposi's sarcoma-associated herpesvirus encodes a functional bcl-2 homologue. Nat Med 3:293–298
Sarid R, Olsen SJ, Moore PS (1999) Kaposi's sarcoma-associated herpesvirus: epidemiology, virology, and molecular biology. Adv Virus Res 52:139–232
Sax JK, Fei P, Murphy ME, et al. (2002) BID regulation by p53 contributes to chemosensitivity. Nat Cell Biol 4:842–849
Scheffner M, Werness BA, Huibregtse JM, et al. (1990) The E6 oncoprotein encoded by human papillomavirus types 16 and 18 promotes the degradation of p53. Cell 63:1129–1136
Sharpless NE, DePinho RA (2002) p53: good cop/bad cop. Cell 110:9–12
Sherr CJ, Weber JD (2000) The ARF/p53 pathway. Curr Opin Genet Dev 10:94–99
Soulier J, Grollet L, Oksenhendler E, et al. (1995) Kaposi's sarcoma-associated herpesvirus-like DNA sequences in multicentric Castleman's disease. Blood 860:1276–1280
Stine JT, Wood C, Hill M, et al. (2000) KSHV-encoded CC chemokine vMIP-III is a CCR4 agonist, stimulates angiogenesis, and selectively chemoattracts TH2 cells. Blood 95:1151–1157
Sun Q, Matta H, Chaudhary PM (2002) The human herpes virus 8 encoded viral FLICE inhibitory protein protects against growth factor withdrawal-induced apoptosis via NF-{kappa}B activation. Blood 24:24
Susin SA, Lorenzo HK, Zamzami N, et al. (1999) Molecular characterization of mitochondrial apoptosis-inducing factor. Nature 397:441–446
Suzuki M, Youle RJ, Tjandra N (2000) Structure of Bax: coregulation of dimer formation and intracellular localization. Cell 103:645–654
Tapon N, Harvey KF, Bell DW, et al. (2002) Salvador promotes both cell cycle exit and apoptosis in *Drosophila* and is mutated in human cancer cell lines. Cell 110:467–478
Thome M, Schneider P, Hofmann K, et al. (1997) Viral FLICE-inhibitory proteins (FLIPs) prevent apoptosis induced by death receptors. Nature 386:517–521
Vanden Abeele F, Skryma R, Shuba Y, et al. (2002) Bcl-2-dependent modulation of Ca(2+) homeostasis and store-operated channels in prostate cancer cells. Cancer Cell 1:169–179
Virgin HWt, Latreille P, Wamsley P, et al. (1997) Complete sequence and genomic analysis of murine gammaherpesvirus 68. J Virol 71:5894–5904
Virgin HWt, Presti RM, Li XY, et al. (1999) Three distinct regions of the murine gammaherpesvirus 68 genome are transcriptionally active in latently infected mice. J Virol 73:2321–2332
Wang GH, Garvey TL, Cohen JI (1999) The murine gammaherpesvirus-68 M11 protein inhibits Fas- and TNF-induced apoptosis. J Gen Virol 80:2737–2740
Wang HW, Sharp TV, Koumi A, et al. (2002) Characterization of an anti-apoptotic glycoprotein encoded by Kaposi's sarcoma-associated herpesvirus which resembles a spliced variant of human survivin. EMBO J 21:2602–2615
Yu J, Zhang L, Hwang PM, et al. (2001) PUMA induces the rapid apoptosis of colorectal cancer cells. Mol Cell 7:673–682

Exploitation of Cell Cycle and Cell Death Controls by Adenoviruses: The Road to a Productive Infection

I. Alasdair Russell[1], J.A. Royds[1] and A.W. Braithwaite[1]

1 Introduction

Adenoviruses have made enormous contributions to our understanding of cell cycle regulation and apoptosis in eukaryotes. Moreover, by virtue of their ability to deregulate these processes, they have contributed significantly to our knowledge of the underlying factors governing oncogenesis. In recent years, their recognized potential as therapeutic agents has led to a dramatic resurgence of interest in Adenoviruses. However, with this has come the realization that knowledge of the basic molecular biology governing adenoviral interactions with the host cell is of vital importance. In this review we aim to address two aspects of Adenovirus/host interactions, those that concern the control of cell cycle progression and those that pertain to the modulation of cell death. We aim to approach this in a teleological manner, focusing upon obstacles presented to the virus throughout its life cycle, thus giving a clear perspective as to why important cellular control mechanisms are altered or subverted by the virus. We therefore begin this chapter with a brief outline of Adenovirus classification, morphology and genome organization, followed by an overview of the major barriers Adenovirus must overcome for successful viral replication. This leads on to an in-depth review of Adenovirus-mediated subversion of cell cycle controls and the perturbation of cell death regulators.

2 Adenovirus and Its Genome

There are at least 49 different Adenovirus (Ad) serotypes which infect man (Shenk 1996; http://www.ncbi.nlm.nih.gov/Taxonomy). The serotypes in turn have been classified into six subgroups (A to F) based on a variety of properties including, for example, their ability to form tumors in newborn rodents. This review will deal mainly with the human group C Ads 2 and 5 (Ad2/5) as the vast majority of literature focuses upon these. Where the data exist, ortholo-

[1]Cell Transformation Group, Department of Pathology, Dunedin School of Medicine, University of Otago, Box 913, Dunedin, New Zealand, e-mail: antony.braithwaite@stonebow.otago.ac.nz

Progress in Molecular and Subcellular Biology
C. Alonso (Ed.): Viruses and Apoptosis
© Springer-Verlag Berlin Heidelberg 2004

gous proteins of other serotypes have been shown to perform equivalent functions to those from Ad2/5.

The Ad virion itself is a non-enveloped icosahedron (70–80 nm in diameter) comprised of 252 capsomeres, which includes 240 capsid hexons and 12 vertex pentons (Fig. 1, inset). Glycoprotein fiber proteins that mediate host cell attachment project from each of the vertex pentons.

The Ad genome is a linear, double-stranded DNA molecule ranging in size from approximately 34 to 36 kb for human Ads, with 100–140 bp inverted terminal repeats that serve as origins of DNA replication. Covalently bound to the 5′ ends are two terminal proteins that serve as primases for DNA replication. The genome is organized into eight discrete transcription units, with each unit giving rise to multiple protein products through alternate splicing and alternative polyadenylation sites (Fig. 2); both of these phenomena were first discovered in Ads (Berget et al. 1977; Chow et al. 1977; Klessig 1977). Transcription involves RNA polymerase II activity, although RNA polymerase III

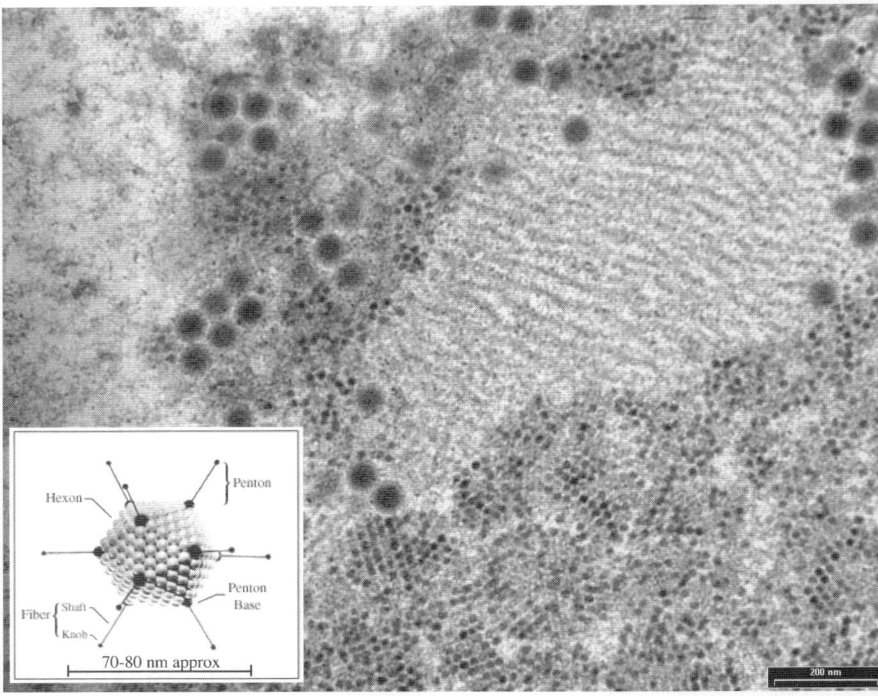

Fig. 1. Electron micrograph of Ad5-infected thyroid cells. Electron micrograph of K1scx cells (thyroid papillary carcinoma expressing a dominant negative mutant p53; Wyllie et al. 1999) infected with Ad5 at 48 h post-infection, contrasted with uranyl acetate and lead citrate. Mature virions can be seen, along with crystalline arrays of hexon capsid protein. *Bar* 200 nm. *Inset* Diagrammatic representation of Ad2/5 with appropriate structures labeled and scale bar (inset adapted from http://hsc.virginia.edu/med-ed/micro/images/viroll/14.jpg)

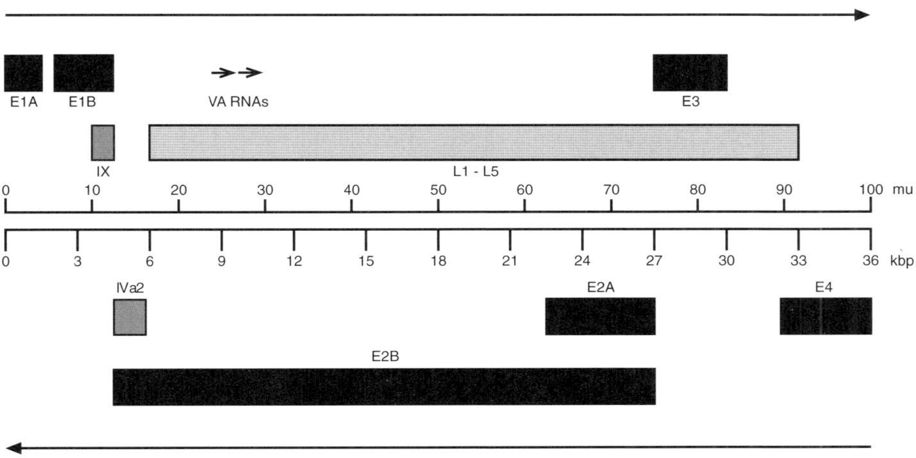

Fig. 2. Schematic representation of the Adenovirus genome. Diagrammatic illustration of the Ad2/5 genome as it is conventionally drawn. The double-stranded genome is represented in both map units (mu) and kilobase pairs (kbp). The *uppermost* and *lowermost arrows* indicate the direction of transcription. The relative positions of coding regions are represented by *shaded boxes*: *black boxes* represent early transcription units; the *dark grey boxes* represent the delayed early polypeptide IX and IVa2 genes; and the *light grey box* represents the late transcription unit. Virus-associated (VA) RNAs are represented by *single lines with small arrowheads*. (Adapted from Braithwaite and Russell 2001)

is necessary for transcription of the virus-associated (VA) RNAs. Historically, these viral transcription units have been divided into early and late genes with respect to the onset of viral DNA replication, although a significant amount of overlap does occur. Each transcription unit appears to have developed through evolution to encode proteins of similar function, with the exception of the E4 region. Briefly, products of the E1 region are primarily concerned with transactivating early viral genes and preparing the cell for DNA synthesis; the E2 region encodes proteins involved directly in the replication of the viral genome, such as the DNA polymerase, primase and DNA binding proteins; products of the E3 region have been shown to play a role in avoiding immunosurveillance of the host; however, in contrast, proteins originating from the E4 region have been implicated in a wide spectrum of interactions with the host, including host cell shutoff, mRNA transport and transcriptional regulation; and finally, proteins from the L1–5 region are principally involved in the structural makeup and assembly of the virion.

The E1, E2 and E3 regions are further broken down into smaller transcription units (E1a, E1b, E2a, E2b, E3a and E3b); however, the E2 and E3 regions are not the main focus of this chapter and thus there will be no additional information regarding their genome organization given. The E1a and E1b transcription units give rise to five detectable protein products each (Boulanger and Blair 1991) that arise by alternate splicing (Fig. 3). Of these, the two most commonly studied E1a proteins are the 243 amino acid residue (R) and 289R species. These are historically referred to by the sedimentation coefficients of their respective mRNA species (12S and 13S). The two products are identical but for an internal stretch of 46 residues, termed the unique region, which is absent in the 243R product (Fig. 3).

Detailed comparison of several Ad serotypes differing in divergence led to the identification of areas of high conservation within the E1a products (van Ormondt et al. 1980; Kimelman et al. 1985; Kimelman 1986). These were termed conserved region (CR) 1, 2 and 3. Recently, a fourth region, CR4, has been described. This lies at the C-terminus of the major E1a proteins and incorporates the extreme C-terminal nuclear localization signal (NLS; Avvakumov et al. 2002; Fig. 3).

Functional assignments were made to these conserved regions through a series of experiments utilizing viruses with mutations in the *E1a* coding region. For example, regions involved in the transactivation of other early viral genes were mapped, although not exclusively, to CR3 (Carlock and Jones 1981; Moran et al. 1986; Zerler et al. 1987), and these were shown to be distinct from those important for DNA synthesis induction (Haley et al. 1984; Moran et al. 1986).

The best-described protein products of the E1b transcription unit are 55- and 19-kDa products which arise from alternate reading frames (these proteins will be discussed in greater detail below). Other transcription units will be discussed in greater detail where necessary. For an excellent review of virion structure and genome organization, see Shenk (1996).

The remainder of this chapter is divided into two major components: an outline of the barriers presented to Ad through the course of a productive infection, and an historical timeline and discussion of current data concerning the interplay between Ad, cell cycle and cell death regulation.

Fig. 3. Expanded Ad5 E1 region showing protein-binding sites. This is an expanded cartoon of the Ad5 E1 region showing the known splice products generated by each transcription unit. In addition, for the better-studied Ad5 E1 products, respective protein binding sites are shown. Coding regions are represented by *boxes* and spliced regions by *interconnecting single lines*. Direction of transcription is in the rightward direction (see *upper arrow*). The 289-residue (13S) E1a protein has been exploded indicating the four conserved regions (CR1–4), two exons and unique region. The binding sites for a selection of known E1a-interacting cellular factors are indicated below. The E1b55- and E1b19-kDa proteins have also been exploded with the putative binding sites for their respective interacting proteins indicated. *Open empty boxes* represent poorly defined or putative binding sites. See text for details. *NES* Nuclear export signal; *RNP* ribonucleoprotein motif; *Zn Finger* putative C_2H_2 zinc finger motif. (Adapted from Braithwaite and Russell 2001; Gallimore and Turnell 2001; Ben-Israel and Kleinberger 2002)

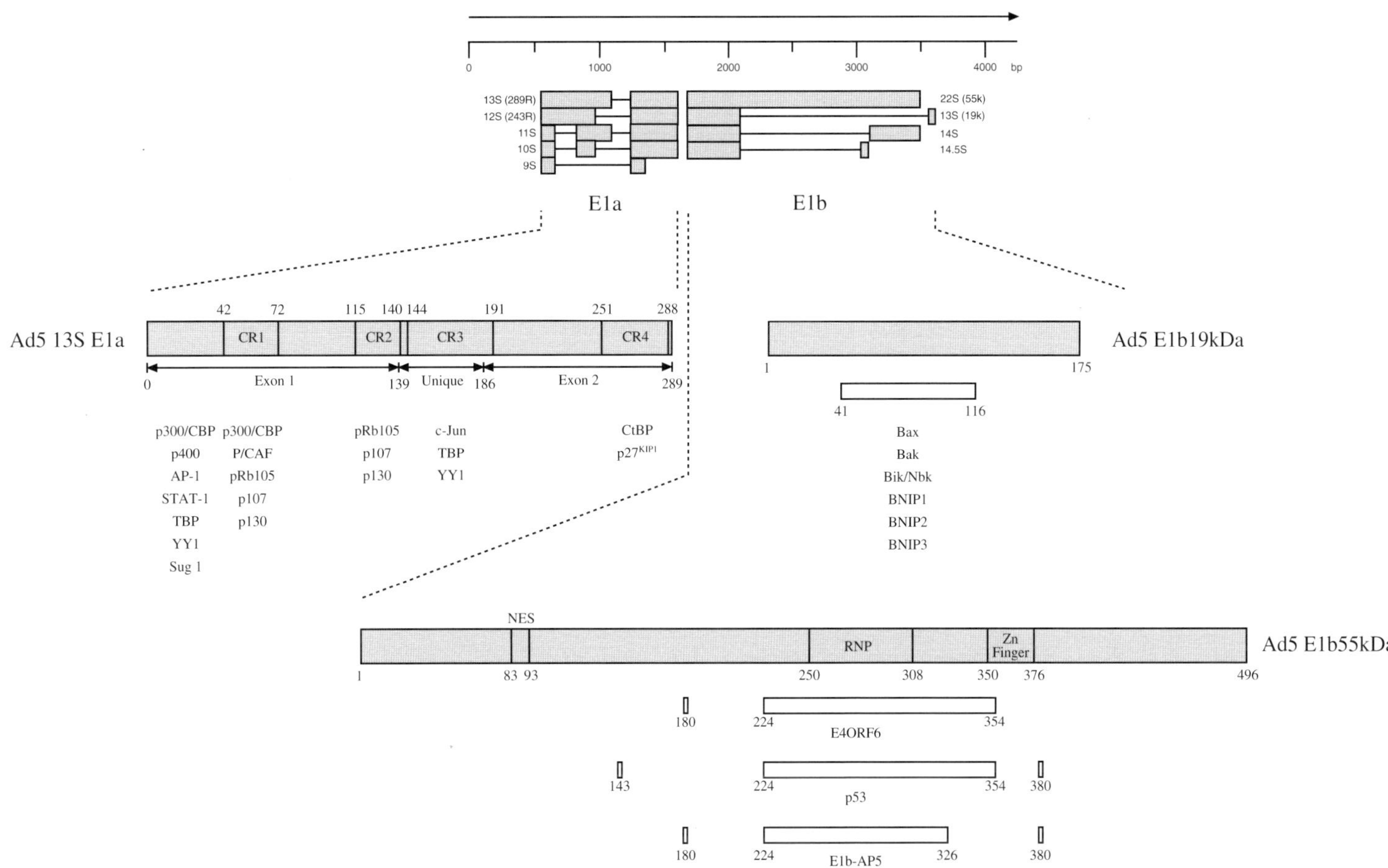

0
1000
2000
3000
4000
bp
13S (289R)
12S (243R)
11S
10S
9S
22S (55k)
13S (19k)
14S
14.5S
E1a
E1b
Ad5 13S E1a
42
72
115
140
144
191
251
288
CR1
CR2
CR3
CR4
0
Exon 1
139
Unique
186
Exon 2
289
p300/CBP
p400
AP-1
STAT-1
TBP
YY1
Sug 1
p300/CBP
P/CAF
pRb105
p107
p130
pRb105
p107
p130
c-Jun
TBP
YY1
CtBP
p27KIP1
Ad5 E1b19kDa
1
175
41
116
Bax
Bak
Bik/Nbk
BNIP1
BNIP2
BNIP3
NES
RNP
Zn Finger
Ad5 E1b55kDa
1
83
93
250
308
350
376
496
180
224
354
E4ORF6
143
224
354
380
p53
180
224
326
380
E1b-AP5

3 Barriers to a Productive Adenovirus Infection

From a viral point of view, the infection process represents a sequence of hurdles or barriers that must be negotiated. How Ad deals with each barrier determines whether the outcome of a particular infection will be successful, producing maximum progeny in an efficient manner, or abortive, with significantly reduced or no virus produced. These barriers are outlined below.

3.1 Binding to the Host Cell Membrane

The first barrier Ad is presented with is the attachment to an appropriate host cell. In Ad2/5 infection, the fiber knobs (Fig. 1, inset, and Fig. 4) are thought to mediate this attachment through binding to a member of the immunoglobulin superfamily, Coxsackie/Adenovirus receptor (CAR; Bergelson et al. 1997; Tomko et al. 1997; Carson 2001). Although the majority of data attest to the importance of CAR in attachment of most subgroups (Roelvink et al. 1998; Hutchin et al. 2000; unpubl. data from our laboratory), this is not the case for group B human Ads (Roelvink et al. 1998), and has not yet been demonstrated in non-human Ads.

3.2 Entry into the Cell

Following attachment Ad must gain entry into the cell. This internalization process is distinct from the attachment described above and involves the interaction of a protruding RGD motif on the penton base (Fig. 1, inset) with specific integrins in the membrane, and subsequent entry via the clathrin pit pathway (Stewart et al. 1997; Meier et al. 2002; Fig. 4). Endocytosis stimulated from this binding has been attributed to activation of the phosphatidylinositol-3-OH kinase (PI3 K) signalling pathway, which in turn leads to massive reorganization of actin polymers (Li et al. 1998; Rauma et al. 1999). Once bound to the integrins, and after stimulation of endocytosis, most of the fiber proteins are shed at the cell surface, with the remainder removed in the early endosome (Greber et al. 1993; Nakano et al. 2000).

3.3 Escaping the Endosome, Viral Uncoating and Nuclear Transport

For the infection process to proceed, it is imperative that the virus escapes the endosome prior to lysosome formation, although the mechanism by which Ad

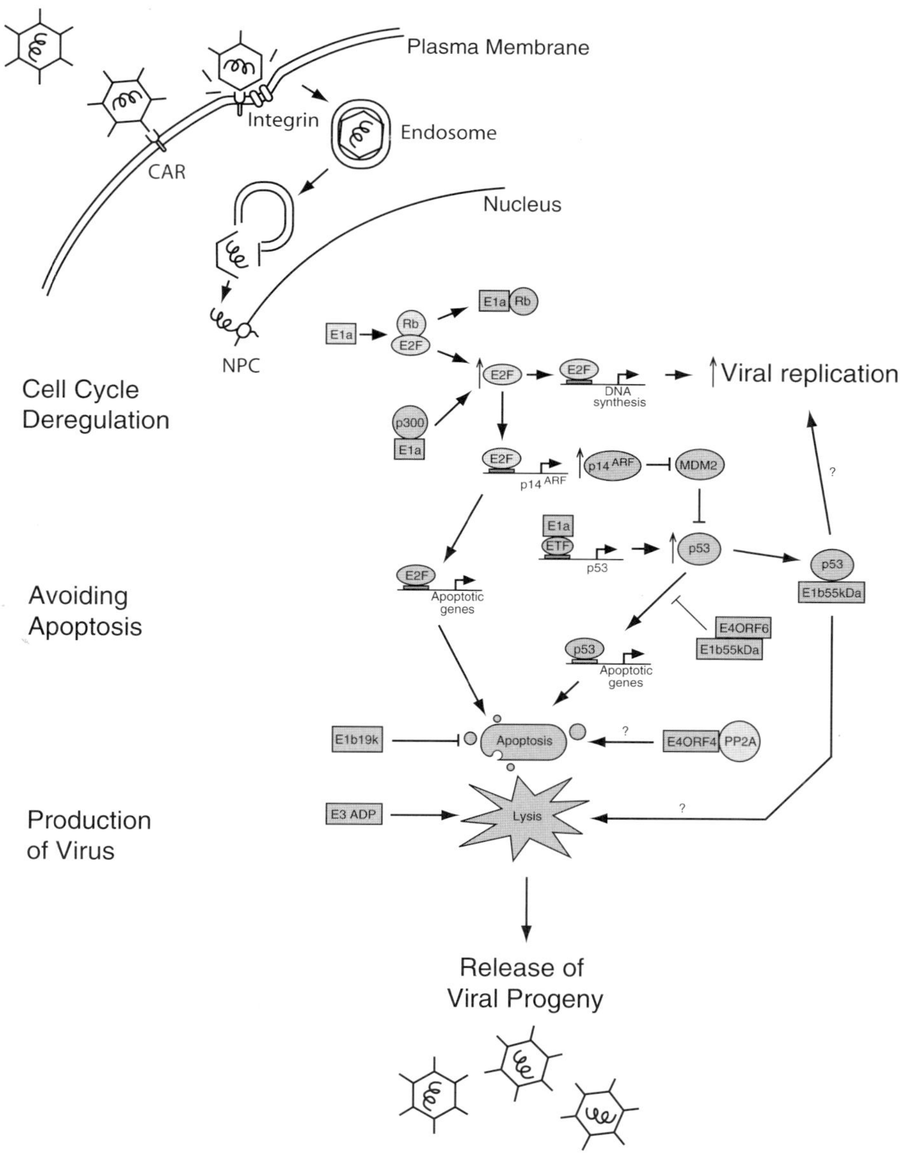

Fig. 4. A productive Adenovirus infection. This diagram represents a simplified overview of the interaction between cell cycle and cell death controls during a productive Adenovirus infection. Viral gene products are indicated by *shaded boxes* while cellular factors are shown by *elliptical shapes*. See text for details. *CAR* Coxsackie/Adenovirus receptor; *NPC* nuclear pore complex; *?* areas of conflicting data. (Adapted from Braithwaite and Russell 2001)

escapes is poorly understood. Recent evidence points to a stepwise disassembly of the virion capsid as playing an essential role in this escape (Greber et al. 1993, 1996; Nakano et al. 2000; Fig. 4). This is thought to be mediated by the Ad-encoded cysteine protease (L3 p23). It has been suggested the shedding of fibers early in infection may allow the penton base access to specific integrins, thereby facilitating penetration of the endosome (Greber et al. 1993; Wickham et al. 1993). More recently, however, binding to the integrin receptors has been shown to increase the rate of macropinocytosis within the cell (Meier et al. 2002). The resultant pinocytes lose 30–50% of their endosomal contents into the cytoplasm, and this has been shown to correlate with viral release into the cytoplasm (Meier et al. 2002).

A continued, ordered disassembly occurs in the cytoplasm (Greber et al. 1993; Greber 1998) as the virion is transported to the nucleus via microtubules (Dales and Chardonnet 1973), and is completed at nuclear pore complexes (NPCs; Greber et al. 1997). The suggestion is that the NPC acts as a regulator of nucleo-cytoplasmic transport, and that the 5′ covalently bound Ad terminal protein may play a role in threading the DNA through the nuclear pore by way of its intrinsic nuclear localization signal (NLS; Zhao and Padmanabhan 1988; Fig. 4). Once inside the nucleus, the Ad genome associates with the nuclear matrix through this terminal protein and awaits transcription (Schaack et al. 1990).

3.4 Early Gene Transcription

The next step in a productive viral lytic cycle is for Ad to express its early genes. The immediate early E1a gene products are the first Ad proteins to be expressed, and are transactivated via a constitutively active cellular factor (Reichel et al. 1988). The major splice variants from the *E1a* gene, 12S and 13S, have been shown to subsequently transactivate Ad's early genes. The role of early genes is generally perceived as being threefold: to create an environment within the cell that is conducive to viral replication by liberating factors and machinery involved in S-phase; to combat death signals that result from the illegitimate push into S-phase, and the host antiviral response; and, finally, to provide the necessary viral replication factors for viral DNA synthesis.

3.5 Countering the Host Antiviral Response

In addition to stimulating early gene transcription, E1a is also responsible for sensitizing the cell to immunological surveillance, the perturbation of which is necessary for establishing a productive infection in the host. Only a brief outline on the mechanisms by which Ad escapes the host immune surveillance will be described here, as it is not the main focus of this chapter (for recent reviews on this subject, see Horwitz 2001; Burgert et al. 2002).

The E3 region encodes at least five proteins that are involved in modulating the host immune response (10.4, 14.5, 14.7, 6.7 and gp19 kDa). The 10.4- and 14.5-kDa proteins exist as a heterodimer, also termed receptor internalization and degradation (RID) complex, which downregulates both cytotoxic T lymphocyte (CTL) activity (Zhang et al. 1991, 1994), and death-receptor-induced apoptosis. The latter is achieved by stimulating endocytosis and degradation of Fas (Shisler et al. 1997; Elsing and Burgert 1998), epidermal growth factor receptor (Tollefson et al. 1991), and tumor necrosis factor (TNF) receptor (Gooding et al. 1991). In addition, the RID complex subverts TNF-induced apoptosis by preventing arachidonic acid (AA) release (Krajcsi et al. 1996), translocation of cytoplasmic phospholipase A_2 to the membrane (Dimitrov et al. 1997), and importantly by inhibiting NF-κB activation (Friedman and Horwitz 2002). It has also been reported that 14.7 kDa can inhibit apoptosis initiated through Fas (Chen et al. 1998), TNF (Gooding et al. 1988) and AA release (Krajcsi et al. 1996). It has been shown that 6.7 kDa acts as an accessory to the RID complex in downregulating TNF-related apoptosis-inducing ligand (TRAIL) receptors (Benedict et al. 2000; Tollefson et al. 2001). More recently, it has been shown to act independently of other E3 proteins to downregulate death-receptor-induced apoptosis and AA release (Moise et al. 2002). The remaining E3 protein, gp19 kDa, is responsible for protecting infected cells against an Ad-specific CTL response. It achieves this in two ways: by binding specific allelotypes of the major histocompatibility complex (MHC) class I antigens and retaining them in the endoplasmic reticulum (ER) lumen (Andersson et al. 1985; Burgert and Kvist 1985; Zhang et al. 1991), and by interfering with the transport of Ad peptides to the maturing MHC class I heterodimer (Bennett et al. 1999). The net effect is to prevent antigen presentation at the cell surface, and subsequent CTL response (Andersson et al. 1987). The combined expression of these five E3 proteins serves to evade the sentinel immune response, thus allowing a productive infection to proceed.

3.6 Late Gene Transcription and Host Cell Shutdown

After the onset of viral DNA synthesis late genes are transcribed from the major late promoter (Shaw and Ziff 1980), producing 18 distinct splice variants, which are further grouped into 5 families (L1–L5; Nevins and Darnell 1978; Ziff and Evans 1978). The majority of late proteins are involved in the capsid makeup and assembly of mature virions.

Although cellular mRNA production still occurs, Ad monopolizes the ribosomes for its own means firstly by blocking the transport of cellular mRNA into the cytoplasm (Beltz and Flint 1979), and subsequently by preferentially translating its own mRNA (Babich et al. 1983; Yoder et al. 1983; Yang et al.

2002). Two Ad2/5 early gene products E1b55 kDa (Babiss and Ginsberg 1984; Pilder et al. 1986) and E4ORF6 (Halbert et al. 1985; Weinberg and Ketner 1986), which exist in a complex (Sarnow et al. 1984), are implicated in this blocking function (Gonzalez and Flint 2002), and are thought to act by hijacking a presently unknown cellular factor important in transporting mRNA to the nuclear pore (Ornelles and Shenk 1991). This inhibition of cellular mRNA accumulation is accompanied by a shutoff of host cell protein synthesis, which is also mediated by the E1b55 kDa protein (Babich et al. 1983; Babiss and Ginsberg 1984), and recent evidence suggests a possible involvement of the tumor suppressor, p53 (Ridgway et al. 1997).

3.7
Packaging and Assembly

Once sufficient quantities of late proteins have accumulated both structural antigen and non-structural virion components, the assembly process begins. Assembly of virion progeny occurs in the nucleus (Fig. 1), although details of the mechanisms involved in the assembly process remain controversial. It is generally accepted that empty capsid shells are constructed first and the Ad DNA is subsequently packaged and condensed into these shells (for reviews on this subject, see D'Halluin 1995; Schmid and Hearing 1995; Greber 1998). Interestingly, at late stages in infection, the major coat component makes up around 10% of the total protein content of the cell.

3.8
Escape from the Nucleus and the Cell

As large numbers of virion progeny increase within the cell, Ads are confronted with the problem of gaining release. Ads are generally thought to be released in a lytic manner. The only implicated gene product in lysis is the Ad E3 11.6 kDa ADP (Ad death protein; Tollefson et al. 1996a,b). Ad mutants lacking the ADP coding region were significantly delayed in release of progeny from infected cells, and nuclei appeared swollen and full of virus (Tollefson et al. 1996b). However, unpublished data from our laboratory suggest otherwise. In our hands, a mutant virus missing the majority of the E3 region, including the *ADP* coding sequence (*dl*327), lyses a range of cell types with wtAd5-like efficiency, suggesting there may be additional routes for escape from the cell. Overcoming this final barrier completes the Ad infectious cycle, and allows for dissemination of progeny to neighboring cells or through aerosol transmission to another host (Fig. 4).

It is clearly evident from the information above that the goal of Ad is not only to simply overcome each barrier, but also to manipulate the cell to its own end as it does so. A crucial component of this manipulation is the alteration of the cell cycle, which is assumed to provide the necessary substrates for replication of the viral genome. Paradoxically, this in turn leads to the induction of cell death. Ad, however, has evolved several approaches to counter this. This interplay between Ad and the regulation of the cell cycle and cell death controls represents the main focus of the following sections.

4 Adenovirus and the Cell Cycle – Defining Milestones

4.1 DNA Synthesis is Stimulated upon Adenovirus Infection

The first data indicating that Ads could interact with cell cycle machinery, and thus induce DNA synthesis, came in 1966 from Takahashi et al. who demonstrated an increase in the number of hamster cells (BHK cells) incorporating ^{3}H-thymidine after Ad12 infection. These observations were consolidated by subsequent reports supporting these findings (Shimojo and Yamashita 1968; Strohl 1973; Harris and Strohl 1980). This research came on the heels of a raft of observations demonstrating a similar induction of DNA synthesis after infection with other DNA viruses such as polyomavirus (Winocour et al. 1965; Sheinin 1966) and simian virus 40 (SV40; Henry et al. 1966). Interest in the induction of cellular DNA synthesis, however, was primarily fuelled by Trentin's discovery in 1962 that Ad12 was able to induce malignant tumors in newborn hamsters (Trentin et al. 1962). This led many investigators to study the effects of Ads on cell cycle regulation, resulting in a rapid shift in Ad research from a virological to a cancer-based focus, and subsequently introducing a "bias" towards systems in which Ad was shown to induce cellular DNA synthesis.

Following these initial observations, Ad was subsequently shown to induce DNA synthesis in BHK cells for a range of serum conditions, and even in its complete absence (Strohl 1969, 1973). This newly synthesized DNA was shown by several different methodologies to be cellular in nature, with little or no evidence of viral DNA (Shimojo and Yamashita 1968; Doerfler 1969; Takahashi et al. 1969). Later, in a variety of human cells and using an array of Ad serotypes, this induction of cellular DNA synthesis was shown to be transient, and followed by a rapid switch to viral DNA synthesis (Hodge and Scharff 1969; Yamashita and Shimojo 1969; Laughlin and Strohl 1976). This apparent discrepancy in the data can be explained by the permissiveness of a given cell to infection by Ad. In a non-permissive infection, as is the case of Ad12 infection of hamster cells, cellular DNA synthesis is stimulated in the absence of viral DNA synthesis. However, in a permissive system, only a transient induction of DNA synthesis exists, which is followed by a rapid switch to viral DNA synthesis. This switch is due to an as yet undefined interaction that occurs in the permissive system. Thus, Ad is able to induce cellular DNA synthesis in both systems, albeit transient in the permissive setting.

In 1981, with the advent of flow cytometry, it was demonstrated that Ad could induce a complete round of DNA replication (Braithwaite et al. 1981). Until this time, incorporation of ^{3}H-thymidine and mitotic indices allowed investigators only to assume that Ad was pushing cells into S-phase.

These findings collectively demonstrated that Ad was able to induce S-phase in situations where this would normally not occur.

4.2 Adenovirus Does Not Progress into S-Phase in a Normal Manner

Although initial observations suggested Ad infection resulted in a normal progression into S-phase, stimulating thymidine kinase and DNA polymerase (Zimmerman et al. 1970), the data soon revealed that this was not the case. In contrast to serum stimulation, Ad was shown to push cells into S-phase in the absence of ribosomal RNA stimulation (Laughlin and Strohl 1976), ornithine decarboxylase induction (Cheetham and Bellett 1982), polyamine accumulation (Cheetham and Bellett 1982; Cheetham et al. 1982), and avoid blockage by dibutyryl cyclic AMP (Braithwaite et al. 1981). In addition to this, Ad was shown to activate only a subset of those cell cycle regulatory genes that are activated by serum (Liu et al. 1985).

In a different set of experiments, Ad2 was shown to induce cellular DNA synthesis at the non-permissive temperature in a panel of BHK-derived cell lines harboring an array of temperature-sensitive (ts) mutations that retard cell cycle progression (Nishimoto et al. 1975; Rossini et al. 1981). One such cell line (tsAF8) harbored a specific block in the transition through G_1 (Nishimoto et al. 1975; Liu et al. 1985), which was later mapped to the catalytic subunit of RNA polymerase II, resulting in decreased activity of the enzyme (Rossini et al. 1980; Shales et al. 1980; Sugaya et al. 2001). This added to the previous data, attesting to the ability of Ad to overcome multiple blocks in the progression through G_1.

From the above data it was apparent that Ad was able to induce DNA synthesis by a mechanism that was distinct from serum, and had evolved to efficiently bypass the normal sequence of events involved in G_1 to S-phase progression. This suggestion was given further weight still by Murray et al. (1982a) who observed that progression times through G_1 to S-phase were significantly shorter in Ad infection (0.5 h) than in serum stimulation (5 h).

4.3 The E1a Gene Products Are Implicated

That Ad could induce cellular DNA synthesis in cells without replicating its own genome (Shimojo and Yamashita 1968; Doerfler 1969; Takahashi et al. 1969), coupled with the finding that incorporation of ^{3}H-thymidine was only seen in cells that expressed so-called "Tumor (T)-antigens" (a marker associated with expression of early virus protein; Strohl 1969), implicated Ad early genes as being responsible for the observed cell cycle deregulation.

After analysis of Ad2 DNA fragments in rat cells, Gallimore (1974) demonstrated the viral genes responsible for transformation were located to the left-hand side (LHS) of the viral genome. In 1977, by exposing human endothelial kidney cells to fractionated Ad5 DNA, Graham et al. (1977) developed the transformed 293 cell line. This not only supported Gallimore's findings implicating the LHS of the Ad genome, but also allowed for the propagation of Ad mutants and thus allowed for their use as tools. The arrival of restriction digest mapping and Southern blot analysis (Southern 1975) allowed these early regions to be characterized and subsequent identification of their respective protein products (Lewis et al. 1976; Halbert et al. 1979).

Viral mutants created by Jones and Shenk (1979) along with other groups were used to show for the first time that early viral proteins, derived from the E1 and E2 regions, were indeed responsible for cellular DNA synthesis induction (Rossini et al. 1981; Shiroki et al. 1981). These findings were subsequently extended, refining the responsible regions initially by the exclusion of the E2 region (Murray et al. 1982b), and further by attributing this function to the E1a region through the demonstration that Ad mutants lacking functional E1a products were defective at inducing DNA synthesis, and that this was not dependent on the ability of E1a to transactivate other early viral genes (Braithwaite et al. 1983; Bellett et al. 1985). Furthermore, the responsible region was mapped to the 289R (13S) transcript of *E1a* (Braithwaite et al. 1983), although the 243R (12S) product was subsequently shown to also play a role (Bellett et al. 1985). Later, Spindler and colleagues (1985) confirmed this work, implicating the E1a region as being responsible, using a variety of human cell lines. Interestingly, Spindler concluded the 12S protein was the responsible element, although the data presented indicate the importance of one transcript over another may well be cell context dependent. Indeed, recently, the respective roles of either the 12S or 13S product *in vivo* have been shown to be heavily dependent on the host cell context (Yang et al. 2002). Also in 1985, Stabel and colleagues neatly complemented the virus work reported above using the corresponding plasmid-based experiments (Stabel et al. 1985). Stabel showed that microinjection of *E1a* into a growth-arrested fibroblastic mouse cell line also resulted in an induction of cellular DNA synthesis. Later, Zerler et al. (1987) showed E1a induces cellular DNA synthesis with no distinction between transcripts, although Quinlan and Grodzicker (1987) demonstrated that infection with an Ad5 mutant expressing only the 12S product was sufficient by itself for DNA synthesis induction. However, the induction of DNA synthesis observed with the 12S product was less than that of the infection control expressing both 12S and 13S products, and no direct comparison was made with the 13S product in this case.

4.4
Summary

Collective interpretation of this data would suggest a model whereby both 12S and 13S Ad2/5 E1a products are able to enact a complete bypass of G_1 and induce subsequent cellular DNA synthesis. Although, neither transcript is sufficient in itself to account for the full induction seen with wtAd5 (Bellett et al. 1985; Spindler et al. 1985; Quinlan and Grodzicker 1987), and the relevance of either the 12S or 13S transcript may indeed depend on the cellular context in which infection takes place (Spindler et al. 1985; Yang et al. 2002).

5
Adenovirus and the Cell Cycle

5.1
Implicating Cellular Counterparts

A range of cellular proteins, known at the time only by their molecular masses, were shown by several groups to bind the E1a proteins (Yee and Branton 1985; Harlow et al. 1986). The most studied of these were polypeptides of 105, 107 and 300 kDa. Mutation of the binding sites for these cellular polypeptides destroyed the ability of Ad to induce DNA synthesis in rat cells, suggesting critical cell cycle regulatory roles for these proteins (Whyte et al. 1988b). Using an extensive panel of *E1a* mutant viruses and immunoprecipitating the complexes formed between E1a major protein products and these cellular proteins, the responsible regions were initially mapped (Egan et al. 1988), and further refined (Whyte et al. 1989; Fig. 3).

5.2
Involvement of pRb

In 1988 the first real insight was gained into the molecular events surrounding Ad cell cycle regulation. Whyte et al. (1988a) purified E1a/cellular protein complexes from 293 cells using E1a antibodies and reacted them with anti-105 kDa antibodies, which they subsequently showed to specifically recognize the retinoblastoma susceptibility product, pRb. These findings represented a defining moment in cancer and cell cycle biology, as this was the first observation of an oncogene directly binding a tumor suppressor. pRb was the best-studied tumor suppressor of that time, being associated with tight regulation or inhibition of cellular proliferation. Whyte postulated that overcoming the growth inhibitory effects of pRb may represent a fundamental step towards a productive infection, and that the mechanism by which Ad may achieve this is by direct binding of pRb with the E1a protein. The other E1a-bound proteins have subsequently been identified as the pRb105 related family member p107, and the histone acetyltransferase p300/CBP (Ben-Israel and Kleinberger 2002). Some of the other E1a-interacting protein products, in addition to others more recently described, have also been identified (see Fig. 3).

This discovery of an interaction between E1a products and pRb shed light on the apparent ability of Ad to overcome multiple blocks in G_1, and helped to clarify the discrepancy in the timing between Ad and serum-induced passage through G_1. By binding pRb, Ad was able to bypass the normal set of events that occur in serum-induced G_1 progression, streamlining transition into S-phase, and consequently activating only a subset of genes that would otherwise be turned on by serum (see below). Obviously, the complement of genes turned on by Ad includes those necessary for its own replication. Other DNA viruses have subsequently been shown to encode proteins that target pRb, thus deregulating the G_1 checkpoint (reviewed in O'Brien 1998). However, this shunting of cells into an unscheduled S-phase is not without its consequences, and these will be discussed later.

Subsequent reports have confirmed the importance of pRb and p300/CBP binding in DNA synthesis induction by Ad (Howe et al. 1990; Barbeau et al. 1992). Although it must be noted that conflicting reports exist as to whether the interactions of both pRb and p300/CBP are necessary for this effect, or whether one product plays a greater role (Whyte et al. 1989; Howe et al. 1990; Chiou and White 1997; Querido et al. 1997). These are curious findings as the same cells were used in some cases with viruses purportedly similar in their ability to selectively bind pRb and/or p300/CBP (Chiou and White 1997; Querido et al. 1997).

5.2.1
Normal Function of pRb

The normal role of pRb in the cell is to act as a regulator of cell cycle progression. The loss of functional pRb, as the name suggests, was initially implicated as playing a causative role in retinoblastoma formation. Inactivation of the pRb pathway, either directly, or indirectly through targeting proteins governing its phosphorylation, is now observed in most human malignancies (for recent reviews, see Sherr 1996; Hickman et al. 2002). This underscores the importance of the pRb family of proteins in cell cycle regulation (reviewed in Lavia and Jansen-Durr 1999). The pRb protein contains two highly conserved domains (A and B) that are critical for its regulatory role (Dyson 1998). These domains form a "pocket" that mediates most of the interactions between pRb and its associated proteins, including the E2F transcription factor family. Upon phosphorylation of the hypophosphorylated form of pRb by cyclin D- and E-dependent kinases, the bound E2F transcription factors are released. E2F in its simplest form is comprised of a heterodimer containing an E2F subunit and a DP subunit. These subunits derive from a pool of at least six E2F genes (*E2F-*

1 to *E2F-6*) and two *DP* genes. Some of the heterodimers act as transcriptional activators and some as repressors (Dyson 1998; Muller et al. 2001). E2F factors play a crucial role in regulating a subset of genes involved in G_1/S-phase transition. More specifically, genes involved in DNA replication (such as enzymes involving nucleotide synthesis and components of the origin of replication complex), DNA repair (such as mismatch repair, *MSH2*, and double-strand break repair, *BRCA1*, genes) and mitosis (such as spindle checkpoint gene, *BUB1b*, and cyclin-dependent kinase, *Cdc2*; Polager et al. 2002). The sequestration of E2F by pRb represents a major G_1 regulatory checkpoint.

5.2.2
E1a Binds pRb to Deregulate the G_1 Checkpoint

E1a products deregulate the G_1 checkpoint by binding pRb (Fig. 4). This is achieved by binding the hypophosphorylated form of pRb (the form in which E2F is bound; Mittnacht et al. 1994), primarily through the LXCXE motif located in CR2, but also secondarily through a portion of CR1 that binds near the E2F-1 binding site on pRb (Fattaey et al. 1993; Ikeda and Nevins 1993). The suggestion is that the pRb complex is brought into close association with E1a via the CR2 binding site, and the competition of CR1 for the E2F-1 binding site on pRb is necessary for the displacement of the bound E2F transcription factors. In addition to liberating these bound E2F factors, E1a also stabilizes the E2F-DP heterodimer, thus increasing the amount of available active E2F (Hateboer et al. 1996).

The liberated E2F transcription factors transcribe both viral and cellular genes, resulting in the coordinated expression of the viral replicative machinery, and presumably a cellular environment conducive to viral replication. E2F, in fact, derives its name (E2 promoter binding factor) from its description as a rate-limiting cellular factor in E1a-mediated transactivation of the Ad E2 promoter (Kovesdi et al. 1987). However, overexpression of E2F-1 in quiescent cells leads not only to induction of S-phase, but can also trigger apoptosis (Shan and Lee 1994; Phillips and Vousden 2001). This induction of apoptosis is usually accompanied by a concomitant increase in p53 tumor suppressor levels and stability (Qin et al. 1994; Wu and Levine 1994; Fig. 4).

5.3
Involvement of p53

There is substantial evidence linking E1a expression and the observed increase in p53 levels and stability. E1a has been shown to increase p53 protein levels by both transcriptional (Liu et al. 1985; Braithwaite et al. 1990) and post-translational (Grand et al. 1993) mechanisms, leading to stabilization of the protein (Lowe and Ruley 1993). The majority of literature is dedicated to post-translational regulation of p53, rather than transcriptional regulation, and, as such, this mechanism is better described.

The mechanism by which E1a upregulates p53 by post-translational regulation is thought to be achieved through E2F transactivation of *p14*^{ARF}, the alternate splice product of the *p16*^{INK4A} locus (Duro et al. 1995; Mao et al. 1995; Fig. 4). E2F-1 has been shown to directly transactivate *p14*^{ARF} (Bates et al. 1998). The p14^{ARF} product is then thought to bind MDM-2 (Pomerantz et al. 1998), neutralizing its activity (Pomerantz et al. 1998; Weber et al. 1999). MDM-2 is a RING-type E3 ubiquitin ligase that aids in the ubiquitination of p53 (Fang et al. 2000; Honda and Yasuda 2000). The normal role of the MDM-2 protein is to maintain p53 at low levels within the cell via an autoregulatory feedback loop (Haupt et al. 1997; Kubbutat et al. 1997). When a cell is stressed, p53 levels rise. p53 in turn activates the expression of the *MDM-2* gene. MDM-2 binds the transcriptional activation domain of p53 and blocks its ability to regulate certain downstream target genes, while signalling for its destruction via the ubiquitin-mediated pathway. The window between p53 induction and destruction is thought to define a time period for p53 to exert its effects. Thus, in the case of E1a expression, MDM-2 downregulation of p53 is lost via the p14^{ARF} pathway, and p53 levels are subsequently seen to rise.

E1a has also been shown to increase p53 levels through indirect transactivation of the *p53* promoter (Hale and Braithwaite 1999; Fig. 4). E1a by itself is unable to directly bind double-stranded DNA in a sequence-specific manner (Ferguson et al. 1985); thus, in order to transactivate its target genes, E1a must utilize other transcription factors to act as a bridge. Hale and Braithwaite (1999) have implicated the transcription factor ETF as an essential bridging protein, allowing E1a to transactivate the *p53* promoter, presumably leading to an increase in p53 protein levels. Importantly, at least two reports have shown transcriptional modulation of p53 levels following treatment with genotoxic agents (Sun et al. 1995; Hellin et al. 1998). These data suggest that, in addition to the well-established post-translational regulation of p53, there is growing evidence of regulation at the transcriptional level.

An increase in p53 protein levels is thought to be a prerequisite for p53-dependent apoptosis, although it must be noted that not all E1a-induced cell death necessarily utilizes the p53 pathway (see below). However, despite a wealth of research, the exact mechanism(s) by which p53 causes cells to die remains unclear (Prives and Hall 1999).

5.4 Significance of Apoptosis to an Adenovirus Infection

The "type" of cell death induced by Ad is critical to the fulfillment of a productive infection. As mentioned earlier, Ad is thought to escape the cell through lysis, allowing the efficient release of maximum numbers of progeny virus. However, we have seen that during the replication cycle a different form of cell death can be induced – apoptosis. Apoptosis, if not efficiently blocked, acts to cripple virus production as viral progeny are condensed into apoptotic bodies and subsequently scavenged by macrophages.

The significance of avoiding this type of death is best demonstrated by a series of early data derived from the development and use of specific Ad mutants. In 1968, Takemori et al. isolated a series of Ad12 mutants termed cytocidal (or cyt) mutants, whose phenotype was characterized by enhanced cellular destruction, low virus yield, and greatly decreased oncogenicity in newborn hamsters. These mutants were subsequently shown to demonstrate extensive degradation of both viral and cellular DNA (Ezoe et al. 1981), and provided some of the first evidence that Ads could induce apoptosis. The *cyt* gene was later mapped to the functional loss of the E1b19 kDa protein (Pilder et al. 1984; Takemori et al. 1984), a Bcl-2 homology (BH) 3 only protein belonging to the Bcl-2 family (see also below; Chiou et al. 1994). Thus, due to the deficiency in E1b19 kDa activity, these mutants were unable to efficiently block apoptosis, resulting in a markedly abortive infection.

As illustrated by the above example, one of the major obstructions to a productive viral infection lies in the fact that unscheduled S-phase is hardwired with premature suicide of the cell. In the event that Ad is unable to prevent this premature death, as in the case of these cyt mutants, the infection becomes abruptly terminated and the result is a non-productive, or abortive infection. The evidence suggests that cell death pathways most inhibitory to productive viral infections are those with p53 as the central mediator. The fact that p53-dependent cell death is an important cellular defence mechanism with respect to viral infection is reflected in the fact that many DNA, and RNA, tumor viruses target p53 (Roulston et al. 1999). It may be that p53 represents a necessary evil and thus something the virus has to deal with, or conversely p53 may be an integral component which aids a productive viral infection (discussed later and in Braithwaite and Russell 2001).

5.5 Summary

It is apparent from the data reported above that Ad can induce an environment conducive to viral replication, freeing up factors and machinery necessary for viral DNA synthesis, in a cellular context where these would otherwise be unavailable. It does this by binding the pRb tumor suppressor through products of the E1a region, thus bypassing the G_1 checkpoint, and allowing progression into an unscheduled S-phase. It is evident also that this plundered S-phase is not without its consequences, as a concomitant increase in p53 levels triggers an apoptotic response; this occurs as the opportunity for a G_1 arrest has been abrogated by E1a. This premature cell suicide would be detrimental to a productive viral infection. Ad however has formulated several strategies to perturb this death induction. This will be discussed in detail below.

6
Regulation and Deregulation of Apoptosis by Adenovirus

Of the five early transcription units, E2 is devoted solely to expressing the replication machinery, whereas all of the remaining four units (E1a, E1b, E3 and E4) code for proteins involved, amongst other things, in overcoming premature death of the host cell. This underscores the importance of the implications premature cell death poses to the virus.

6.1
Strategies to Counter Increased p53 Function

Highlighting the importance of p53 to the virus, Ad has developed a complex arsenal of proteins directed at various levels of p53 regulation, acting to manipulate its activity and to perturb its normal stress response. The major E1a proteins have been shown to bind the transcriptional co-activators of p53 [p300/CREB-binding protein (CBP)], downregulating the ability of p53 to activate some of its normal regulatory promoters. In addition, two Ad proteins (E1b55 kDa and E4ORF6) are able to directly bind p53, also inactivating the transcriptional regulation of a selection of promoters, and target it for destruction via the proteasome. And, finally, the Bcl-2 homologue, E1b19 kDa, acts downstream of p53, inhibiting the execution of p53-dependent apoptosis, as well as other forms of cell death. These approaches are discussed in greater detail below.

6.1.1
E1a Binds p300/CBP

One strategy utilized by Ad that is thought to specifically counteract the transactivational activities of p53 is the binding and sequestering of p300/CBP and its associated factor P/CAF by E1a. p300 and CBP are highly related co-activators of transcription, and are recruited to many promoters through their association with DNA-binding transcription factors, such as p53 (for a recent review, see Goodman and Smolik 2000). Numerous studies have demonstrated that an interaction between p53 and p300/CBP augments the transcriptional activation of the p53-responsive *MDM-2*, $p21^{WAF1/CIP1}$ and *Bax* promoters, genes critical for the response of p53 to DNA damage (Avantaggiati et al. 1997; Gu and Roeder 1997; Lill et al. 1997). Although the majority of the literature is focused on the ability of p300/CBP to augment p53-dependent transcription, it must be noted these co-activators have also been shown to play a major role in p53-dependent repression of transcription (reviewed in Goodman and Smolik 2000). It is thought that, once recruited, p300/CBP activates transcription by acetylating the N-terminal tails of the core histones or by acetylating specific lysines in other transcription factors. They do this by using their intrinsic histone acetyltransferase (HAT)/acetyltransferase activity and that of the associated HAT co-activator p300/CBP-associated factor (P/CAF).

The binding of E1a to p300/CBP has been shown to mediate the ability of Ad to induce DNA synthesis (Whyte et al. 1989; Stein et al. 1990). The residues of E1a important for this interaction were shown to lie in the non-conserved N-terminus and within the CR1 region (Stein et al. 1990; O'Connor et al. 1999; see Fig. 3). The N-terminus of E1a binds p300/CBP directly at the transcriptional-adaptor motif (TRAM) domain (O'Connor et al. 1999). The popular suggestion is that this sequestration of p300/CBP by E1a inhibits the HAT activities of p300/CBP (Chakravarti et al. 1999; Hamamori et al. 1999), thus repressing p53-dependent transactivation (Somasundaram and El-Deiry 1997) and subsequent apoptosis (Lill et al. 1997). It must be noted, however, that little data have been accumulated from the large number of p53-responsive promoters not directly involved in the DNA damage pathway.

Although it is generally held that E1a inhibits p300/CBP function there is a conflicting report demonstrating that E1a increases the HAT activity of p300/CBP (Ait-Si-Ali et al. 1998). In addition, p300/CBP has been shown to possess transactivational activity despite high levels of E1a in Ad-infected 293 cells (Nakajima et al. 1997). In fact, there is a suggestion that the interaction with E1a redirects the acetyltransferase activities of p300/CBP to other regulatory promoters, presumably those involved in ensuring a productive Ad infection. It is also of note that p300/CBP can activate transcription through direct interactions with RNA polymerase II (Nakajima et al. 1997), and can associate with E2F-5, and to a lesser degree with E2F-1, enhancing transcription of E2F target genes (Martinez-Balbas et al. 2000; Morris et al. 2000). Interestingly, the SV40 and polyomavirus LT antigens, HPV E6, and the Ad E1a proteins have all been demonstrated to bind the same region of p300/CBP, attesting to the importance of this interaction during infection (reviewed in Goodman and Smolik 2000).

6.1.2
E1a Binds P/CAF

P/CAF exists within a multisubunit complex with p300/CBP and also possesses HAT activity. Following DNA damage, P/CAF has been shown to enhance the binding of p53 to the promoters of its target genes, such as $p21^{WAF1/CIP1}$ (Sakaguchi et al. 1998; Liu et al. 1999), promoting cell cycle arrest.

E1a, through its N-terminal 60 amino acids, binds P/CAF directly at its HAT domain (Reid et al. 1998). This is thought to inhibit its HAT activity (Chakravarti et al. 1999) and repress P/CAF-mediated activation of transcription (Reid et al. 1998). Interestingly, P/CAF can also acetylate E2F-1 (Martinez-Balbas et al. 2000). This acetylation extends the half-life of E2F-1, and increases its DNA-binding and transactivational abilities. pRb, through its association with histone deacetylases, has been shown to counter the activity of P/CAF (Brehm et al. 1998; Luo et al. 1998). However, the significance of this during an Ad infection is unclear, as pRb is sequestered by E1a.

6.1.3
Summary

In general the majority of data implicate the binding of E1a to p300/CBP, and P/CAF, as neutralizing the ability of p53 to transcriptionally activate a subset of its normal target promoters predominantly involved in the DNA damage response. In addition, there is also a small body of data suggesting E1a may bind these acetyltransferases so as to utilize them for its own purposes (discussed below).

6.2
Strategies to Counter Increased p53 Levels

6.2.1
The E1b55 kDa and E4ORF6 Proteins

Ad encodes a further two proteins that, when expressed alone, act to counter the ability of p53 to transactivate a subset of its responsive promoters and, when expressed together, lead to the degradation of p53 (Fig. 4). The responsible proteins are the 55 kDa product of the E1b region and the 34 kDa product of the E4 region (E4ORF6). In addition to binding p53, E1b55 kDa and E4ORF6 can also bind each other (Sarnow et al. 1984).

These proteins play different roles in a productive Ad infection with respect to the timing of their expression. In the latter stages of infection, after the onset of viral DNA replication, these proteins play a major role in the selective nucleo-cytoplasmic trafficking of viral (and certain cellular) mRNA transcripts. In conjunction with this activity, E1b55 kDa and E4ORF6 are also required for shutoff of host protein synthesis (Babich et al. 1983; Babiss and Ginsberg 1984; Halbert et al. 1985), and for efficient DNA replication (reviewed in Shenk 1996). In the early stages of infection, however, these proteins are thought to combat the effects of increased p53 induced by Ad infection.

As mentioned above, E1b55 kDa and E4ORF6 can bind p53 separately or together, neutralizing p53 function, and signalling its degradation. The E1b55 kDa protein inhibits p53 activity through binding to the N-terminal transcriptional activation domain of the p53 protein (Sarnow et al. 1982; Kao et al. 1990; Yew and Berk 1992; Martin and Berk 1998), whereas the E4ORF6 protein binds at the C-terminus (Dobner et al. 1996), also inhibiting p53's transactivation ability. When both Ad proteins are bound, they have been shown to assist in the ubiquitination of p53 and its subsequent degradation via the proteasome (Ridgway et al. 1997; Steegenga et al. 1998; Wienzek et al. 2000).

6.3
Adenovirus Induces p53-Independent Apoptosis

6.3.1
Involvement of E4ORF4

It is evident from the data reported above that Ad's exploitation of the cell cycle results in the initiation of apoptotic pathways. However, it must be noted that not all apoptotic pathways induced by Ad gene products are centered around p53. In fact, an Ad gene product from the E4 region, E4ORF4, has been shown to cooperate with E1a to induce apoptosis independent of p53 (Marcellus et al. 1996; Lavoie et al. 1998). Furthermore, E4ORF4 expression has been shown to be sufficient in itself to induce this apoptosis (Lavoie et al. 1998; Marcellus et al. 1998; Shtrichman and Kleinberger 1998). The suggestion is E4ORF4 binds the regulatory subunit of protein phosphatase (PP)2A, presumably in order to redirect its activity, and this provides the necessary impetus for induction of p53-independent apoptosis (Shtrichman and Kleinberger 1998; Shtrichman et al. 2000; Fig. 4). However, these results have only been demonstrated by two groups, and, in our hands, E4ORF4 does not lead to cell death when expressed alone, nor in combination with E1a expression (unpubl. data).

6.3.2
Involvement of E4ORF6/7

It is clearly evident that Ad goes to great lengths to liberate and stabilize sufficient free E2F to grant the virus access to the replication machinery (see above). Another Ad gene product, E4ORF6/7, has been shown to facilitate the binding of E2F-1 to the Ad *E2* promoter (Huang and Hearing 1989; Marton et al. 1990; Neill et al. 1990; Obert et al. 1994), thus stabilizing the E2F-1/DP-1 heterodimer (Cress and Nevins 1994; Obert et al. 1994). In addition to playing a role in Ad gene transcription and S-phase induction, E2F-1 overexpression has also been demonstrated to induce apoptosis in a variety of cells by a mechanism that does not necessarily involve p53. This has been shown in rat, mouse and monkey cells using transient transfection of cloned *E2F-1*, tetracycline-inducible *E2F-1* transformed cells and infection with recombinant Ads expressing *E2F-1* cDNA (Mymryk et al. 1994; Qin et al. 1994; Shan and Lee 1994; Wu and Levine 1994; DeGregori et al. 1997). Thus the data are strong for the existence of a p53-independent apoptotic pathway. Ad, however, has a potent mechanism whereby premature death of the host cell via p53-independent pathways can be avoided (see below).

6.4 E1b19 kDa Blocks p53-Dependent and Independent Apoptosis

We have already mentioned that the smaller E1b product, E1b19 kDa, plays a major role in averting apoptosis. Ad mutants unable to express a functional E1b19 kDa protein (cyt mutants) culminate in an abortive infection due to premature cell death (see above). Accumulation of recent data suggests this protein acts to block both p53-dependent and p53-independent pathways (Fig. 4). The suggestion is that E1b19 kDa accomplishes this in a manner analogous to the cellular anti-apoptotic protein Bcl-2. There are several lines of evidence that support this suggestion, namely the presence of the BH3 domain in E1b19 kDa (see above), its ability to associate with other Bcl-2 family members (Boyd et al. 1994; Chen et al. 1996; Han et al. 1996a,b), and the ability of Bcl-2 to substitute for E1b19 kDa during Ad infection (Chiou et al. 1994; Subramanian et al. 1995).

The exact mechanism by which E1b19 kDa prevents apoptosis during an infection, however, is less clear. In addition to preventing p53-dependent (Rao et al. 1992) and p53-independent (Putzer et al. 2000) apoptosis, E1b19 kDa has been shown to suppress a diverse array of apoptotic signals from death receptor pathways, such as TNF (White et al. 1992), Fas (Huang et al. 1997) and TRAIL (Routes et al. 2000). The popular suggestion is that E1b19 kDa acts at a downstream point of convergence of several apoptotic pathways, and indeed E1b19 kDa has been reported to form complexes with Bcl-2 family members that regulate mitochondrial membrane integrity, such as Bak and Bax (Farrow et al. 1995; Han et al. 1996a; Lomonosova et al. 2002). These proteins appear to be required for the execution of apoptosis derived from diverse death stimuli (Wei et al. 2001; Zong et al. 2001). A recent report using isogenic *Bax*$^{+/-}$ and *Bax*$^{-/-}$ cell lines has shown that Bax is essential for efficient Ad-induced apoptosis to occur (Lomonosova et al. 2002), and that E1b19 kDa expression prevents N-terminal processing of Bax, an event that appears to be responsible for mitochondrial permeabilization, cytochrome-C release and apoptosis (Gao and Dou 2000).

Presumably, indicated merely by its presence in the Ad genome, E1b19 kDa must have evolved functions distinct from those of Bcl-2. This is evidenced in part by the observation that Bcl-2 can only partially complement for E1b19 kDa in infection (Rao et al. 1992). At the present time the additional functions of E1b19 kDa which relate to the inhibition of apoptosis are unclear.

What is clear, however, is that E1b19 kDa serves to prevent premature cell death generated from the deregulation of the cell cycle machinery and from external death signals, thus maintaining host cell viability. In accordance with this, E1b19 kDa is produced at higher levels later in infection (Spector et al. 1978; Wilson and Darnell 1981; Montell et al. 1984), presumably as more death pathways are triggered, allowing the virus time to maximize its progeny.

6.5 Summary

The interplay between life and death signals during a productive Ad infection is complex. The above data indicate at least two pathways through which premature death of the host cell can be triggered. Both of these pathways stem from the deregulation of the G_1 checkpoint and subsequent increase in E2F levels and its stabilization. The best-described pathway is that involving p53, where E2F-1, via the $p14^{ARF}$ pathway, brings about increased levels of p53 and subsequent p53-dependent apoptosis. The alternate pathway described here acts independently of p53, but also stems from E2F-1 overexpression. The data for E4ORF4 cooperating in this pathway are less convincing. Irrespective of the pathway through which Ad induces apoptosis, it is blocked by the mainstay, if not the only, anti-apoptotic protein encoded by Ad, E1b19 kDa. E1b19 kDa functions to destroy the cell's ability to commit itself to an early apoptotic fate, and subsequently affords the virus ample time to complete a productive replicative cycle.

7 Evaluation of the Role of p53 in an Adenoviral Infection

The data for the role of the E1b19 kDa protein in the inhibition of both p53-dependent and p53-independent apoptosis induced by viral gene products are strong. Inhibition has been demonstrated from many different laboratories in an array of cell lines using both viruses and cloned genes (see above). This, however, begs the question as to why Ad devotes such a large proportion of its genome to coding multiple gene products that act upstream of E1b19 kDa to supposedly knock out p53 function and to target it for degradation.

Perhaps one explanation for this apparent redundancy of Ad gene products lies in the notion that, at least during the early stages of infection, p53 provides functions to the virus that are necessary for a productive infection to occur. Although this appears to contradict current understanding of virus/p53 interactions, there is in fact substantial evidence to support such a claim whereby Ad manipulates p53 function to its own benefit instead of simply promoting its destruction. In fact, even in viral systems with well-established dogma surrounding p53 degradation, as with the E6-mediated p53 degradation in HPV infection, there is considerable evidence that p53 levels and stabilization are first stimulated by the E7 protein (Munger et al. 2001; Eichten et al. 2002).

It is apparent that early in infection Ad devotes great effort to controlling p53 levels and function. Not only does E1a stabilize p53 levels, it also specifically transactivates the p53 promoter (Fig. 4). This seems somewhat surprising given the apparent toxicity of p53 to the cell when present at high levels. Even more surprisingly, as p53 levels are seen to increase, instead of simply degrading p53, Ad encodes two proteins, E1b55 kDa and E4ORF6, which modulate its activity, but can only degrade the protein when jointly bound to p53. The complexity of this scenario suggests at least some of the functions of p53 may be useful for Ad, if properly regulated.

Looking closer at the roles of E1b55 kDa and E4ORF6 early in infection, it is somewhat surprising, given the strength of current dogma, that there exists very little direct evidence demonstrating the ability of E1b55 kDa to actually inhibit p53-dependent apoptosis. In fact, the evidence for inhibition of p53-dependent apoptosis by E1b55 kDa is weak. Only partial inhibition has been demonstrated in mouse cells after infection with a 12S expressing, E1b19 kDa mutant virus (Teodoro and Branton 1997), and in transiently transfected and stably transformed rat cells (Rao et al. 1992). In addition, an E1b55 kDa transformed murine cell line has been shown to downregulate apoptosis from growth-factor deprivation; however, this effect too proved transient (BenJilani et al. 2002). In human cells, however, E1b55 kDa has not been shown to demonstrate any inhibitory effects towards p53-dependent apoptosis (Dix et al. 2000). The inhibition of p53-dependent or -independent apoptosis by E4ORF6 has yet to be demonstrated.

Recently, several E1-transformed cell lines, including 293s, have been shown to actually possess transcriptionally active p53 despite the presence of E1b55 kDa (Lober et al. 2002). In fact, the complex between E1b55 kDa and p53 has even been demonstrated to facilitate viral replication (Ridgway et al. 1997; Hall et al. 1998; Fig. 4). Furthermore, Ad mutants lacking the E1b55 kDa protein are highly attenuated for cell killing compared with wtAd5 due, at least in part, to a lack of E1b55 kDa/p53 complex formation (Grand et al. 1996; Goodrum and Ornelles 1997; Turnell et al. 1999; Dix et al. 2000). These latter data point to a positive role for the E1b55 kDa/p53 complex during the infection process. Indeed, recent work from our laboratory supports this positive role for p53 in a productive infection. This is demonstrated by two lines of evidence: infection experiments using isogenic cell lines have shown that, in the absence of wtp53, viral replicative lysis is inhibited compared with wtp53 counterparts, and, secondly, transient transfection of p53 into a functionally p53 null cell line prior to infection increases viral recovery yields by approximately ten-fold. The possibility that p53 may indeed be necessary for a productive infection, at least in complex with Ad proteins, raises the question as to the role of p53 in the infection process.

One scenario might involve p53 acting in much the same way as pRb does. When E1b55 kDa and E4ORF6 proteins bind p53 they displace other bound factors, such as a component of transcription factor IID (part of the transcription initiation complex; Dobner et al. 1996). It may be possible that these liberated factors are necessary for an early event in the infection process, and that the binding of these Ad products to p53 may result in their displacement.

In contrast to this passive role of p53 alluded to above, p53 may in fact play a more active part in Ad infection. It may be that Ad requires a specific subset of p53 functions, and, as such, hijacks p53 for its own purposes. This is not without precedence, as viruses commonly alter or redirect proteins rather than simply inhibiting their function. Clearly more work is needed to address these issues that are not consistent with current dogma regarding the binding and inhibition of p53 during Ad infection.

8 Conclusions

For Ad, the ultimate goal of an infection is to replicate its genome. In order to complete this task Ad must overcome a series of barriers presented to it by the cell. To effectively overcome these barriers and allow for successful propagation of progeny virus, Ad must utilize host cell functions and machinery. From the very beginning of the infection process, Ad deceives the cell into allowing viral attachment and internalization. It achieves this by presumably acting under the guise of a normal extracellular ligand and by exploiting the macropinocytotic ability of the host cell. This theme of deception and exploitation is carried throughout the infection process.

A core component of this exploitation is through manipulation of cell cycle regulation. Disruption of the G_1 regulatory checkpoint allows Ad to increase levels of free E2F, thus promoting the transcription of viral and cellular factors necessary for Ad replication. Interestingly, E2F, through $p14^{ARF}$, also leads to increased levels of stabilized p53. This accumulation of p53 may occur as p53 is hardwired with cellular factors involved in S-phase, or alternately as some component of p53 activity may play a necessary role in a productive infection. In fact, there is growing data supporting the latter notion that E1a along with E1b55 kDa and E4ORF6 may actually manipulate, or hijack, p53. It may be that certain functions of p53, if properly regulated, benefit Ad in the early stages of infection. However, in the later stages of infection, it appears either p53 becomes less important to Ad, or it becomes too difficult to control. In either case, uncontrolled or prolonged p53 levels result in premature cell death unless efficiently blocked. Ad combats this death, as well as premature death from other signals, by inhibiting downstream disruption of the mitochondria. The responsible product, E1b19 kDa, affords Ad time to produce sufficient progeny, and to lyse the cell through E3 ADP expression, thus completing a productive Ad infection.

Half a century of dedicated research has allowed investigators to better understand some of the complexities involved in the exploitation of host cell processes by Ad. However, in a number of areas limited and/or weak data combined with reductionist approaches have led to the establishment of dogma that may not fairly reflect the data. In addition, it cannot be overlooked that Ad2 and Ad5 have evolved to infect differentiated epithelial cells in the nasopharynx. Thus, it is highly likely some of the barriers that these Ads have evolved to overcome *in vivo* may not be relevant in an *in vitro* setting. With this in mind, it is clear that certain areas need to be further explored to allow better insight into virus/host interactions. This insight will in turn allow for the development of Ads as therapeutic tools for use in the clinical setting.

Acknowledgements. We would like to extend warm thanks to members of the Cell Transformation Group, Department of Pathology, Otago Medical School, both past and present, for spirited discussion and helpful advice. In addition we would also like to acknowledge the University of Otago Electron Microscopy Unit for their technical assistance. Most importantly, we would like to thank our colleagues whose work is cited here for their contributions and apologize to those whose references were omitted due to space constraints. This work was funded by the Royal Society of New Zealand Marsden Fund and by the Health Research Council of New Zealand.

References

Ait-Si-Ali S, Ramirez S, Barre FX, Dkhissi F, Magnaghi-Jaulin L, Girault JA, Robin P, Knibiehler M, Pritchard LL, Ducommun B, Trouche D, Harel-Bellan A (1998) Histone acetyltransferase activity of CBP is controlled by cycle-dependent kinases and oncoprotein E1a. Nature 396:184–186

Andersson M, Paabo S, Nilsson T, Peterson PA (1985) Impaired intracellular transport of class I MHC antigens as a possible means for adenoviruses to evade immune surveillance. Cell 43:215–222

Andersson M, McMichael A, Peterson PA (1987) Reduced allorecognition of adenovirus-2 infected cells. J Immunol 138:3960–3966

Avantaggiati ML, Ogryzko V, Gardner K, Giordano A, Levine AS, Kelly K (1997) Recruitment of p300/CBP in p53-dependent signal pathways. Cell 89:1175–1184

Avvakumov N, Wheeler R, D'Halluin JC, Mymryk JS (2002) Comparative sequence analysis of the largest E1a proteins of human and simian adenoviruses. J Virol 76:7968–7975

Babich A, Feldman LT, Nevins JR, Darnell JE Jr, Weinberger C (1983) Effect of adenovirus on metabolism of specific host mRNAs: transport control and specific translational discrimination. Mol Cell Biol 3:1212–1221

Babiss LE, Ginsberg HS (1984) Adenovirus type 5 early region 1b gene product is required for efficient shutoff of host protein synthesis. J Virol 50:202–212

Barbeau D, Marcellus RC, Bacchetti S, Bayley ST, Branton PE (1992) Quantitative analysis of regions of adenovirus E1a products involved in interactions with cellular proteins. Biochem Cell Biol 70:1123–1134

Bates S, Phillips AC, Clark PA, Stott F, Peters G, Ludwig RL, Vousden KH (1998) $p14^{ARF}$ links the tumour suppressors Rb and p53. Nature 395:124–125

Bellett AJ, Li P, David ET, Mackey EJ, Braithwaite AW, Cutt JR (1985) Control functions of adenovirus transformation region E1a gene products in rat and human cells. Mol Cell Biol 5:1933–1939

Beltz GA, Flint SJ (1979) Inhibition of HeLa cell protein synthesis during adenovirus infection. Restriction of cellular messenger RNA sequences to the nucleus. J Mol Biol 131:353–373

Ben-Israel H, Kleinberger T (2002) Adenovirus and cell cycle control. Front Biosci 7: d1369–1395

Benedict CA, Norris PS, Prigozy TI, Bodmer JL, Mahr JA, Garnett CT, Martinon F, Tschopp J, Gooding LR, Ware CF (2000) Three adenovirus E3 proteins cooperate to evade apoptosis by TRAIL receptor-1 and 2. J Biol Chem 24:24

BenJilani KE, Gaillard JP, Petit F, Arnoult D, Roumier AS, Labalette M, Ameisen JC, Estaquier J (2002) A suppressive effect of the adenovirus 5 protein E1b 55 K on apoptosis induced by IL-3 deprivation and gamma-irradiation. Biol Cell 94:77–89

Bennett EM, Bennink JR, Yewdell JW, Brodsky FM (1999) Cutting edge: adenovirus E19 has two mechanisms for affecting class I MHC expression. J Immunol 162:5049–5052

Bergelson JM, Cunningham JA, Droguett G, Kurt-Jones EA, Krithivas A, Hong JS, Horwitz MS, Crowell RL, Finberg RW (1997) Isolation of a common receptor for Coxsackie B viruses and adenoviruses 2 and 5. Science 275:1320–1323
Berget SM, Moore C, Sharp PA (1977) Spliced segments at the 5′ terminus of adenovirus 2 late mRNA. Proc Natl Acad Sci USA 74:3171–3175
Boulanger PA, Blair GE (1991) Expression and interactions of human adenovirus oncoproteins. Biochem J 275:281–299
Boyd JM, Malstrom S, Subramanian T, Venkatesh LK, Schaeper U, Elangovan B, D'Sa-Eipper C, Chinnadurai G (1994) Adenovirus E1b 19 kDa and Bcl-2 proteins interact with a common set of cellular proteins. Cell 79:341–351
Braithwaite AW, Russell IA (2001) Induction of cell death by adenoviruses. Apoptosis 6:359–370
Braithwaite AW, Murray JD, Bellett AJ (1981) Alterations to controls of cellular DNA synthesis by adenovirus infection. J Virol 39:331–340
Braithwaite AW, Cheetham BF, Li P, Parish CR, Waldron-Stevens LK, Bellett AJ (1983) Adenovirus-induced alterations of the cell growth cycle: a requirement for expression of E1a but not of E1b. J Virol 45:192–199
Braithwaite A, Nelson C, Skulimowski A, McGovern J, Pigott D, Jenkins J (1990) Transactivation of the p53 oncogene by E1a gene products. Virology 177:595–605
Brehm A, Miska EA, McCance DJ, Reid JL, Bannister AJ, Kouzarides T (1998) Retinoblastoma protein recruits histone deacetylase to repress transcription. Nature 391:597–601
Burgert HG, Kvist S (1985) An adenovirus type 2 glycoprotein blocks cell surface expression of human histocompatibility class I antigens. Cell 41:987–997
Burgert HG, Ruzsics Z, Obermeier S, Hilgendorf A, Windheim M, Elsing A (2002) Subversion of host defense mechanisms by adenoviruses. Curr Top Microbiol Immunol 269:273–318
Carlock LR, Jones NC (1981) Transformation-defective mutant of adenovirus type 5 containing a single altered E1a mRNA species. J Virol 40:657–664
Carson SD (2001) Receptor for the group B coxsackieviruses and adenoviruses: CAR. Rev Med Virol 11:219–226
Chakravarti D, Ogryzko V, Kao HY, Nash A, Chen H, Nakatani Y, Evans RM (1999) A viral mechanism for inhibition of p300 and PCAF acetyltransferase activity. Cell 96:393–403
Cheetham BF, Bellett AJ (1982) A biochemical investigation of the adenovirus-induced G1 to S-phase progression: thymidine kinase, ornithine decarboxylase, and inhibitors of polyamine biosynthesis. J Cell Physiol 110:114–122
Cheetham BF, Shaw DC, Bellett AJ (1982) Adenovirus type 5 induces progression of quiescent rat cells into S-phase without polyamine accumulation. Mol Cell Biol 2:1295–1298
Chen G, Branton PE, Yang E, Korsmeyer SJ, Shore GC (1996) Adenovirus E1b 19-kDa death suppressor protein interacts with Bax but not with Bad. J Biol Chem 271:24221–24225
Chen P, Tian J, Kovesdi I, Bruder JT (1998) Interaction of the adenovirus 14.7-kDa protein with FLICE inhibits Fas ligand-induced apoptosis. J Biol Chem 273:5815–5820
Chiou SK, White E (1997) p300 binding by E1a cosegregates with p53 induction but is dispensable for apoptosis. J Virol 71:3515–3525
Chiou SK, Tseng CC, Rao L, White E (1994) Functional complementation of the adenovirus E1b 19-kilodalton protein with Bcl-2 in the inhibition of apoptosis in infected cells. J Virol 68:6553–6566
Chow LT, Gelinas RE, Broker TR, Roberts RJ (1977) An amazing sequence arrangement at the 5′ ends of adenovirus 2 messenger RNA. Cell 12:1-8
Cress WD, Nevins JR (1994) Interacting domains of E2F1, DP1, and the adenovirus E4 protein. J Virol 68:4213–4219
Dales S, Chardonnet Y (1973) Early events in the interaction of adenoviruses with HeLa cells. IV. Association with microtubules and the nuclear pore complex during vectorial movement of the inoculum. Virology 56:465–483
DeGregori J, Leone G, Miron A, Jakoi L, Nevins JR (1997) Distinct roles for E2F proteins in cell growth control and apoptosis. Proc Natl Acad Sci USA 94:7245–7250
D'Halluin JC (1995) Virus assembly. Curr Top Microbiol Immunol 199:47–66

Dimitrov T, Krajcsi P, Hermiston TW, Tollefson AE, Hannink M, Wold WS (1997) Adenovirus E3-10.4 K/14.5 K protein complex inhibits tumor necrosis factor-induced translocation of cytosolic phospholipase A2 to membranes. J Virol 71:2830–2837
Dix BR, O'Carroll SJ, Myers CJ, Edwards SJ, Braithwaite AW (2000) Efficient induction of cell death by adenoviruses requires binding of E1b55 k and p53. Cancer Res 60:2666–2672
Dobner T, Horikoshi N, Rubenwolf S, Shenk T (1996) Blockage by adenovirus E4ORF6 of transcriptional activation by the p53 tumor suppressor. Science 272:1470–1473
Doerfler W (1969) Nonproductive infection of baby hamster kidney cells (BHK21) with adenovirus type 12. Virology 38:587–606
Duro D, Bernard O, Della Valle V, Berger R, Larsen CJ (1995) A new type of p16INK4/MTS1 gene transcript expressed in B-cell malignancies. Oncogene 11:21–29
Dyson N (1998) The regulation of E2F by pRb-family proteins. Genes Dev 12:2245–2262
Egan C, Jelsma TN, Howe JA, Bayley ST, Ferguson B, Branton PE (1988) Mapping of cellular protein-binding sites on the products of early-region 1a of human adenovirus type 5. Mol Cell Biol 8:3955–3959
Eichten A, Westfall M, Pietenpol JA, Munger K (2002) Stabilization and functional impairment of the tumor suppressor p53 by the human papillomavirus type 16 E7 oncoprotein. Virology 295:74–85
Elsing A, Burgert HG (1998) The adenovirus E3/10.4K-14.5 K proteins down-modulate the apoptosis receptor Fas/Apo-1 by inducing its internalization. Proc Natl Acad Sci USA 95:10072–10077
Ezoe H, Fatt RB, Mak S (1981) Degradation of intracellular DNA in KB cells infected with cyt mutants of human adenovirus type 12. J Virol 40:20–27
Fang S, Jensen JP, Ludwig RL, Vousden KH, Weissman AM (2000) MDM2 is a RING finger-dependent ubiquitin protein ligase for itself and p53. J Biol Chem 275:8945–8951
Farrow S, White J, Martinou I, Raven T, Pun K-T, Grinham C, Martinou J-C, Brown R (1995) Cloning of a Bcl-2 homologue by interaction with adenovirus E1b 19 K. Nature 374:731–733
Fattaey AR, Harlow E, Helin K (1993) Independent regions of adenovirus E1a are required for binding to and dissociation of E2F-protein complexes. Mol Cell Biol 13:7267–7277
Ferguson B, Krippl B, Andrisani O, Jones N, Westphal H, Rosenberg M (1985) E1a 13S and 12S mRNA products made in *Escherichia coli* both function as nucleus-localized transcription activators but do not directly bind DNA. Mol Cell Biol 5:2653–2661
Friedman JM, Horwitz MS (2002) Inhibition of tumor necrosis factor alpha-induced NF-kappa B activation by the adenovirus E3–10.4/14.5 K complex. J Virol 76:5515–5521
Gallimore PH (1974) Viral DNA in transformed cells. II. A study of the sequences of adenovirus 2 DNA in nine lines of transformed rat cells using specific fragments of the viral genome. J Mol Biol 89:49–72
Gallimore PH, Turnell AS (2001) Adenovirus E1a: remodelling the host cell, a life or death experience. Oncogene 20:7824–7835
Gao G, Dou QP (2000) N-terminal cleavage of Bax by calpain generates a potent proapoptotic 18-kDa fragment that promotes Bcl-2-independent cytochrome C release and apoptotic cell death. J Cell Biochem 80:53–72
Gonzalez RA, Flint SJ (2002) Effects of mutations in the adenoviral E1b 55-kilodalton protein coding sequence on viral late mRNA metabolism. J Virol 76:4507–4519
Gooding LR, Elmore LW, Tollefson AE, Brady HA, Wold WS (1988) A 14,700 MW protein from the E3 region of adenovirus inhibits cytolysis by tumor necrosis factor. Cell 53:341–346
Gooding LR, Ranheim TS, Tollefson AE, Aquino L, Duerksen-Hughes P, Horton TM, Wold WS (1991) The 10,400- and 14,500-dalton proteins encoded by region E3 of adenovirus function together to protect many but not all mouse cell lines against lysis by tumor necrosis factor. J Virol 65:4114–4123
Goodman RH, Smolik S (2000) CBP/p300 in cell growth, transformation, and development. Genes Dev 14:1553–1577
Goodrum FD, Ornelles DA (1997) The early region 1b 55-kilodalton oncoprotein of adenovirus relieves growth restrictions imposed on viral replication by the cell cycle. J Virol 71:548–561

Graham FL, Smiley J, Russell WC, Nairn R (1977) Characteristics of a human cell line transformed by DNA from human adenovirus type 5. J Gen Virol 36:59–72
Grand RJ, Lecane PS, Roberts S, Grant ML, Lane DP, Young LS, Dawson CW, Gallimore PH (1993) Overexpression of wild-type p53 and c-Myc in human fetal cells transformed with adenovirus early region 1. Virology 193:579–591
Grand RJ, Owen D, Rookes SM, Gallimore PH (1996) Control of p53 expression by adenovirus 12 early region 1a and early region 1b 54 K proteins. Virology 218:23–34
Greber UF (1998) Virus assembly and disassembly: the adenovirus cysteine protease as a trigger factor. Rev Med Virol 8:213–222
Greber UF, Willetts M, Webster P, Helenius A (1993) Stepwise dismantling of adenovirus 2 during entry into cells. Cell 75:477–486
Greber UF, Webster P, Weber J, Helenius A (1996) The role of the adenovirus protease on virus entry into cells. EMBO J 15:1766–1777
Greber UF, Suomalainen M, Stidwill RP, Boucke K, Ebersold MW, Helenius A (1997) The role of the nuclear pore complex in adenovirus DNA entry. EMBO J 16:5998–6007
Gu W, Roeder RG (1997) Activation of p53 sequence-specific DNA binding by acetylation of the p53 C-terminal domain. Cell 90:595–606
Halbert DN, Spector DJ, Raskas HJ (1979) In vitro translation products specified by the transforming region of adenovirus type 2. J Virol 31:621–629
Halbert DN, Cutt JR, Shenk T (1985) Adenovirus early region 4 encodes functions required for efficient DNA replication, late gene expression, and host cell shutoff. J Virol 56:250–257
Hale TK, Braithwaite AW (1999) The adenovirus oncoprotein E1a stimulates binding of transcription factor ETF to transcriptionally activate the p53 gene. J Biol Chem 274:23777–23786
Haley KP, Overhauser J, Babiss LE, Ginsberg HS, Jones NC (1984) Transformation properties of type 5 adenovirus mutants that differentially express the E1a gene products. Proc Natl Acad Sci USA 81:5734–5738
Hall AR, Dix BR, O'Carroll SJ, Braithwaite AW (1998) p53-dependent cell death/apoptosis is required for a productive adenovirus infection. Nat Med 4:1068–1072
Hamamori Y, Sartorelli V, Ogryzko V, Puri PL, Wu HY, Wang JY, Nakatani Y, Kedes L (1999) Regulation of histone acetyltransferases p300 and PCAF by the bHLH protein twist and adenoviral oncoprotein E1a. Cell 96:405–413
Han J, Sabbatini P, Perez D, Rao L, Modhas D, White E (1996a) The E1b 19 K protein blocks apoptosis by interacting with and inhibiting the p53-inducible and death promoting Bax protein. Genes Dev 10:461–470
Han J, Sabbatini P, White E (1996b) Induction of apoptosis by human Nbk/Bik, a BH3-containing protein that interacts with E1b 19 K. Mol Cell Biol 16:5857–5864
Harlow E, Whyte P, Franza BR Jr, Schley C (1986) Association of adenovirus early-region 1a proteins with cellular polypeptides. Mol Cell Biol 6:1579–1589
Harris MS, Strohl WA (1980) The induction of cellular DNA synthesis in G1-arrested BHK21 cell infected with adenovirus type 2. J Gen Virol 46:455–466
Hateboer G, Kerkhoven RM, Shvarts A, Bernards R, Beijersbergen RL (1996) Degradation of E2F by the ubiquitin-proteasome pathway: regulation by retinoblastoma family proteins and adenovirus transforming proteins. Genes Dev 10:2960–2970
Haupt Y, Maya R, Kazaz A, Oren M (1997) MDM2 promotes the rapid degradation of p53. Nature 387:296–299
Hellin AC, Calmant P, Gielen J, Bours V, Merville MP (1998) Nuclear factor-kappa B-dependent regulation of p53 gene expression induced by daunomycin genotoxic drug. Oncogene 16:1187–1195
Henry P, Black PH, Oxman MN, Weissman SM (1966) Stimulation of DNA synthesis in mouse cell line 3T3 by Simian virus 40. Proc Natl Acad Sci USA 56:1170–1176
Hickman ES, Moroni MC, Helin K (2002) The role of p53 and pRb in apoptosis and cancer. Curr Opin Genet Dev 12:60–66

Hodge LD, Scharff MD (1969) Effect of adenovirus on host cell DNA synthesis in synchronized cells. Virology 37:554–564
Honda R, Yasuda H (2000) Activity of MDM2, a ubiquitin ligase, toward p53 or itself is dependent on the RING finger domain of the ligase. Oncogene 19:1473–1476
Horwitz MS (2001) Adenovirus immunoregulatory genes and their cellular targets. Virology 279:1–8
Howe JA, Mymryk JS, Egan C, Branton PE, Bayley ST (1990) Retinoblastoma growth suppressor and a 300-kDa protein appear to regulate cellular DNA synthesis. Proc Natl Acad Sci USA 87:5883–5887
Huang DC, Cory S, Strasser A (1997) Bcl-2, Bcl-XL and adenovirus protein E1b19kD are functionally equivalent in their ability to inhibit cell death. Oncogene 14:405–414
Huang MM, Hearing P (1989) The adenovirus early region 4 open reading frame 6/7 protein regulates the DNA binding activity of the cellular transcription factor, E2F, through a direct complex. Genes Dev 3:1699–1710
Hutchin ME, Pickles RJ, Yarbrough WG (2000) Efficiency of adenovirus-mediated gene transfer to oropharyngeal epithelial cells correlates with cellular differentiation and human coxsackie and adenovirus receptor expression. Hum Gene Ther 11:2365–2375
Ikeda MA, Nevins JR (1993) Identification of distinct roles for separate E1a domains in disruption of E2F complexes. Mol Cell Biol 13:7029–7035
Jones N, Shenk T (1979) Isolation of adenovirus type 5 host range deletion mutants defective for transformation of rat embryo cells. Cell 17:683–689
Kao CC, Yew PR, Berk AJ (1990) Domains required for in vitro association between the cellular p53 and the adenovirus 2 E1b 55 K proteins. Virology 179:806–814
Kimelman D (1986) A novel general approach to eukaryotic mutagenesis functionally identifies conserved regions within the adenovirus 13S E1a polypeptide. Mol Cell Biol 6:1487–1496
Kimelman D, Miller JS, Porter D, Roberts BE (1985) E1a regions of the human adenoviruses and of the highly oncogenic simian adenovirus 7 are closely related. J Virol 53:399–409
Klessig DF (1977) Two adenovirus mRNAs have a common 5′ terminal leader sequence encoded at least 10 kb upstream from their main coding regions. Cell 12:9-21
Kovesdi I, Reichel R, Nevins JR (1987) Role of an adenovirus E2 promoter binding factor in E1a-mediated coordinate gene control. Proc Natl Acad Sci USA 84:2180–2184
Krajcsi P, Dimitrov T, Hermiston TW, Tollefson AE, Ranheim TS, Vande Pol SB, Stephenson AH, Wold WS (1996) The adenovirus E3-14.7 K protein and the E3-10.4 K/14.5 K complex of proteins, which independently inhibit tumor necrosis factor (TNF)-induced apoptosis, also independently inhibit TNF-induced release of arachidonic acid. J Virol 70:4904–4913
Kubbutat MH, Jones SN, Vousden KH (1997) Regulation of p53 stability by MDM2. Nature 387:299–303
Laughlin C, Strohl WA (1976) Factors regulating cellular DNA synthesis induced by adenovirus infection. II. The effects of actinomycin D on productive virus-cell systems. Virology 74:44–56
Lavia P, Jansen-Durr P (1999) E2F target genes and cell-cycle checkpoint control. Bioessays 21:221–230
Lavoie JN, Nguyen M, Marcellus RC, Branton PE, Shore GC (1998) E4ORF4, a novel adenovirus death factor that induces p53-independent apoptosis by a pathway that is not inhibited by zVAD-fmk. J Cell Biol 140:637–645
Lewis JB, Atkins JF, Baum PR, Solem R, Gesteland RF, Anderson CW (1976) Location and identification of the genes for adenovirus type 2 early polypeptides. Cell 7:141–151
Li E, Stupack D, Klemke R, Cheresh DA, Nemerow GR (1998) Adenovirus endocytosis via alpha(v) integrins requires phosphoinositide-3-OH kinase. J Virol 72:2055–2061
Lill NL, Grossman SR, Ginsberg D, DeCaprio J, Livingston DM (1997) Binding and modulation of p53 by p300/CBP coactivators. Nature 387:823–827
Liu HT, Baserga R, Mercer WE (1985) Adenovirus type 2 activates cell cycle dependent genes that are a subset of those activated by serum. Mol Cell Biol 5:2936–2942

Liu L, Scolnick DM, Trievel RC, Zhang HB, Marmorstein R, Halazonetis TD, Berger SL (1999) p53 sites acetylated in vitro by PCAF and p300 are acetylated in vivo in response to DNA damage. Mol Cell Biol 19:1202–1209

Lober C, Lenz-Stoppler C, Dobbelstein M (2002) Adenovirus E1-transformed cells grow despite the continuous presence of transcriptionally active p53. J Gen Virol 83:2047–2057

Lomonosova E, Subramanian T, Chinnadurai G (2002) Requirement of Bax for efficient adenovirus-induced apoptosis. J Virol 76:11283–11290

Lowe SW, Ruley HE (1993) Stabilization of the p53 tumour suppressor is induced by adenovirus 5 E1a and accompanies apoptosis. Genes Dev 7:535–545

Luo RX, Postigo AA, Dean DC (1998) Rb interacts with histone deacetylase to repress transcription. Cell 92:463–473

Mao L, Merlo A, Bedi G, Shapiro GI, Edwards CD, Rollins BJ, Sidransky D (1995) A novel p16INK4A transcript. Cancer Res 55:2995–2997

Marcellus RC, Teodoro JG, Wu T, Brough DE, Ketner G, Shore GC, Branton PE (1996) Adenovirus type 5 early region 4 is responsible for E1a-induced p53-independent apoptosis. J Virol 70:6207–6215

Marcellus RC, Lavoie JN, Boivin D, Shore GC, Ketner G, Branton PE (1998) The early region 4 ORF4 protein of human adenovirus type 5 induces p53-independent cell death by apoptosis. J Virol 72:7144–7153

Martin ME, Berk AJ (1998) Adenovirus E1b 55 K represses p53 activation in vitro. J Virol 72:3146–3154

Martinez-Balbas MA, Bauer UM, Nielsen SJ, Brehm A, Kouzarides T (2000) Regulation of E2F1 activity by acetylation. EMBO J 19:662–671

Marton MJ, Baim SB, Ornelles DA, Shenk T (1990) The adenovirus E4 17-kilodalton protein complexes with the cellular transcription factor E2F, altering its DNA-binding properties and stimulating E1a-independent accumulation of E2 mRNA. J Virol 64:2345–2359

Meier O, Boucke K, Hammer SV, Keller S, Stidwill RP, Hemmi S, Greber UF (2002) Adenovirus triggers macropinocytosis and endosomal leakage together with its clathrin-mediated uptake. J Cell Biol 158:1119–1131

Mittnacht S, Lees JA, Desai D, Harlow E, Morgan DO, Weinberg RA (1994) Distinct subpopulations of the retinoblastoma protein show a distinct pattern of phosphorylation. EMBO J 13:118–127

Moise AR, Grant JR, Vitalis TZ, Jefferies WA (2002) Adenovirus E3-6.7 K maintains calcium homeostasis and prevents apoptosis and arachidonic acid release. J Virol 76:1578–1587

Montell C, Fisher EF, Caruthers MH, Berk AJ (1984) Control of adenovirus E1b mRNA synthesis by a shift in the activities of RNA splice sites. Mol Cell Biol 4:966–972

Moran E, Zerler B, Harrison TM, Mathews MB (1986) Identification of separate domains in the adenovirus E1a gene for immortalization activity and the activation of virus early genes. Mol Cell Biol 6:3470–3480

Morris L, Allen KE, La Thangue NB (2000) Regulation of E2F transcription by cyclin E-Cdk2 kinase mediated through p300/CBP co-activators. Nat Cell Biol 2:232–239

Muller H, Bracken AP, Vernell R, Moroni MC, Christians F, Grassilli E, Prosperini E, Vigo E, Oliner JD, Helin K (2001) E2Fs regulate the expression of genes involved in differentiation, development, proliferation, and apoptosis. Genes Dev 15:267–285

Munger K, Basile JR, Duensing S, Eichten A, Gonzalez SL, Grace M, Zacny VL (2001) Biological activities and molecular targets of the human papillomavirus E7 oncoprotein. Oncogene 20:7888–7898

Murray JD, Bellett AJ, Braithwaite A, Waldron LK, Taylor IW (1982a) Altered cell cycle progression and aberrant mitosis in adenovirus-infected rodent cells. J Cell Physiol 111:89–96

Murray JD, Braithwaite AW, Taylor IW, Bellett AJ (1982b) Adenovirus-induced alterations of the cell growth cycle: effects of mutations in early regions E2a and E2b. J Virol 44:1072–1075

Mymryk JS, Shire K, Bayley ST (1994) Induction of apoptosis by adenovirus type 5 E1a in rat cells requires a proliferation block. Oncogene 9:1187–1193

Nakajima T, Uchida C, Anderson SF, Lee CG, Hurwitz J, Parvin JD, Montminy M (1997) RNA helicase A mediates association of CBP with RNA polymerase II. Cell 90:1107–1112

Nakano MY, Boucke K, Suomalainen M, Stidwill RP, Greber UF (2000) The first step of adenovirus type 2 disassembly occurs at the cell surface, independently of endocytosis and escape to the cytosol. J Virol 74:7085–7095

Neill SD, Hemstrom C, Virtanen A, Nevins JR (1990) An adenovirus E4 gene product trans-activates E2 transcription and stimulates stable E2F binding through a direct association with E2F. Proc Natl Acad Sci USA 87:2008–2012

Nevins JR, Darnell JE (1978) Groups of adenovirus type 2 mRNA's derived from a large primary transcript: probable nuclear origin and possible common 3′ ends. J Virol 25:811–823

Nishimoto T, Raskas HJ, Basilico C (1975) Temperature-sensitive cell mutations that inhibit adenovirus 2 replication. Proc Natl Acad Sci USA 72:328–332

O'Brien V (1998) Viruses and apoptosis. J Gen Virol 79:1833–1845

O'Connor MJ, Zimmermann H, Nielsen S, Bernard HU, Kouzarides T (1999) Characterization of an E1a-CBP interaction defines a novel transcriptional adapter motif (TRAM) in CBP/p300. J Virol 73:3574–3581

Obert S, O'Connor RJ, Schmid S, Hearing P (1994) The adenovirus E4-6/7 protein transactivates the E2 promoter by inducing dimerization of a heteromeric E2F complex. Mol Cell Biol 14:1333–1346

Ornelles DA, Shenk T (1991) Localization of the adenovirus early region 1b 55-kilodalton protein during lytic infection: association with nuclear viral inclusions requires the early region 4 34-kilodalton protein. J Virol 65:424–429

Phillips AC, Vousden KH (2001) E2F-1 induced apoptosis. Apoptosis 6:173–182

Pilder S, Logan J, Shenk T (1984) Deletion of the gene encoding the adenovirus 5 early region 1b 21,000-molecular-weight polypeptide leads to degradation of viral and host cell DNA. J Virol 52:664–671

Pilder S, Moore M, Logan J, Shenk T (1986) The adenovirus E1b-55 K transforming polypeptide modulates transport or cytoplasmic stabilization of viral and host cell mRNAs. Mol Cell Biochem 6:470–476

Polager S, Kalma Y, Berkovich E, Ginsberg D (2002) E2Fs up-regulate expression of genes involved in DNA replication, DNA repair and mitosis. Oncogene 21:437–446

Pomerantz J, Schreiber-Agus N, Liegeois NJ, Silverman A, Alland L, Chin L, Potes J, Chen K, Orlow I, Lee HW, Cordon-Cardo C, DePinho RA (1998) The INK4a tumor suppressor gene product, p19^{ARF}, interacts with MDM2 and neutralizes MDM2's inhibition of p53. Cell 92:713–723

Prives C, Hall PA (1999) The p53 pathway. J Pathol 187:112–126

Putzer BM, Stiewe T, Parssanedjad K, Rega S, Esche H (2000) E1a is sufficient by itself to induce apoptosis independent of p53 and other adenoviral gene products. Cell Death Differ 7:177–188

Qin XQ, Livingston DM, Kaelin WG Jr, Adams PD (1994) Deregulated transcription factor E2F-1 expression leads to S-phase entry and p53-mediated apoptosis. Proc Natl Acad Sci USA 91:10918–10922

Querido E, Teodoro JG, Branton PE (1997) Accumulation of p53 induced by the adenovirus E1a protein requires regions involved in the stimulation of DNA synthesis. J Virol 71:3526–3533

Quinlan MP, Grodzicker T (1987) Adenovirus E1a 12S protein induces DNA synthesis and proliferation in primary epithelial cells in both the presence and absence of serum. J Virol 61:673–682

Rao L, Debbas M, Sabbatini P, Hockenbery D, Korsmeyer S, White E (1992) The adenovirus E1a proteins induce apoptosis, which is inhibited by the E1b 19-kDa and Bcl-2 proteins. Proc Natl Acad Sci USA 89:7742–7746

Rauma T, Tuukkanen J, Bergelson JM, Denning G, Hautala T (1999) rab5 GTPase regulates adenovirus endocytosis. J Virol 73:9664–9668

Reichel R, Kovesdi I, Nevins JR (1988) Activation of a preexisting cellular factor as a basis for adenovirus E1a-mediated transcription control. Proc Natl Acad Sci USA 85:387–390

Reid JL, Bannister AJ, Zegerman P, Martinez-Balbas MA, Kouzarides T (1998) E1a directly binds and regulates the P/CAF acetyltransferase. EMBO J 17:4469–4477
Ridgway PJ, Hall AR, Myers CJ, Braithwaite AW (1997) p53/E1b58 kDa complex regulates adenovirus replication. Virology 237:404–413
Roelvink PW, Lizonova A, Lee JG, Li Y, Bergelson JM, Finberg RW, Brough DE, Kovesdi I, Wickham TJ (1998) The coxsackievirus-adenovirus receptor protein can function as a cellular attachment protein for adenovirus serotypes from subgroups A, C, D, E, and F. J Virol 72:7909–7915
Rossini M, Baserga S, Huang CH, Ingles CJ, Baserga R (1980) Changes in RNA polymerase II in a cell cycle-specific temperature-sensitive mutant of hamster cells. J Cell Physiol 103:97–103
Rossini M, Jonak GJ, Baserga R (1981) Identification of adenovirus 2 early genes required for induction of cellular DNA synthesis in resting hamster cells. J Virol 38:982–986
Roulston A, Marcellus RC, Branton PE (1999) Viruses and apoptosis. Annu Rev Microbiol 53:577–628
Routes JM, Ryan S, Clase A, Miura T, Kuhl A, Potter TA, Cook JL (2000) Adenovirus E1a oncogene expression in tumor cells enhances killing by TNF-related apoptosis-inducing ligand (TRAIL). J Immunol 165:4522–4527
Sakaguchi K, Herrera JE, Saito S, Miki T, Bustin M, Vassilev A, Anderson CW, Appella E (1998) DNA damage activates p53 through a phosphorylation-acetylation cascade. Genes Dev 12:2831–2841
Sarnow P, Sullivan CA, Levine AJ (1982) A monoclonal antibody detecting the adenovirus type 5-E1b-58Kd tumor antigen: characterization of the E1b-58Kd tumor antigen in adenovirus-infected and -transformed cells. Virology 120:510–517
Sarnow P, Hearing P, Anderson CW, Halbert DN, Shenk T, Levine AJ (1984) Adenovirus early region 1b 58,000-dalton tumor antigen is physically associated with an early region 4 25,000-dalton protein in productively infected cells. J Virol 49:692–700
Schaack J, Ho WY, Freimuth P, Shenk T (1990) Adenovirus terminal protein mediates both nuclear matrix association and efficient transcription of adenovirus DNA. Genes Dev 4:1197–1208
Schmid SI, Hearing P (1995) Selective encapsidation of adenovirus DNA. Curr Top Microbiol Immunol 199:67–80
Shales M, Bergsagel J, Ingles CJ (1980) Defective RNA polymerase II in the G1 specific temperature sensitive hamster cell mutant tsAF8. J Cell Physiol 105:527–532
Shan B, Lee W-H (1994) Deregulated expression of E2F-1 induces S-phase entry and leads to apoptosis. Mol Cell Biol 14:8166–8173
Shaw AR, Ziff EB (1980) Transcripts from the adenovirus-2 major late promoter yield a single early family of 3′ coterminal mRNAs and five late families. Cell 22:905–916
Sheinin R (1966) Deoxyribonucleic acid synthesis in cells replicating polyoma virus. Virology 28:621–632
Shenk T (1996) Adenoviridae: the viruses and their replication. In: Fields BN, Knipe DM, Howley PM, et al (eds). Fields virology, vol 2. Lippincott-Raven, Philadelphia, pp 2111–2148
Sherr CJ (1996) Cancer cell cycles. Science 274:1672–1677
Shimojo H, Yamashita T (1968) Induction of DNA synthesis by adenoviruses in contact-inhibited hamster cells. Virology 36:422–433
Shiroki K, Maruyama K, Saito I, Fukui Y, Shimojo H (1981) Incomplete transformation of rat cells by a deletion mutant of adenovirus type 5. J Virol 38:1048–1054
Shisler J, Yang C, Walter B, Ware CF, Gooding LR (1997) The adenovirus E3-10.4 K/14.5 K complex mediates loss of cell surface Fas (CD95) and resistance to Fas-induced apoptosis. J Virol 71:8299–8306
Shtrichman R, Kleinberger T (1998) Adenovirus type 5 E4 open reading frame 4 protein induces apoptosis in transformed cells. J Virol 72:2975–2982
Shtrichman R, Sharf R, Kleinberger T (2000) Adenovirus E4ORF4 protein interacts with both Balpha and B′ subunits of protein phosphatase 2A, but E4ORF4-induced apoptosis is mediated only by the interaction with Balpha. Oncogene 19:3757–3765

Somasundaram K, El-Deiry WS (1997) Inhibition of p53-mediated transactivation and cell cycle arrest by E1a through its p300/CBP-interacting region. Oncogene 14:1047–1057
Southern EM (1975) Detection of specific sequences among DNA fragments separated by gel electrophoresis. J Mol Biol 98:503–517
Spector DJ, McGrogan M, Raskas HJ (1978) Regulation of the appearance of cytoplasmic RNAs from region 1 of the adenovirus 2 genome. J Mol Biol 126:395–414
Spindler KR, Eng CY, Berk AJ (1985) An adenovirus early region 1a protein is required for maximal viral DNA replication in growth-arrested human cells. J Virol 53:742–750
Stabel S, Argos P, Philipson L (1985) The release of growth arrest by microinjection of adenovirus E1a DNA. EMBO J 4:2329–2336
Steegenga WT, Riteco N, Jochemsen AG, Fallaux FJ, Bos JL (1998) The large E1b protein together with the E4ORF6 protein target p53 for active degradation in adenovirus infected cells. Oncogene 16:349–357
Stein RW, Corrigan M, Yaciuk P, Whelan J, Moran E (1990) Analysis of E1a-mediated growth regulation functions: binding of the 300-kilodalton cellular product correlates with E1a enhancer repression function and DNA synthesis-inducing activity. J Virol 64:4421–4427
Stewart PL, Chiu CY, Huang S, Muir T, Zhao Y, Chait B, Mathias P, Nemerow GR (1997) Cryo-EM visualization of an exposed RGD epitope on adenovirus that escapes antibody neutralization. EMBO J 16:1189–1198
Strohl WA (1969) The response of BHK21 cells to infection with type 12 adenovirus. II. Relationship of virus-stimulated DNA synthesis to other viral functions. Virology 39:653–665
Strohl WA (1973) Alterations in hamster cell regulatory mechanisms resulting from abortive infection with an oncogenic adenovirus. Prog Exp Tumor Res 18:199–239
Subramanian T, Tarodi B, Chinnadurai G (1995) p53-independent apoptotic and necrotic cell deaths induced by adenovirus infection: suppression by E1b 19 K and Bcl-2 proteins. Cell Growth Differ 6:131–137
Sugaya K, Sasanuma S, Cook PR, Mita K (2001) A mutation in the largest (catalytic) subunit of RNA polymerase II and its relation to the arrest of the cell cycle in G(1) phase. Gene 274:77–81
Sun XF, Shimizu H, Yamamoto K (1995) Identification of a novel p53 promoter element involved in genotoxic stress-inducible p53 gene expression. Mol Cell Biol 15:4489–4496
Takahashi M, Van Hoosier GL, Jr., Trentin JJ (1966) Stimulation of DNA synthesis in human and hamster cells by human adenovirus types 12 and 5. Proc Soc Exp Biol Med 122:740–746
Takahashi M, Ogino T, Baba K, Onaka M (1969) Synthesis of deoxyribonucleic acid in human and hamster kidney cells infected with human adenovirus types 5 and 12. Virology 37:513–520
Takemori N, Riggs JL, Aldrich C (1968) Genetic studies with tumorigenic adenoviruses.1. Isolation of cytocidal (cyt) mutants of adenovirus type 12. Virology 36:575–586
Takemori N, Cladaras C, Bhat B, Conley AJ, Wold WS (1984) Cyt gene of adenoviruses 2 and 5 is an oncogene for transforming function in early region E1b and encodes the E1b 19,000-molecular-weight polypeptide. J Virol 52:793–805
Teodoro JG, Branton PE (1997) Regulation of p53-dependent apoptosis, transcriptional repression, and cell transformation by phosphorylation of the 55-kilodalton E1b protein of human adenovirus type 5. J Virol 71:3620–3627
Tollefson AE, Stewart AR, Yei SP, Saha SK, Wold WS (1991) The 10,400- and 14,500-dalton proteins encoded by region E3 of adenovirus form a complex and function together to down-regulate the epidermal growth factor receptor. J Virol 65:3095–3105
Tollefson AE, Ryerse JS, Scaria A, Hermiston TW, Wold WS (1996a) The E3-11.6-kDa adenovirus death protein (ADP) is required for efficient cell death: characterization of cells infected with adp mutants. Virology 220:152–162
Tollefson AE, Scaria A, Hermiston TW, Ryerse JS, Wold LJ, Wold WSM (1996b) The adenovirus death protein (E3-11.6 k) is required at very late stages of infection for efficient cell lysis and release of adenovirus from infected cells. J Virol 70:2296–2306

Tollefson AE, Toth K, Doronin K, Kuppuswamy M, Doronina OA, Lichtenstein DL, Hermiston TW, Smith CA, Wold WS (2001) Inhibition of TRAIL-induced apoptosis and forced internalization of TRAIL receptor 1 by adenovirus proteins. J Virol 75:8875–8887
Tomko RP, Xu R, Philipson L (1997) HCAR and MCAR: the human and mouse cellular receptors for subgroup C adenoviruses and group B coxsackieviruses. Proc Natl Acad Sci USA 94:3352–3356
Trentin JJ, Yabe Y, Taylor G (1962) The quest for human cancer viruses. Science 137:835–849
Turnell AS, Grand RJ, Gallimore PH (1999) The replicative capacities of large E1b-null group A and group C adenoviruses are independent of host cell p53 status. J Virol 73:2074–2083
van Ormondt H, Maat J, Dijkema R (1980) Comparison of nucleotide sequences of the early E1a regions for subgroups A, B and C of human adenoviruses. Gene 12:63–76
Weber JD, Taylor LJ, Roussel MF, Sherr CJ, Bar-Sagi D (1999) Nucleolar ARF sequesters MDM2 and activates p53. Nat Cell Biol 1:20–26
Wei MC, Zong WX, Cheng EH, Lindsten T, Panoutsakopoulou V, Ross AJ, Roth KA, MacGregor GR, Thompson CB, Korsmeyer SJ (2001) Proapoptotic Bax and Bak: a requisite gateway to mitochondrial dysfunction and death. Science 292:727–730
Weinberg DH, Ketner G (1986) Adenoviral early region 4 is required for efficient viral DNA replication and for late gene expression. J Virol 57:833–838
White E, Sabbatini P, Debbas M, Wold W, Kusher D, Gooding L (1992) The 19-Kilodalton adenovirus E1b transforming protein inhibits programmed cell death and prevents cytolysis by tumour necrosis factor α. Mol Cell Biol 12:2570–2580
Whyte P, Buchkovich KJ, Horowitz JM, Friend SH, Raybuck M, Weinberg RA, Harlow E (1988a) Association between an oncogene and an anti-oncogene: the adenovirus E1a proteins bind to the retinoblastoma gene product. Nature 334:124–129
Whyte P, Ruley HE, Harlow E (1988b) Two regions of the adenovirus early region 1a proteins are required for transformation. J Virol 62:257–265
Whyte P, Williamson NM, Harlow E (1989) Cellular targets for transformation by the adenovirus E1a proteins. Cell 56:67–75
Wickham TJ, Mathias P, Cheresh DA, Nemerow GR (1993) Integrins alpha v beta 3 and alpha v beta 5 promote adenovirus internalization but not virus attachment. Cell 73:309–319
Wienzek S, Roth J, Dobbelstein M (2000) E1b 55-kilodalton oncoproteins of adenovirus types 5 and 12 inactivate and relocalize p53, but not p51 or p73, and cooperate with E4ORF6 proteins to destabilize p53. J Virol 74:193–202
Wilson MC, Darnell JE Jr (1981) Control of messenger RNA concentration by differential cytoplasmic half-life. Adenovirus messenger RNAs from transcription units 1a and 1b. J Mol Biol 148:231–251
Winocour E, Kaye AM, Stollar V (1965) Synthesis and transmethylation of DNA in polyoma-infected cultures. Virology 27:156–169
Wu X, Levine AJ (1994) p53 and E2F-1 cooperate to mediate apoptosis. Proc Natl Acad Sci USA 91:3602–3606
Wyllie FS, Haughton MF, Rowson JM, Wynford-Thomas D (1999) Human thyroid cancer cells as a source of iso-genic, iso-phenotypic cell lines with or without functional p53. Br J Cancer 79:1111–1120
Yamashita T, Shimojo H (1969) Induction of cellular DNA synthesis by adenovirus 12 in human embryo kidney cells. Virology 38:351–355
Yang Y, McKerlie C, Borenstein SH, Lu Z, Schito M, Chamberlain JW, Buchwald M (2002) Transgenic expression in mouse lung reveals distinct biological roles for the adenovirus type 5 E1a 243- and 289-amino-acid proteins. J Virol 76:8910–8919
Yee SP, Branton PE (1985) Detection of cellular proteins associated with human adenovirus type 5 early region 1a polypeptides. Virology 147:142–153
Yew PR, Berk AJ (1992) Inhibition of p53 transactivation required for transformation by adenovirus early 1b protein. Nature 357:82–85

Yoder SS, Robberson BL, Leys EJ, Hook AG, Al-Ubaidi M, Yeung CY, Kellems RE, Berget SM (1983) Control of cellular gene expression during adenovirus infection: induction and shut-off of dihydrofolate reductase gene expression by adenovirus type 2. Mol Cell Biol 3:819–828
Zerler B, Roberts RJ, Mathews MB, Moran E (1987) Different functional domains of the adenovirus E1a gene are involved in regulation of host cell cycle products. Mol Cell Biol 7:821–829
Zhang XL, Bellett AJ, Hla RT, Braithwaite AW, Mullbacher A (1991) Adenovirus type 5 E3 gene products interfere with the expression of the cytolytic T cell immunodominant E1a antigen. Virology 180:199–206
Zhang X, Bellett AJ, Hla RT, Voss T, Mullbacher A, Braithwaite AW (1994) Down-regulation of human adenovirus E1a by E3 gene products: evidence for translational control of E1a by E3 14.5 K and/or E3 10.4 K products. J Gen Virol 75:1943–1951
Zhao LJ, Padmanabhan R (1988) Nuclear transport of adenovirus DNA polymerase is facilitated by interaction with preterminal protein. Cell 55:1005–1015
Ziff EB, Evans RM (1978) Coincidence of the promoter and capped 5′ terminus of RNA from the adenovirus 2 major late transcription unit. Cell 15:1463–1475
Zimmerman JE Jr, Raska K Jr, Strohl WA (1970) The response of BHK21 cells to infection with type 12 adenovirus. IV. Activation of DNA-synthesizing apparatus. Virology 42:1147–1150
Zong WX, Lindsten T, Ross AJ, MacGregor GR, Thompson CB (2001) BH3-only proteins that bind pro-survival Bcl-2 family members fail to induce apoptosis in the absence of Bax and Bak. Genes Dev 15:1481–1486

Induction of Transformed Cell-Specific Apoptosis by the Adenovirus E4orf4 Protein

T. Kleinberger[1]

1 Introduction

1.1 Mechanisms of Apoptosis

Apoptosis is a genetically regulated cellular process, which plays an important role in conservation of homeostasis and protection from tumorigenesis. Apoptosis can be induced by various signals that activate the evolutionarily conserved core cell death apparatus, committing the cells to die. The commitment to cell death triggers a cascade of events that results in acquisition of morphological and biochemical features unique to apoptotic cells. These include alterations in the cytoskeleton leading to cell rounding, cell shrinkage and membrane blebbing; chromatin condensation; and DNA fragmentation. At the final stages, the cell fragments into apoptotic bodies that are engulfed by neighboring cells.

The core death machinery includes Bcl-2 family members and cysteine proteases called caspases. Caspases are constitutively expressed as catalytically inactive proenzymes, and are activated by proteolytic processing (reviewed in Cryns and Yuan 1998). Currently, two main caspase-activating pathways that regulate apoptosis are well characterized. The "extrinsic pathway" is initiated from cell surface death receptors, and the "intrinsic pathway" is triggered by various stress signals and involves changes in mitochondrial integrity (reviewed in Budihardjo et al. 1999). The death receptors are a family of transmembrane proteins, including Fas/APO-1/CD95 and the TNF receptor, which share a region of homology at the cytoplasmic face, termed the death domain. Upon binding of ligands to the receptors and subsequent receptor trimerization, the death domains recruit adaptor proteins, which contain a death domain and a death effector domain. The death domain of Fas, for example, recruits an adaptor protein called FADD/MORT1, and the FADD/MORT1 death effector domain is critical for binding an upstream procaspase, such as procaspase-8 or -10, to form the "DISC", death-inducing signaling

[1]The Gonda Center of Molecular Microbiology, The Bruce Rappaport Faculty of Medicine, Technion – Israel Institute of Technology, 31096 Haifa , Israel, e-mail: tamark@tx.technion.ac.il

Progress in Molecular and Subcellular Biology
C. Alonso (Ed.): Viruses and Apoptosis
© Springer-Verlag Berlin Heidelberg 2004

complex (Peter and Krammer 1998; Wallach et al. 1999). Immediately following recruitment, the procaspase is proteolytically processed into the active caspase form. Activation of caspase-8 can lead to a direct cleavage and activation of downstream caspases, such as caspase-3, -6, and -7, or can activate the mitochondrial apoptotic pathway (Wallach et al. 1999).

The second caspase activation cascade involves mitochondrial changes induced by several stress signals, such as serum starvation, oncogene activation, DNA damaging agents, kinase inhibitors, etc., as well as activation of cell surface death receptors (Yang et al. 1997; Scaffidi et al. 1998). Cytochrome c is released from the mitochondria and associates with two cytosolic proteins, Apaf-1 and procaspase-9. Caspase-9 is activated in this complex, and is released to further activate downstream caspases (reviewed in Budihardjo et al. 1999). Bcl-2 family members regulate cytochrome c release from the mitochondria. Overexpression of Bcl-2 or Bcl-xL blocks the release, whereas the proapoptotic Bcl-2 proteins, including Bax and Bid, promote it.

The phosphorylation status of various apoptotic factors has been shown to affect their activity as cell death regulators (Ito et al. 1997; Datta et al. 1999; Yamamoto et al. 1999). Thus kinases and phosphatases may well influence the apoptotic process.

Various recent reports indicate the existence of caspase-independent apoptosis (Green and Kroemer 1998; Monney et al. 1998; Quignon et al. 1998; Elliott et al. 2000). Thus, whereas in some cases inhibition of caspases prevented cell death (Tewari et al. 1995), in others death was delayed, but not abolished (Xiang et al. 1996).

1.2
Adenoviruses and Apoptosis

Virus infection results in many host responses that may lead to death of the virally infected cells. These responses include activation of the immune system, interferon production, and induction of apoptosis. Often, viruses infect quiescent cells, which do not provide an optimal environment for viral DNA synthesis due to rate-limiting levels of deoxynucleotides and low levels of proteins involved in DNA synthesis. Various viruses, including adenoviruses, have evolved different means to overcome this obstacle, by deregulating the cell cycle machinery (reviewed in Ben-Israel and Kleinberger 2002). This interference with normal cell cycle control may lead to induction of apoptosis. To prevent premature cell death, which would be deleterious for virus replication, these viruses have also evolved several mechanisms to hinder the apoptotic process. Thus, adenoviruses, as well as other viruses, contain several genes whose products either induce, or protect against, apoptosis. The most-studied adenovirus inducer of apoptosis is the E1A gene product. E1A overrides normal regulatory constraints at the G1/S border by acting on at least three levels: (1) by inactivating pRb-family proteins and releasing active E2F transcription factor; (2) by modulating the function of chromatin remodeling factors; and (3) by targeting additional cellular proteins, including downstream targets of cdk2 and transcription factors involved in regulation of genes that participate in cell cycle control. The first two mechanisms lead to stabilization and activation of the tumor suppressor p53 protein, which acts to induce cell cycle arrest or apoptosis, depending on the cellular circumstances. In addition, E2F can induce p53-independent apoptosis. At the same time, adenovirus expresses two products of the E1B transcription unit, which counteract the unwanted consequences of E1A action. The E1B-55 kDa protein utilizes several mechanisms to inactivate p53, including binding p53 and repressing its ability to transactivate transcription, and, together with another adenovirus protein, E4orf6, targeting p53 for degradation by the proteasome (reviewed in Ben-Israel and Kleinberger 2002). The E1B-19 kDa protein is a Bcl-2 family homologue and acts similarly to its homologues to inhibit both p53-dependent and -independent apoptosis (White 2001).

This review will discuss another early adenovirus protein, E4orf4, whose activities lead to induction of multiple cellular mechanisms, resulting in apoptosis. The finding (described in Sect. 2.2) that oncogenic transformation sensitizes cells to the apoptotic effect of E4orf4 adds an exciting clinical potential to E4orf4 research.

2
E4orf4 – a Multifunctional Viral Regulator

The adenovirus genome contains five transcription units (E1A, E1B, E2, E3, E4), which are transcribed early during infection. The E4 transcription unit encodes seven open reading frames (orfs), one of which is E4orf4 (Shenk 1996). Although transcription of the early units ceases during the late phase of viral infection, the E4orf4 protein is presumably stable and can still be detected in cells late in the infectious cycle (Boivin et al. 1999). In HeLa cells, commonly used in adenovirus research, deletion of E4orf4 is only marginally deleterious to viral growth, reducing viral titer by only ten-fold (Halbert et al. 1985; unpubl. results). However, in an untransformed immortalized rat fibroblast cell line, CREF, which is non-permissive for adenovirus replication, a deletion of E4orf4 led to increased toxicity, resulting in a significantly reduced ability of the infected cells to form colonies (Müller et al. 1992). These results suggest that in situations of persistent adenovirus infection, normally occurring in its human host, the E4orf4 protein may play an important role in supporting virus replication.

Like many other viral regulators, the 14-kDa E4orf4 protein has been shown to perform multiple tasks during adenovirus infection. Previous work has shown that the adenovirus E1A proteins cooperate with cAMP in the S49 mouse T cell lymphoma cell line to induce the accumulation of AP-1 transcription factor. Elevated AP-1 levels result from activated transcription of the genes encoding the AP-1 components, c-fos and junB. Increased AP-1 levels activate transcription of early adenovirus genes through AP-1 and ATF sites in adenoviral promoters (Engel et al. 1988; Müller et al. 1989). One of the activated genes is the E4 transcription unit, encoding E4orf4. Experiments using viral mutants lacking E4orf4 have indicated that increased E4orf4 levels result in downregulation of AP-1 due to inhibition of *junB* and *c-fos* transcription and repression of c-fos translation (Müller et al. 1992). As a result, E4orf4 expression leads to downregulation of most of the early viral promoters, including E2 and E4 itself. It appears, therefore, that E4orf4 functions to counterbalance transactivation of cellular and viral genes by E1A and cAMP. In addition to the effect exerted by E4orf4 through AP-1, transfection experiments have suggested that it reduces E4 expression in HeLa cells through other transcription factors, which may include ATF-2 and E4F (Bondesson et al. 1996). Moreover, E4orf4 inhibits E2 expression in HeLa and SAOS-2 cells, possibly by reducing the stability of E2F/DNA complexes binding to the E2 promoter (Mannervik et al. 1999). Decreased E2 expression, attributed to E4orf4, has also been shown to occur during adenovirus infection, and it is much enhanced when the E4orf3 and E4orf6 gene products are absent (Medghalchi et al. 1997). A strong temporal control of viral gene expression, as seems to be exerted by the adenovirus E4orf4 protein, has been shown to contribute to efficient infections by several other viruses as well.

In addition to its effects on transcription and translation, the E4orf4 gene product has been shown to be responsible for alternative splicing of late viral mRNAs (Kanopka et al. 1998).

2.1
E4orf4 Interacts with Protein Phosphatase 2A

During early studies of E4orf4, it was observed that E4orf4 expression correlates with the appearance of hypophosphorylated viral and cellular proteins, including E1A and c-fos (Müller et al. 1992). It was soon discovered that E4orf4 associates with the cellular protein phosphatase 2A (PP2A) by binding directly to the regulatory Bα subunit in a complex that contains all three subunits (A, B, and C) of PP2A (Kleinberger and Shenk 1993).

PP2A is one of the major protein serine/threonine phosphatases in the cell, playing a role in several cellular processes, including metabolism, transcription, RNA splicing, translation, cell cycle progression, morphogenesis, signal transduction, development, and transformation (Mumby and Walter 1993; Wera and Hemmings 1995). The predominant form of PP2A in cells is a heterotrimer consisting of three subunits. Two of them, the 36 kDa catalytic C subunit and the 63 kDa scaffolding A subunit (PR65), form the core enzyme, and the regulatory B subunit binds the core enzyme to form the holoenzyme. The A and C subunits both exist as two isoforms (α and β), which are closely related. The B subunit is variable, and its multiple isoforms belong to at least three unrelated gene families, B/B55/PR55 (including 4 isoforms: α, β, γ, δ), B′/B56/PR61 (including α, β, γ, δ, ε), and B″ (including PR48, PR59, PR72 and PR130); (Kamibayashi et al. 1994; McCright and Virshup 1995; Csortos et al. 1996). Viral proteins, such as the SV40 small t antigen and the polyomavirus small and middle T antigens, can replace the cellular B subunits (Mumby 1995). The various cellular PP2A B subunits target the PP2A holoenzyme to different substrates and dictate its subcellular localization (Virshup 2000).

The association of E4orf4 with PP2A was shown to be required for transcriptional repression by the viral protein, since the addition of okadaic acid, a PP2A inhibitor, reversed E4orf4-induced downregulation of transcription from the junB, E2, and E4 promoters (Kleinberger and Shenk 1993; Bondesson et al. 1996; Mannervik et al. 1999). The sites of E4orf4-induced hypophosphorylation in E1A have been identified, and include Ser-185 and Ser-188. However, although in CHO cells enhanced phosphorylation of these sites correlated with enhanced E4 promoter transactivation by E1A (Whalen et al. 1997), no correlation was found in HeLa cells between the ability of E4orf4 to cause hypophosphorylation of E1A mutants and its ability to repress transactivation by these mutants (Bondesson et al. 1996). Thus E1A hypophosphorylation may not be crucial for E4orf4 inhibition of transcription.

An interaction with PP2A also contributes to the regulation of alternative splicing of adenoviral mRNAs by E4orf4. The splicing factors, SR proteins, bind to an intronic splicing repressor element present upstream of the 3′ splice site of the adenovirus IIIa mRNA, preventing the usage of this splice-site (Kanopka et al. 1996). E4orf4 induces dephosphorylation of SR proteins, reducing their ability to bind the splicing repressor element, thus allowing splicing and generation of IIIa mRNA late in virus infection. Okadaic acid can reverse the E4orf4-induced dephosphorylation of SR proteins, indicating the involvement of PP2A (Kanopka et al. 1998). It has been further shown that E4orf4 binds a subset of SR proteins (SF2/ASF and SRp30c), and the E4orf4-SR complexes preferentially contain the hyperphosphorylated form of the SR proteins. Since E4orf4 mutant proteins that fail to bind either SR proteins or PP2A are unable to alter splicing of the IIIa mRNA, it has been suggested that a ternary complex containing all three proteins is required for E4orf4-induced dephosphorylation of SR proteins and the ensuing differential splicing (Estmer Nilsson et al. 2001).

2.2
E4orf4 Induces Transformed Cell-Specific Apoptosis

The finding that E4orf4 downregulates a signaling pathway initiated by E1A and cAMP suggested the possibility that E4orf4 may interfere with E1A-induced cellular proliferation. When this notion was tested, an exciting new role for the E4orf4 protein was revealed. Introduction of E4orf4 into 293 cells, which express the adenovirus E1A and E1B proteins, resulted in a strongly diminished ability of these cells to form colonies. The morphology of the E4orf4-expressing cells suggested that they were undergoing apoptosis, and this was confirmed by several tests (Shtrichman and Kleinberger 1998). Interestingly, E4orf4-induced apoptosis was not confined to E1A-expressing cells but occurred in many types of transformed cells (Lavoie et al. 1998; Marcellus et al. 1998; Shtrichman and Kleinberger 1998). Furthermore, E4orf4-induced apoptosis was shown to be p53-independent, since it occurred in p53-deficient cells, and, unlike E1A, E4orf4 expression was not accompanied by accumulation of p53 (Lavoie et al. 1998; Marcellus et al. 1998; Shtrichman and Kleinberger 1998). These results were consistent with work showing that E1A proteins from viral mutants lacking the anti-apoptotic E1B gene products induced both p53-dependent and -independent apoptosis (Subramanian et al. 1995; Teodoro et al. 1995). The p53-independent apoptosis was shown to require the adenovirus E4 gene region, whose transcription is activated by E1A (Marcellus et al. 1996).

Since E4orf4 induces apoptosis in many transformed cell types and reduces colony formation by transformed cells, it was predicted that E4orf4 would prevent oncogenic transformation of primary cells. Indeed, focus formation by baby rat kidney cells transformed with various combinations of oncogenes could be dramatically reduced by co-transfection with E4orf4, but not with a non-apoptogenic E4orf4 mutant (Shtrichman et al. 1999). However, the crucial question was whether E4orf4 specifically killed transformed cells, or whether its effect was universal. To address this issue, primary rat cells were either transfected with E4orf4 alone, or co-transfected with E4orf4 and various combinations of oncogenes, such as E1A and Ras, or Myc and Ras. The number of apoptotic cells was measured 48 h post-transfection. The results indicated that expression of oncogenes sensitized the cells to killing by E4orf4 (Shtrichman et al. 1999). An unpublished work, reported in a recent review, appears to confirm these results by claiming that E4orf4 induced cell death in 40 human cancer cell lines, but not in several primary human cell types derived from various tissues (Branton and Roopchand 2001). This review further reported that injection of adenovirus vectors expressing E4orf4 from an inducible promoter could, upon induction, reduce the size of tumors generated by human tumor xenografts in mice. The E4orf4-expressing adenovirus vectors were claimed to be more effective than p53-expressing vectors. The finding that E4orf4-induced apoptosis is cancer-cell-specific indicates that E4orf4 research may have exciting clinical implications. However, to understand the molecular basis of this specificity, the details of the E4orf4 apoptotic pathway have to be uncovered. It has been suggested that activation of the oncogenic state leads to induction of latent apoptotic signals that are uncoupled from the basic apoptotic machinery, thus providing a lower threshold for activation of apoptosis by additional signals (Fearnhead et al. 1998). Other hypotheses suggested that E4orf4 downstream targets or even E4orf4 itself may be differentially modified in cancer vs. normal cells leading to a differential response to the E4orf4 signal (Kleinberger 2000). This issue is presently unresolved. The following sections will summarize our current understanding of the E4orf4 apoptotic pathway.

3 Mechanisms of E4orf4-Induced Apoptosis

3.1 Is Caspase Activation Dispensable for E4orf4-Induced Apoptosis?

Early work showed that induction of E4orf4 expression in CHO cells was not accompanied by activation of caspase-3, common to many other apoptosis-inducing pathways. Furthermore, the broad-range caspase inhibitor zVAD-fmk neither inhibited, nor slowed down the appearance of apoptotic morphologies associated with E4orf4-induced apoptosis (Lavoie et al. 1998). These results suggest that E4orf4 induces a caspase-independent apoptotic pathway. However, it was later discovered that induction of caspase activation by E4orf4 was cell line-dependent (Livne et al. 2001). Thus, caspase activity, measured in an enzymatic assay using the caspase substrate Ac-DEVD-pNA, could be induced in two human cell lines, H1299 and 293T. The picture was further complicated by the finding that, in H1299 cells, caspase activation was required for accumulation of a sub-G1 cell population, representing cells with fragmented DNA, but was dispensable for nuclear condensation and cell killing (measured by colony formation). In contrast, in 293T cells, inhibition of caspase activation partially relieved nuclear condensation and cell death (Livne et al. 2001). To complicate matters even further, it was later reported that, in 293T cells grown in another laboratory, no caspase activation was detected, using a specific antibody against the cleaved form of caspase-3, and two kinds of broad-spectrum caspase inhibitors (zVAD-fmk and BocD-fmk) did not inhibit nuclear condensation (Robert et al. 2002). The same laboratory showed that the human C-33A cells behaved in a manner similar to that described above for H1299 cells. Contradictory results were also presented regarding the pathway involved in caspase activation. Livne et al. (2001) showed that the death receptor pathway was activated in 293T cells. Co-transfection of E4orf4 with dominant-negative mutants of caspase-8 or the death adaptor molecule FADD/MORT1, both downstream effectors of the death receptor pathway, resulted in inhibition of E4orf4-induced nuclear condensation and trypan blue uptake in 293T cells. Co-transfection of E4orf4 with a dominant-negative mutant of caspase-9, a downstream effector of the mitochondrial pathway, did not inhibit nuclear condensation, although E4orf4 was shown to cause cytochrome c release into the cytoplasm. Furthermore, no increase in caspase-9 activity was detected in the E4orf4-expressing 293T cells, as measured by western blot, using antibodies recognizing both precursor and cleaved caspase-9. In addition, E4orf4 was shown to enhance the levels of reactive oxygen species (ROS) in 293T cells, and this elevation in ROS levels could be inhibited by CrmA, a caspase-8 inhibitor, or by the dominant-negative FADD/MORT1 mutant. Reducing ROS levels with an antioxidant inhibited E4orf4-induced nuclear condensation. In contrast to these results, Robert et al. (2002) showed that, in C-33A cells, a dominant-negative mutant of caspase-9 as well as Bcl-2, but not a caspase-8 inhibitor (zIETD-fmk), inhibited the DNA fragmentation phenotype. These results implicate the mitochondria-apoptosome pathway in caspase-mediated DNA fragmentation in C-33A cells. Interestingly, it has been previously reported that E4orf4-induced cell killing, measured in CHO and HyA4 cells by a toxicity assay based on inhibition of RSV promoter-driven expression of luciferase, was reduced significantly by Bcl-2 and Bcl-$_x$L (Lavoie et al. 1998).

To make sense of the various conflicting results, we suggest that the physiological state of the cells, as well as their genetic content, may affect caspase activation by E4orf4 and its outcome. For example, unpublished results from our laboratory indicate that caspase-induced DNA fragmentation in H1299 and 293T cells varied with the serum used to feed the cells. The fact that under some circumstances E4orf4 can engage the death receptor pathway, whereas under other conditions the mitochondria pathway is recruited, may reflect variations in the cellular proteome, including targets of E4orf4 and other regulators of the E4orf4-induced pathway. However, the fact that E4orf4 can induce a unique, caspase-independent mode of apoptosis, in addition to any other type of apoptosis, is undisputed.

3.2
An Interaction with PP2A is Required for Induction of Apoptosis by E4orf4

As described in Section 2.1, an interaction with PP2A has been shown to be important for several of the E4orf4 functions. Thus, when the ability of E4orf4 to induce apoptosis was revealed, it was only natural to test whether association with PP2A was required for this function as well. Two approaches were taken to address this question. First, E4orf4 mutants were generated by site-directed mutagenesis of amino acids that are conserved in the E4orf4 proteins of several adenovirus serotypes, or by random PCR mutagenesis, using amplification conditions that reduce the fidelity of Taq polymerase (Shtrichman et al. 1999; Marcellus et al. 2000). Generation of deletion mutants was mostly unfruitful due to the unstable nature of the mutated proteins. The ability of the various mutants to bind an active PP2A heterotrimer was assessed by immunoprecipitation of the E4orf4 mutant proteins, followed by western blots to detect the presence of the Bα or C subunits of PP2A, and by enzymatic assays to measure the amount of active phosphatase present in the immune complexes. Shtrichman et al. (1999) demonstrated that there was a very good correlation between the ability of mutants to bind an active PP2A and their ability to induce apoptosis. One mutant in particular, A3 (S95P), was completely deficient in both functions, whereas other mutants showed diminished

PP2A binding and a similar reduced apoptogenic function. Some mutants showed only a slightly reduced ability to induce apoptosis, compared with a much more impaired ability to bind an active PP2A, indicating that, although PP2A may be required for induction of apoptosis, even low levels or a short duration of interaction are sufficient to initiate the pathway. In addition, there were mutants that physically associated with PP2A quite efficiently, although a relatively low phosphatase activity was associated with the complex. These mutants could suggest that the interaction of PP2A with E4orf4 must be positioned precisely to maintain the phosphatase activity, or that the interaction of E4orf4 with PP2A may affect PP2A activity. Another class of E4orf4 mutants was described by Marcellus et al. (2000). These mutants could bind PP2A but were unable to induce apoptosis, thus they may have lost their ability to bind the relevant PP2A substrate, while retaining the activity towards the phosphopeptide used in the assay. It is also possible that an additional PP2A-independent mechanism contributes to cell killing by E4orf4, and these mutants lost the ability to bind another mediator of E4orf4-induced apoptosis. However, among 60 mutants analyzed, substituting altogether 59 out of 114 residues of the E4orf4 protein (Shtrichman et al. 1999; Marcellus et al. 2000; Afifi et al. 2001), there were no mutants that lost completely the ability to bind an active PP2A while retaining the ability to induce apoptosis. These findings strongly suggest that binding to PP2A is necessary, if not sufficient, for induction of apoptosis by E4orf4.

The conclusion that association of E4orf4 with an active form of PP2A, rather than the B subunit alone, is required for induction of apoptosis is supported by the results of a closer scrutiny of the A3 (S95P) mutant of E4orf4 (Shtrichman et al. 2000). This mutant has been found to associate with the regulatory B subunit, but not with the A and C subunits. The A3 mutant was found to act in a dominant-negative fashion and repress wild-type E4orf4 activity when co-transfected into cells. This finding suggests that the A3 mutant competes with wild-type E4orf4 for binding holoenzymes containing the B subunit, but the AC subunits are released from this complex, which is thus unable to mediate apoptosis.

Identification of the non-apoptotic E4orf4 mutants described above, and of additional mutants of a similar nature obtained in yeast (Sect. 3.2.2; Afifi et al. 2001), indicated that mutations decreasing the binding of E4orf4 to PP2A and reducing apoptosis by two-fold or more were localized to several regions of the protein (Fig. 1). These regions include clusters of conserved amino acids: a.a. 4–10, 21–29, 47–62, and 78–95. However, until the E4orf4 structure can be solved, it is difficult to predict the structural features underlying this interaction.

The second approach used to investigate the role of PP2A in E4orf4-induced apoptosis involved the use of an antisense Bα construct. When co-transfected into H1299 cells together with E4orf4, this construct decreased Bα levels in the cells and significantly inhibited apoptosis, whereas Bα in the sense orientation elevated E4orf4-induced apoptosis. This effect was specific to E4orf4 since co-

```
          10         20         30         40
MVLPALPAPP VCDSQNECV  GWLGVAYSAVV DVIRAAAHEG VYIEPE
   ______            __________
     I                   II

 50          60         70         80         90
ARGR LDALREWIYY NYYTERAKRR DRRRRSVCHA RTWFCFRKYD YVRRS
______________          ________   __________________
     III                  Basic            IV

 100        110
IWHDT TTNTISVVSA HSVQ
```

Fig. 1. E4orf4 binding sites for PP2A, determined by mutant analysis. Mutations in the residues marked in *bold* caused at least 50% reduction in PP2A binding and induction of apoptosis. Residues that are conserved among several adenovirus serotypes are *underlined.* Clusters of amino acids that appear to contribute to PP2A binding are marked with *Roman numerals.* There are contradicting reports (Shtrichman et al. 1999; Marcellus et al. 2000) regarding the contribution of the basic region of E4orf4 (residues R73–75) to PP2A binding, and therefore it is marked separately

transfection of the Bα antisense construct did not inhibit p53-induced apoptosis (Shtrichman et al. 1999). Thus the interaction with PP2A appears to be indispensable for induction of apoptosis by E4orf4.

3.2.1
A Specific Subpopulation of PP2A Mediates E4orf4-Induced Apoptosis

Early work mentioned above (Sect. 2.1; Kleinberger and Shenk 1993) identified Bα as the PP2A subunit mediating the interaction of the phosphatase with E4orf4. However, a later work, assaying co-immunoprecipitation of E4orf4 with epitope-tagged PP2A B′ subunits, revealed an interaction of E4orf4 with B′family members as well (Shtrichman et al. 2000). The E4orf4 mutants described by Shtrichman et al. (1999) were used to compare the binding of Bα and B′β to the viral protein. The results indicated that E4orf4 binding sites required for the interaction with the two B subunits overlapped, but were not identical. Recent unpublished work showed that E4orf4 could also bind epitope-tagged B″/PR59 and B″/PR72 subunits (Ben-Israel and Kleinberger, unpubl.). Although there is no apparent sequence homology between the various B subunit families, all of them bind the A subunit of PP2A. Recent work identified conserved amino acids shared by members of all three families of PP2A B subunits, which may contribute to A subunit binding (Li and Virshup 2002). Similarly, different B subunits may share conserved amino acids that contribute to E4orf4 binding. Work in yeast, described in Section 3.3.6, has been undertaken to identify residues in the Bα subunit required for its interaction with E4orf4 and with the core PP2A-AC heterodimer.

Since PP2A contributes to different functions of E4orf4, it is tempting to suggest that interactions with different B subunits may mediate individual E4orf4 functions. If so, it could be predicted that only one of the B subunits interacting with E4orf4 would mediate induction of apoptosis. Indeed, Shtrichman et al. (2000) showed that co-transfection of the Bα subunit with E4orf4 increased the levels of apoptosis, whereas co-transfection with any of the B′ subunits did not. Furthermore, one B′ subunit that was tested, B′β, acted in a dominant-negative fashion: addition of B′β to a transfection mixture containing Bα and E4orf4 resulted in reduced levels of Bα found in association with the C subunit, and in a diminished degree of apoptosis. The exclusive role of the Bα-containing PP2A heterotrimers in mediating cell death was further supported by results in yeast, described later in Section 3.3.2.

3.3
A Genetic Search in Yeast for Effectors of the E4orf4-PP2A-Initiated Pathway Reveals an Interaction of E4orf4 with the Cell Cycle Machinery

3.3.1
Apoptosis Research in Yeast

Once it had been shown that PP2A was an important partner mediating E4orf4 functions, including induction of apoptosis, it became necessary to identify downstream components of the pathway initiated by the E4orf4-PP2A complex. When searching for a genetic approach to address this issue, the yeast *Saccharomyces cerevisiae* was considered. Although many components of the metazoan core machinery of cell death, such as caspases and Bcl-2 family members, are absent in yeast, it has been previously shown that this organism could serve as a powerful tool for apoptosis research (Matsuyama et al. 1999). For example, a screen for mutant *S. cerevisiae* strains resistant to Bax-induced death demonstrated that F_0F_1-ATPase is required for killing by Bax in yeast and mammalian cells (Matsuyama et al. 1998). A second complementary screen for mammalian proteins that can suppress Bax-induced killing of *S. cerevisiae* yielded the gene BI-1, which encodes a conserved integral membrane protein that can suppress apoptosis when expressed in mammalian cells (Xu and Reed 1998). Furthermore, yeast and mammalian PP2A subunits share an extensive homology, suggesting that E4orf4 may be able to bind yeast PP2A and direct it to targets similar to its targets in mammalian cells.

3.3.2
E4orf4 Induces PP2A-Dependent Loss of Viability in Yeast

To test the feasibility of using yeast to study E4orf4-induced apoptosis, E4orf4 was cloned into a yeast vector downstream of the GAL1,10 promoter, and was introduced into yeast cells (Kornitzer et al. 2001). It soon became apparent that E4orf4, but not the previously described A3 mutant, inhibited yeast cell growth on galactose. Furthermore, Cdc55, the yeast homologue of the PP2A-B subunit, was required for growth inhibition by E4orf4, whereas Rts1, the yeast B′ homologue, was dispensable for E4orf4-induced toxicity. High levels of Cdc55 expression accentuated the E4orf4 toxic effect. Loss of yeast cell viability was irreversible, since transferring the cells back to glucose medium, thus inhibiting E4orf4 expression, did not rescue their ability to form colonies. These results were highly consistent with previous findings in mammalian cells, and were further confirmed by Roopchand et al. (2001). Tpd3, the yeast homologue of the A subunit, was also required for growth inhibition of E4orf4-expressing cells. The binding of E4orf4 to the PP2A holoenzyme in yeast was shown to require the presence of Cdc55 (Roopchand et al. 2001). E4orf4 was also shown to increase ROS levels in yeast cells (Kornitzer et al. 2001), again indicating that E4orf4-induced pathways in yeast and mammalian cells share common characteristics.

To further demonstrate the relevance of the yeast system to E4orf4-induced apoptosis in mammalian cells, E4orf4 was randomly mutagenized in vitro using chemical mutagenesis, and the yeast system was used to select non-toxic E4orf4 mutants (Afifi et al. 2001). These mutants were shown to be non-apoptotic in mammalian cells, and demonstrated a reduced ability to associate with an active PP2A. These results support the suggestion that the yeast system can be used to investigate the E4orf4-PP2A-regulated pathway.

3.3.3
E4orf4 Interacts with Cell Cycle Regulators in Yeast to Induce Growth Arrest

Microscopic examination of yeast cells expressing E4orf4 suggested that E4orf4 may be interfering with normal cell cycle regulation. FACS analysis further demonstrated that these cells contained 2n DNA content, typical of cells in G_2/M. Using several known yeast cell cycle mutants, it was shown that E4orf4 is synthetically lethal with a reduction in Cdc28 (the yeast Cdc2 kinase) activity. In other words, reduced Cdc28 activity and E4orf4 expression acted in synergy to arrest yeast growth. On the other hand, E4orf4 increased Cdc28 activity in yeast cells in a manner dependent on Mih1, the yeast homologue of the Cdc25 phosphatase (Kornitzer et al. 2001). Thus E4orf4 may partially counteract its own cell cycle inhibitory effect by stimulating dephosphorylation of Cdc28-Y19 by Mih1. The E4orf4-induced increase in Cdc28 activity in yeast cells was shown to require Cdc55 (Roopchand et al. 2001).

In addition to its functional interaction with Cdc28, E4orf4 was also shown to genetically interact with the anaphase-promoting complex (APC)/Cyclosome (Kornitzer et al. 2001). APC/C is a mitotic ubiquitin ligase complex responsible for the degradation of several mitotic substrates, including Cyclin B and Pds1/securin. The degradation of these mitotic regulators allows progression through mitosis (Peters 2002). E4orf4 expression led to inhibition of degradation of two APC/C substrates: Pds1/securin, a substrate of APC/C complexed with the activating Cdc20 subunit and active at the metaphase to anaphase transition; and Ase1, a substrate of APC/C complexed with the activating Cdh1/Hct1 subunit, active at the end of mitosis. Co-immunoprecipitation experiments demonstrated that E4orf4 associated with the APC/C, and, furthermore, PP2A-APC/C complexes could be detected in the presence of E4orf4, but not in its absence. It is well established that APC/C phosphorylation by Cdk1 is required for its activation in both metazoans (Lahav-Baratz et al. 1995; Shteinberg et al. 1999) and yeast cells (Rudner and Murray 2000). Moreover, genetic interactions between Cdc55, the PP2A-B subunit, and Cdc20, the APC/C activating subunit, have been reported in the absence of E4orf4 (Wang and Burke 1997; Kornitzer et al. 2001). Thus we suggested that one of the normal functions of PP2A is to inhibit the APC/C, and E4orf4 may be stabilizing a transient physiological interaction between APC/C and PP2A, enhancing APC/C inhibition and subsequent mitotic arrest. The specific phosphorylation sites on the APC/C targeted by the E4orf4-PP2A complex remain unidentified.

3.3.4 *E4orf4 Can Induce G_2/M Arrest in Mammalian Cells*

How relevant are the findings in yeast for mammalian cells? In a 293-derived cell line, expressing E4orf4 from a tetracycline-regulated promoter, G_2/M arrest was observed 24 h post-induction with doxycycline. No such arrest was observed upon treatment of a parallel control cell line. Forty-eight hours post-induction, the G_2/M arrest was released and apoptotic cells with a "sub-G1" DNA content started to accumulate (Kornitzer et al. 2001). These results indicate that, under certain physiological conditions, E4orf4 can initially induce cell cycle arrest, from which the cells escape eventually and undergo apoptosis. Other studies further indicate that unidentified adenovirus E4 gene products (excluding E4orf6) cause a delay in cell cycle progression, associated with elevated levels of cyclin B (a substrate of APC/C) and partial G_2 growth arrest. However, this effect is seen only in cells infected with adenovirus mutants lacking the E1 region (Wersto et al. 1998; Brand et al. 1999; Steinwaerder et al. 2000). Many cellular modulators were shown to induce G_2/M arrest, followed by apoptosis, and other studies demonstrated that induction of apoptosis by various stimuli required activation of Cdc2 or Cdk2 kinases (reviewed in Kornitzer et al. 2001). Thus, cell cycle deregulation may lead to apoptosis. More specifically, it has been shown that, in Jurkat cells induced to undergo apoptosis by Fas, APC/C was inactivated by caspase-mediated proteolysis of its Cdc27 subunit, resulting in stabilization of cyclins A and B. At the same time, Cdc2 was independently activated by caspase-mediated destruction of its inhibitor kinase Wee1 (Zhou et al. 1998). The similarity between these events and the E4orf4-induced events in yeast suggests that activation of cyclin-dependent kinases and inactivation of their downstream effectors may cause serious perturbations in the cell cycle and may result in induction of apoptosis.

3.3.5
A Non-PP2A-dependent E4orf4-Induced Pathway May Exist in Yeast

Using E4orf4 mutants in yeast cells, it has been shown that mutants that did not bind PP2A retained a residual growth inhibitory effect (Roopchand et al. 2001). These results raise the possibility that E4orf4 may initiate more than one growth inhibitory pathway in yeast. A yeast genetic screen performed in our laboratory has identified a protein that contributes to E4orf4-mediated toxicity, physically associates with E4orf4, and, as judged by epistasis analysis, is not required for E4orf4-Cdc55-mediated toxicity (Maoz and Kleinberger, in preparation). The relevance of this protein to E4orf4-induced apoptosis in mammalian cells remains to be determined.

3.3.6
Use of E4orf4-Induced Toxicity in Yeast for Analysis of Cdc55 Interactions

The E4orf4-yeast system is a powerful tool that can be used not only to identify elements of the E4orf4-initiated pathways, but can also be used to obtain information on the E4orf4 partner, PP2A. In vitro mutagenesis of *CDC55* and selection for Cdc55 mutant proteins that lost the ability to transduce the E4orf4 signal has yielded several mutants with defects in binding E4orf4 or the PP2A-A subunit (Tpd3) or both (Rainis et al., submitted). Sequence analysis of these mutants revealed that the domains of interaction between Cdc55 and its binding partners are not limited to a focused region in the primary sequence of the protein. Furthermore, some point mutations inhibited binding of both E4orf4 and Tpd3, suggesting that they may affect the conformation of the protein, or that the binding of one Cdc55 partner may affect binding of another partner.

3.4
Association of E4orf4 with Src Family Kinases Modulates Src Signaling Pathways, Induces Morphological Changes in the Cells, and Augments Apoptosis

When E4orf4 cDNA is transfected into cells, the protein is found initially in the nucleus, and it later accumulates in the cytoplasm and plasma membrane (Lavoie et al. 2000; Shtrichman et al. 2000; Gingras et al. 2002). Furthermore, at early times after the onset of E4orf4 expression in transformed cells, the cells undergo morphological changes, including changes in the actin cytoskeleton, rounding up, and intense membrane blebbing (Lavoie et al. 2000). These cytosolic changes are typical of "extranuclear apoptosis" (Mills et al. 1999) and precede nuclear condensation. Inhibitors of actin polymerization, such as cytochalasin D, were shown to reduce not only E4orf4-induced membrane blebbing, but also nuclear condensation and cell death. These results suggest that the early morphological changes occurring in the cytoplasm and plasma membrane provide important signals that contribute to the cell death process. Investigation of this phenomenon led to the discovery that E4orf4 associates with Src family kinases, co-localizes with them in membrane blebs, and modulates Src-dependent phosphorylation of specific cellular proteins (Lavoie et al. 2000). E4orf4 contains a proline-rich motif at its N-terminus (MVLPALPAPP), which resembles an SH3-binding motif, and could potentially mediate binding to Src. Interestingly, some of the residues in this region are also important for the E4orf4 interaction with PP2A (Sect. 3.2, Fig. 1). If this region is indeed involved in both interactions, they may be mutually exclusive.

E4orf4 expression was shown to be linked with reduced Tyr-phosphorylation of the Src substrate focal adhesion kinase (FAK), leading to decreased FAK activity, and reduced phosphorylation of its likely substrate paxillin. At the same time, Tyr-phosphorylation of cortactin, an F-actin binding protein that is a known substrate of Src family kinases, was elevated. In addition, the state of phosphorylation of other unidentified proteins was also altered in the presence of E4orf4 (Lavoie et al. 2000). The changes in phosphorylation correlated with redistribution of Tyr-phosphorylated proteins at the periphery of the cells, misassembley of focal adhesions, and reorganization of the cytoskeleton. Using constitutively activated and kinase-dead Src mutants, as well as pharmacological agents that inhibit Src kinases, it was shown that Src activity contributed to accumulation of E4orf4 in the cytoplasm and plasma membrane, to E4orf4-induced membrane blebbing, and to nuclear condensation (Lavoie et al. 2000).

3.4.1 Tyrosine-Phosphorylation of E4orf4 by Src Kinases is Required for Membrane Translocation of E4orf4 and Induction of a Cytoplasmic Signal for Apoptosis

One of the consequences of the association between Src kinases and E4orf4 is the Src-mediated Tyr-phosphorylation of E4orf4 itself on Tyr 26, 42, and 59. Mutations of these tyrosines to phenylalanines inhibited Src-induced phosphorylation of E4orf4 in vitro and in vivo, but did not affect the ability of the two proteins to interact (Gingras et al. 2002). These mutations did interfere, however, with the ability of E4orf4 to modulate Src-mediated phosphorylation of its target proteins. Furthermore, Src substrates such as cortactin and p62dok associated with wild-type E4orf4, but not with the non-phosphorylatable variant. Thus, it has been suggested that Tyr-phosphorylation of E4orf4 is required for formation of an active E4orf4-Src signaling complex. Furthermore, the non-phosphorylatable mutant was not translocated to the cytoplasm or plasma membrane, but remained concentrated in the nucleus, and was highly impaired in its ability to cause membrane blebbing. However, it retained a residual ability to induce nuclear condensation. In contrast with this mutant, a mutation of Tyr42 to glutamic acid, replacing Tyr42 with a negative charge, increased E4orf4 ability to induce membrane blebbing, nuclear condensation, and cell death, while efficiently shifting E4orf4 from a nuclear localization to the cytoplasm and the plasma membrane (Gingras et al. 2002). Based on these results it was suggested that Src-induced phosphorylation of E4orf4 is important for maintaining E4orf4 in the cytoplasm and plasma membrane. Phosphorylation of E4orf4 may change its conformation and unmask domains involved in protein–protein interactions that allow cytoplasmic and membrane translocation. Furthermore, E4orf4 phosphorylation may be required for recruitment of Src substrates to a signaling complex, where they are phosphorylated and participate in actin remodeling and cell death. It is currently not clear how Src-mediated morphological changes in the cells induce apoptosis. Gingras et al. (2002) suggest that these events become proapoptotic when uncoupled from Src-induced survival signals that may normally be mediated by activation of FAK or other Src effectors.

3.5
Nuclear Versus Cytoplasmic/Membranal Pathways of E4orf4-Induced Apoptosis

The finding that non-phosphorylatable E4orf4 or E4orf4 in the presence of kinase-dead Src can still induce apoptosis (Lavoie et al. 2000; Gingras et al. 2002) indicates that Src signaling contributes to E4orf4-induced apoptosis, but that other signaling pathways may also be involved. Furthermore, the presence of E4orf4 in the nucleus, prior to its accumulation in the cytoplasm and in membranes, suggests that nuclear localization could also be important for E4orf4-initiated apoptosis. Using E4orf4 constructs targeting E4orf4 accumulation to the nucleus (E4orf4-GFP-NLS), to the cytoplasm (E4orf4-GFP-NES), or to membranes (E4orf4-GFP-CAAX, Myr-GFP-E4orf4), it was demonstrated that nuclear accumulation of 94% of E4orf4 molecules could still lead to induction of apoptosis by E4orf4, although it was delayed, as compared with wild-type E4orf4-induced apoptosis. Moreover, targeting E4orf4 outside the nucleus, or into membranes, facilitated wild-type-like apoptosis, measured by the appearance of membrane blebbing and nuclear condensation (Robert et al. 2002). It was further demonstrated that, whereas Src-mediated E4orf4 phosphorylation was required for the cytoplasmic cell death program, it was dispensable for the nuclear program. Moreover, calpastatin, an inhibitor of calpains, was shown to inhibit the Src-mediated cytoplasmic pathway, but not the nuclear pathway. Calpains are Ca^{2+}-dependent proteases that cleave substrate proteins localized near membranes and the cytoskeleton (Glading et al.

2002). Some of the substrates of calpains were shown to be modulated by the E4orf4-Src pathway (FAK, cortactin; Lavoie et al. 2000; Tatosyan and Mizenina 2000). Changes in tyrosine-phosphorylation of some calpain substrates can modify their susceptibility to calpains, raising the possibility that alterations in Src-phosphorylation and calpain activation could together play a role in the cytoplasmic cell death pathway.

Based on the results presented by Robert et al. (2002) it was suggested that E4orf4 triggers two cell death programs that are mediated through distinct death effectors. It is quite possible, however, that the early nuclear presence of E4orf4 is required to initiate the Src-mediated as well as the nuclear pathway. Robert et al. (2002) suggested that the presence of E4orf4 in the cytoplasm and membranes was both necessary and sufficient to trigger the Src-mediated pathway. However, the possibility that a small nuclear population of E4orf4 molecules was enough to initiate a nucleus-to-cytoplasm signaling, which is required to trigger Src association with E4orf4, cannot be ruled out at this stage. Thus, the nuclear and cytoplasmic pathways could act in parallel, the contribution of each determined by the availability of targets and signaling pathways in the cells, or the nuclear pathway could be responsible for both Src-dependent and -independent signaling.

4
Summary: Unresolved Questions and Future Directions

Much progress has been made towards elucidation of the underlying mechanisms of E4orf4-induced apoptosis, and a current model summarizing the available information is presented in Fig. 2. However, many questions remain unanswered.

It is clear that E4orf4 targets two important cellular regulators, PP2A and Src, which mediate its proapoptotic activities. However, the relevant downstream effectors of the E4orf4-PP2A and E4orf4-Src complexes have not been identified yet. Although several Src targets, whose phosphorylation is altered in the presence of E4orf4, have been described, the contribution of each of these targets to the apoptotic process has yet to be determined. Moreover, yet unidentified targets of Src may also be important for induction of apoptosis. The nature of the signaling pathway leading back from the cytoplasm and the membrane to the nucleus must also be addressed.

Relevant E4orf4-PP2A downstream effectors have been suggested in yeast, but not yet in mammalian cells, and the direct phosphorylation targets have not been identified. Furthermore, the immediate consequences of the E4orf4-PP2A interaction are not yet understood. It is possible that E4orf4 targets PP2A to new substrates, or that it changes PP2A activity towards specific substrates, increasing or decreasing it, depending on the substrate. Thus, it will be important to identify and characterize the direct targets and other downstream effectors of E4orf4-PP2A complexes. This can be done either by means of

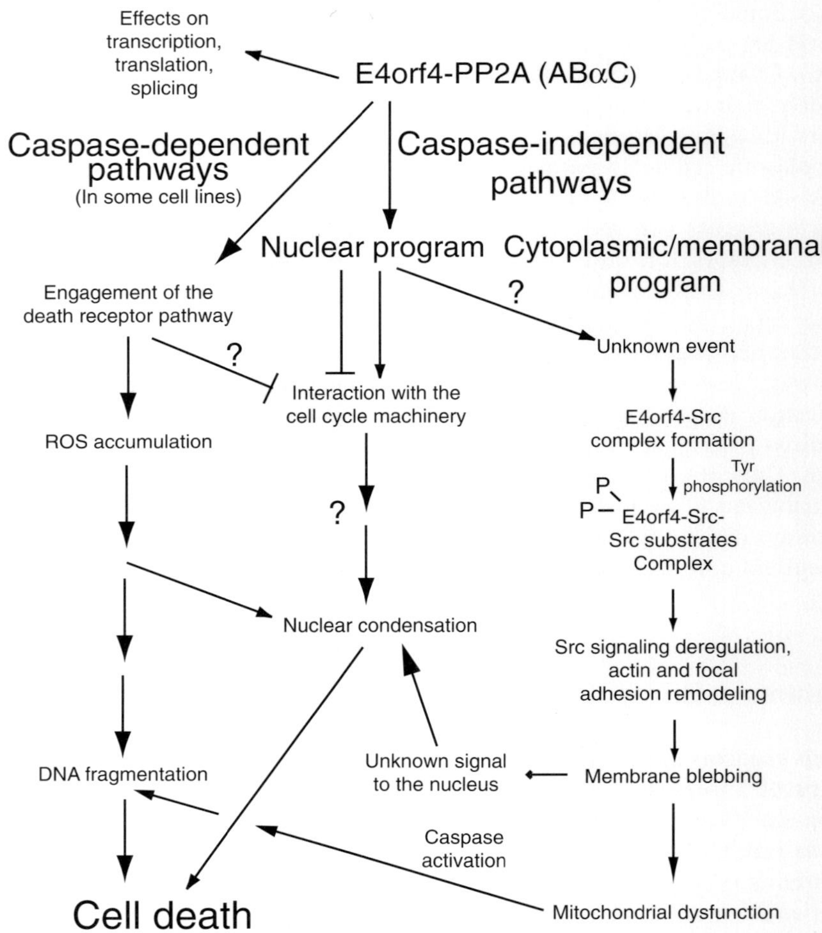

Fig. 2. A model describing the various pathways reported to contribute to E4orf4-induced apoptosis. See text for details

genetic screens in yeast and mammalian cells, or using proteomics methods to uncover the immediate targets of PP2A.

As mentioned above (Sect. 3.5), it is also not clear whether the cytoplasmic and nuclear pathways contributing to E4orf4-induced apoptosis are linked, or whether they are totally independent of each other. Thus, it is unclear why the interaction of E4orf4 with Src kinase and the ensuing Tyr-phosphorylation of E4orf4 do not happen immediately upon introduction of E4orf4 into the cells, but, rather, a nuclear accumulation is seen during the early phase, while the shift to cytoplasm and membranes occurs later. This finding could suggest the requirement for an initial event occurring in the nucleus, which permits the association of E4orf4 with Src. It has also not been demonstrated yet whether the interaction of E4orf4 with PP2A is required for both nuclear and cytoplasmic pathways, or just for one of them. However, since inhibiting the interaction with PP2A (using E4orf4 mutants or the PP2A-Bα antisense construct) leads to inhibition of apoptosis, PP2A may interact with both pathways, or, more simply, it may be required for the initial triggering of apoptosis, followed by divergence of the nuclear and cytoplasmic pathways.

In addition to PP2A and Src, there may be other E4orf4-associating proteins that contribute to induction of apoptosis. Based on work in yeast, it has been suggested that there are Cdc55-independent mechanisms mediating E4orf4-induced toxicity (Roopchand et al. 2001). Since yeast does not encode Src family kinases, other pathways may be involved, as suggested in Section 3.3.5. Furthermore, identification of additional E4orf4-interacting proteins may reveal immediate targets of PP2A or Src, as well as other mediators of E4orf4-induced apoptosis.

E4orf4 has been shown to affect transcription, alternative splicing, and translation (Sect. 2), and these activities of E4orf4 may indirectly affect the apoptotic process. Thus, identification of genes whose expression patterns are altered when E4orf4 expression is induced may provide additional information on the E4orf4-induced pathways. Furthermore, effects of E4orf4 on splicing patterns of apoptosis-related proteins may also contribute to the apoptotic process.

The question whether E4orf4-initiated apoptosis contributes to the adenovirus replication cycle is also not resolved at present. It has been suggested that E4orf4-induced apoptosis contributes to viral spread at the end of the infectious cycle, allowing the virus to evade the immune system (Roulston et al. 1999); however, no data have been provided to support this hypothesis. It may also be possible that E4orf4-induced apoptosis is a consequence of its other functions, which can be avoided with the help of the adenovirus anti-apoptotic gene products, or which may not present a problem in normal cells that are not highly susceptible to E4orf4-induced cell killing. In any case, expression of E4orf4 early during adenovirus infection does not result in premature apoptosis of the host cell, and a negative-feedback mechanism, described above (Sect. 2), is employed to prevent enhanced accumulation of E4orf4 in the cells.

Despite the progress made in understanding E4orf4-induced pathways, the basis for the cancer-cell-specificity of E4orf4-induced apoptosis is not yet obvious. Src kinases appear to be dispensable for the process, although they accelerate it and contribute to its efficiency. The possible inhibitory effect of E4orf4 on progression of cells through the cell cycle may conflict with growth stimulatory signals in transformed cells, triggering apoptosis, whereas these conflicts could be less pronounced in normal cells. More insights into E4orf4-initiated pathways have to be gained to elucidate this point. However, it is clear that E4orf4-induced apoptosis is a novel and complicated process and its study may lead to the uncovering of new apoptotic signaling pathways and to the discovery of new targets for cancer therapy.

Acknowledgements. I thank R. Sharf and H. Ben-Israel for their comments on the manuscript. Research conducted in our laboratory on E4orf4 is supported by grants from the Israel Science Foundation founded by the Israel Academy of Sciences and Humanities, by the German-Israeli Foundation for Scientific Research and Development, by the Israel Ministry of Science, and by the Association for International Cancer Research (UK).

References

Afifi R, Sharf R, Shtrichman R, Kleinberger T (2001) Selection of apoptosis-deficient adenovirus E4orf4 mutants in *S. cerevisiae*. J Virol 75:4444–4447
Ben-Israel H, Kleinberger T (2002) Adenovirus and cell cycle control. Front Biosci 7:d1369–1395
Boivin D, Morrison MR, Marcellus RC, Querido E, Branton PE (1999) Analysis of synthesis, stability, phosphorylation, and interacting polypeptides of the 34-kilodalton product of open reading frame 6 of the early region 4 protein of human adenovirus type 5. J Virol 73:1245–1253
Bondesson M, Ohman K, Mannervik M, Fan S, Akusjarvi G (1996) Adenovirus E4 open reading 4 protein autoregulates E4 transcription by inhibiting E1A transactivation of the E4 promoter. J Virol 70:3844–3851
Brand K, Klocke R, Possling A, Paul D, Strauss M (1999) Induction of apoptosis and G2/M arrest by infection with replication-deficient adenovirus at high multiplicity of infection. Gene Ther 6:1054–1063
Branton PE, Roopchand DE (2001) The role of adenovirus E4orf4 protein in viral replication and cell killing. Oncogene 20:7855–7865
Budihardjo I, Oliver H, Lutter M, Luo X, Wang X (1999) Biochemical pathways of caspase activation during apoptosis. Annu Rev Cell Dev Biol 15:269–290
Cryns V, Yuan J (1998) Proteases to die for. Genes Dev 12:1551–1570
Csortos C, Zolnierowicz S, Bako E, Durbin SD, DePaoli-Roach AA (1996) High complexity in the expression of the B′ subunit of protein phosphatase 2A0. J Biol Chem 271:2578–2588
Datta SR, Brunet A, Greenberg ME (1999) Cellular survival: a play in three Akts. Genes Dev 13:2905–2927
Elliott K, Ge K, Du W, Prendergast GC (2000) The c-Myc-interacting adaptor protein Bin1 activates a caspase-independent cell death program. Oncogene 19:4669–4684
Engel DA, Hardy S, Shenk T (1988) cAMP acts in synergy with E1A protein to activate transcription of the adenovirus early genes E4 and E1A. Genes Dev 2:1517–1528
Estmer Nilsson C, Petersen-Mahrt S, Durot C, Shtrichman R, Krainer AR, Kleinberger T, Akusjarvi G (2001) The adenovirus E4-ORF4 splicing enhancer protein interacts with a subset of phosphorylated SR proteins. EMBO J 20:864–871
Fearnhead HO, Rodriguez J, Govek EE, Guo W, Kobayashi R, Hannon G, Lazebnik YA (1998) Oncogene-dependent apoptosis is mediated by caspase-9. Proc Natl Acad Sci USA 95:13664–13669
Gingras MC, Champagne C, Roy M, Lavoie JN (2002) Cytoplasmic death signal triggered by SRC-mediated phosphorylation of the adenovirus E4orf4 protein. Mol Cell Biol 22:41–56
Glading A, Lauffenburger DA, Wells A (2002) Cutting to the chase: calpain proteases in cell motility. Trends Cell Biol 12:46–54
Green D, Kroemer G (1998) The central executioners of apoptosis: caspases or mitochondria? Trends Cell Biol 8:267–271
Halbert DN, Cutt JR, Shenk T (1985) Adenovirus early region 4 encodes functions required for efficient DNA replication, late gene expression, and host cell shutoff. J Virol 56:250–257
Ito T, Deng X, Carr B, May WS (1997) Bcl-2 phosphorylation required for anti-apoptotic function. J Biol Chem 272:11671–11673
Kamibayashi C, Estes R, Lickteig RL, Yang S-I, Craft C, Mumby M (1994) Comparison of heterotrimeric protein phosphatase 2A containing different B subunits. J Biol Chem 269:20139–20148

Kanopka A, Muhlemann O, Akusjarvi G (1996) Inhibition by SR proteins of splicing of a regulated adenovirus pre-mRNA. Nature 381:535–538

Kanopka A, Muhlemann O, Petersen-Mahrt S, Estmer C, Ohrmalm C, Akusjarvi G (1998) Regulation of adenovirus alternative RNA splicing by dephosphorylation of SR proteins. Nature 393:185–187

Kleinberger T (2000) Induction of apoptosis by adenovirus E4orf4 protein. Apoptosis 5:211–215

Kleinberger T, Shenk T (1993) Adenovirus E4orf4 protein binds to protein phosphatase 2A, and the complex down regulates E1A-enhanced junB transcription. J Virol 67:7556–7560

Kornitzer D, Sharf R, Kleinberger T (2001) Adenovirus E4orf4 protein induces PP2A-dependent growth arrest in *S. cerevisiae* and interacts with the anaphase promoting complex/cyclosome. J Cell Biol 154:331–344

Lahav-Baratz S, Sudakin V, Ruderman JV, Hershko A (1995) Reversible phosphorylation controls the activity of cyclosome-associated cyclin-ubiquitin ligase. Proc Natl Acad Sci USA 92:9303–9307

Lavoie JN, Nguyen M, Marcellus RC, Branton PE, Shore GC (1998) E4orf4, a novel adenovirus death factor that induces p53-independent apoptosis by a pathway that is not inhibited by zVAD-fmk. J Cell Biol 140:637–645

Lavoie JN, Champagne C, Gingras M-C, Robert A (2000) Adenovirus E4 open reading frame 4-induced apoptosis involves dysregulation of Src family kinases. J Cell Biol 150:1037–1055

Li X, Virshup DM (2002) Two conserved domains in regulatory B subunits mediate binding to the A subunit of protein phosphatase 2A. Eur J Biochem 269:546–552

Livne A, Shtrichman R, Kleinberger T (2001) Caspase activation by adenovirus E4orf4 protein is cell line-specific and is mediated by the death receptor pathway. J Virol 75:789–798

Mannervik M, Fan S, Strom AC, Helin K, Akusjarvi G (1999) Adenovirus E4 open reading frame 4-induced dephosphorylation inhibits E1A activation of the E2 promoter and E2F-1-mediated transactivation independently of the retinoblastoma tumor suppressor protein. Virology 256:313–321

Marcellus RC, Teodoro JG, Wu T, Brough DE, Ketner G, Shore G, Branton PE (1996) Adenovirus type 5 early region 4 is responsible for E1A-induced p53-independent apoptosis J Virol 70:6207–6215

Marcellus RC, Lavoie JN, Boivin D, Shore GC, Ketner G, Branton PE (1998) The early region 4 orf4 protein of human adenovirus type 5 induces p53-independent cell death by apoptosis. J Virol 72:7144–7153

Marcellus RC, Chan H, Paquette D, Thirlwell S, Boivin D, Branton PE (2000) Induction of p53-independent apoptosis by the adenovirus E4orf4 protein requires binding to the Balpha subunit of protein phosphatase 2A. J Virol 74:7869–7877

Matsuyama S, Xu Q, Velours J, Reed JC (1998) The mitochondrial F0F1-ATPase proton pump is required for function of the proapoptotic protein Bax in yeast and mammalian cells. Mol Cell 1:327–336

Matsuyama S, Nouraini S, Reed JC (1999) Yeast as a tool for apoptosis research. Curr Opin Microbiol 2:618–623

McCright B, Virshup DM (1995) Identification of a new family of protein phosphatase 2A regulatory subunits. J Biol Chem 270:26123–26128

Medghalchi S, Padmanabhan R, Ketner G (1997) Early region 4 modulates adenovirus DNA replication by two genetically separable mechanisms. Virology 236:8–17

Mills JC, Stone NL, Pittman RN (1999) Extranuclear apoptosis. The role of the cytoplasm in the execution phase. J Cell Biol 146:703–708

Monney L, Otter I, Olivier R, Ozer HL, Haas AL, Omura S, Borner C (1998) Defects in the ubiquitin pathway induce caspase-independent apoptosis blocked by Bcl-2. J Biol Chem 273:6121–6131

Müller U, Roberts MP, Engel DA, Doerfler W, Shenk T (1989) Induction of transcription factor AP-1 by adenovirus E1A protein and cAMP. Genes Dev 3:1991–2002

Müller U, Kleinberger T, Shenk T (1992) Adenovirus E4orf4 protein reduces phosphorylation of c-fos and E1A proteins while simultaneously reducing the level of AP-1. J Virol 66:5867–5878

Mumby M (1995) Regulation by tumor antigens defines a role for PP2A in signal transduction. Semin Cancer Biol 6:229–237
Mumby MC, Walter G (1993) Protein serine/threonine phosphatases: structure, regulation, and functions in cell growth. Physiol Rev 73:673–699
Peter ME, Krammer PH (1998) Mechanisms of CD95 (APO-1/Fas)-mediated apoptosis. Curr Opin Immunol 10:545–551
Peters JM (2002) The anaphase-promoting complex: proteolysis in mitosis and beyond. Mol Cell 9:931–943
Quignon F, De Bels F, Koken M, Feunteun J, Ameisen J-C, de The H (1998) PML induces a novel caspase-independent death process. Nat. Genet 20:259–265
Robert A, Miron MJ, Champagne C, Gingras MC, Branton PE, Lavoie JN (2002) Distinct cell death pathways triggered by the adenovirus early region 4 ORF 4 protein. J Cell Biol 158:519–528
Roopchand DE, Lee JM, Shahinian S, Paquette D, Bussey H, Branton PE (2001) Toxicity of human adenovirus E4orf4 protein in *Saccharomyces cerevisiae* results from interactions with the Cdc55 regulatory B subunit of PP2A. Oncogene 20:5279–5290
Roulston A, Marcellus RC, Branton PE (1999) Viruses and apoptosis. Annu Rev Microbiol 53:577–628
Rudner AD, Murray AW (2000) Phosphorylation by Cdc28 activates the Cdc20-dependent activity of the anaphase-promoting complex. J Cell Biol 149:1377–1390
Scaffidi C, Fulda S, Srinivasan A, Friesen C, Li F, Tomaselli KJ, Debatin K-M, Krammer PH, Peter ME (1998) Two CD95(APO-1/Fas) signalling pathways. EMBO J 17:1675–1687
Shenk T (1996). Adenoviridae: the viruses and their replication In: Fields B, Howley P, Knipe D (eds) Virology. Raven Press, New York, pp 2111–2148
Shteinberg M, Protopopov Y, Listovsky T, Brandeis M, Hershko A (1999) Phosphorylation of the cyclosome is required for its stimulation by Fizzy/cdc20. Biochem Biophys Res Commun 260:193–198
Shtrichman R, Kleinberger T (1998) Adenovirus type 5 E4 open reading frame 4 protein induces apoptosis in transformed cells. J Virol 72:2975–2982
Shtrichman R, Sharf R, Barr H, Dobner T, Kleinberger T (1999) Induction of apoptosis by adenovirus E4orf4 protein is specific to transformed cells and requires an interaction with protein phosphatase 2A. Proc Natl Acad Sci USA 96:10080–10085
Shtrichman R, Sharf R, Kleinberger T (2000) Adenovirus E4orf4 protein interacts with both Bα and B′ subunits of protein phosphatase 2A, but E4orf4-induced apoptosis is mediated only by the interaction with Bα. Oncogene 19:3757–3765
Steinwaerder DS, Carlson CA, Lieber A (2000) DNA replication of first-generation adenovirus vectors in tumor cells. Hum Gene Ther 11:1933–1948
Subramanian T, Tarodi B, Chinnadurai G (1995) p53-independent apoptotic and necrotic cell deaths induced by adenovirus infection: suppression by E1B 19 K and Bcl-2 proteins. Cell Growth Differ 6:131–137
Tatosyan AG, Mizenina OA (2000) Kinases of the Src family: structure and functions. Biochemistry (Mosc) 65:49–58
Teodoro JG, Shore GC, Branton PE (1995) Adenovirus E1A proteins induce apoptosis by both p53-dependent and p53-independent mechanisms. Oncogene 11:467–474
Tewari M, Quan LT, O'Rourke K, Desnoyers S, Zeng Z, Beidler DR, Poirier GG, Salvesen GS, Dixit VM (1995) Yama/CPP32 beta, a mammalian homolog of CED-3, is a CrmA-inhibitable protease that cleaves the death substrate poly(ADP-ribose) polymerase. Cell 81:801–809
Virshup DM (2000) Protein phosphatase 2A: a panoply of enzymes. Curr Opin Cell Biol 12:180–185
Wallach D, Varfolomeev EE, Malinin NL, Goltsev YV, Kovalenko AV, Boldin MP (1999) Tumor necrosis factor receptor and Fas signaling mechanisms. Annu Rev Immunol 17:331–367
Wang Y, Burke DJ (1997) Cdc55p, the B-type regulatory subunit of protein phosphatase 2A, has multiple functions in mitosis and is required for the kinetochore/spindle checkpoint in *Saccharomyces cerevisiae*. Mol Cell Biol 17:620–626
Wera S, Hemmings BA (1995) Serine/threonine protein phosphatases. Biochem J 311:17–29

Wersto RP, Rosenthal ER, Seth PK, Eissa NT, Donahue RE (1998) Recombinant, replication-defective adenovirus gene transfer vectors induce cell cycle dysregulation and inappropriate expression of cyclin proteins. J Virol 72:9491–9502
Whalen SG, Marcellus RC, Whalen A, Ahn NG, Ricciardi RP, Branton PE (1997) Phosphorylation within the transactivation domain of adenovirus E1A protein by mitogen-activated protein kinase regulates expression of early region 4. J Virol 71:3545–3553
White E (2001) Regulation of the cell cycle and apoptosis by the oncogenes of adenovirus. Oncogene 20:7836–7846
Xiang J, Chao DT, Korsmeyer SJ (1996) Bax-induced cell death may not require interleukin 1 beta-converting enzyme-like proteases. Proc Natl Acad Sci USA 93:14559–14563
Xu Q, Reed JC (1998) Bax inhibitor-1, a mammalian apoptosis suppressor identified by functional screening in yeast. Mol Cell 1:337–346
Yamamoto K, Ichijo H, Korsmeyer SJ (1999) BCL-2 is phosphorylated and inactivated by an ASK1/Jun N-terminal protein kinase pathway normally activated at G(2)/M. Mol Cell Biol 19:8469–8478
Yang J, Liu X, Bhalla K, Kim CN, Ibrado AM, Cai J, Peng T-I, Jones DP, Wang X (1997) Prevention of apoptosis by Bcl-2: release of cytochrome c from mitochondria blocked. Science 275:1129–1132
Zhou B-B, Li H, Yuan J, Kirschner MW (1998) Caspase-dependent activation of cyclin-dependent kinases during Fas-induced apoptosis in Jurkat cells. Proc Natl Acad Sci USA 95:6785–6790

Epstein-Barr Virus Signal Transduction and B-Lymphocyte Growth Transformation

K.M. Izumi[1]

1 Introduction

Epstein-Barr virus (EBV) is a human herpesvirus that persistently infects its host for a lifetime through a strategy of growth transforming resting B-cells into continuously proliferating lymphoblastoid cell lines (LCLs) while maintaining itself as an incomplete or "latent" virus in the infected cell nucleus. Immune surveillance is critical to maintaining this symbiosis as perturbations to T-cell immunity reveal the malignant potential of latent EBV-transformed B-cells. Epidemiological evidence further associates EBV infection with the development of other malignancies including Hodgkin's lymphoma, Burkitt's lymphoma, and nasopharyngeal carcinoma. Genetic and biochemical analyses have revealed that five viral proteins are essential for transformation by commandeering signal transduction pathways or intervening in transcriptional control. Latent infection membrane protein 1 (LMP1), a principal effector of transformation signaling, is a constitutively active receptor that exploits the tumor necrosis factor receptor-signaling pathway. This chapter reviews current models regarding the molecular mechanisms through which LMP1 enables EBV to transform B-cell growth.

2 EBV Infection and Disease

Epidemiological data indicate that most humans are asymptomatic EBV carriers who shed virus into saliva. Primary infections typically occur early in life, and most individuals are infected by early adulthood. Transmitted in saliva, EBV lytically infects the oropharyngeal epithelium. EBV spreads from this site via blood or lymph and gains access to lymphoid tissues where it infects B-lymphocytes. In these cells, EBV does not usually undergo lytic infection proteins. Instead, EBV expresses three latent infection membrane proteins (LMPs), six EBV nuclear antigens (EBNAs), a family of complementary strand

[1]Department of Microbiology and Immunology, University of Texas Health Science Center at San Antonio, 7703 Floyd Curl Drive, San Antonio, Texas 78229-3900, USA, e-mail: izumi@uthscsa.edu

Progress in Molecular and Subcellular Biology
C. Alonso (Ed.): Viruses and Apoptosis
© Springer-Verlag Berlin Heidelberg 2004

transcripts (CSTs), and two EBV-encoded RNAs (EBERs). These gene products transform resting B-cells into proliferating cell lines and maintain "latent" EBV as a circular DNA in the cell nucleus. An anti-viral T-cell response principally restricts further lytic infection or progression of EBV-transformed B-cells into a malignant proliferation. Thus, most childhood infections are either asymptomatic or mild febrile illnesses. Although these individuals recover, they become life-long carriers. Infrequently, primary EBV infection is delayed to adolescence or early adulthood, and a more clinically sustained infectious disease mononucleosis may result. Despite the severity of symptoms, most individuals recover fully. These individuals also become life-long carriers (Cohen 1999; Kieff and Rickinson 2001; Rickinson and Kieff 2001).

Because EBV latency proteins are immunogenic, there is strong selection to downregulate protein expression to evade host immunity. Protein expression cannot be completely extinguished since B-cells may undergo apoptosis or terminal differentiation. To resolve this dilemma, EBV will infect most B-cells and by chance infect a memory B-cell (Babcock et al. 1998, 1999; Thorley-Lawson and Babcock 1999). These differentiated B-cells no longer express immunoglobulin D (IgD) but have undergone Ig gene rearrangements and express surface Ig receptors. Most critically, these cells are long-lived. Thus, latent EBV protein expression is not absolutely required to enable host cell survival. EBV continues to express EBNA1 to insure that in cells undergoing mitosis that EBV DNA is replicated and transmitted to the daughter cells. EBNA1 is less antigenic due to repetitive stretches of glycines and alanines that interfere with antigen presentation (Levitskaya et al. 1995, 1997; Dantuma et al. 2000; Sharipo et al. 2001). While downregulated protein expression and EBNA1 effects on antigen presentation help reduce T-cell recognition, infected cells are limited to about 1 to 50 per million lymphocytes (Miyashita et al. 1995; Khan et al. 1996). Because the status of an infected memory B-cell may change, EBV has other survival strategies. EBV may re-express transforming proteins to expand the pool of infected cells, and then downregulate gene expression. Expansion may be afforded by reactivation to lytic infection. Virus that infects epithelial cells of the oropharynx may be shed into saliva to infect a new host, whereas virus that infects another B-cell may recapitulate the process of growth transformation. Secondary immune responses will restrict lytic infection or transformed cell proliferation. However, an expansion of EBV-transformed memory B-cells will likely have developed (Rickinson and Kieff 2001).

Transplant recipients treated with immunosuppressive drugs and patients with AIDS reveal the importance of T-cell immunity in controlling EBV. In these individuals, EBV infection initially causes an infectious mononucleosis, and then infected B-cells progress to a malignant cell proliferation. Post-transplant lymphoproliferative disease is a serious cause of morbidity and mortality in allograft recipients. AIDS-associated lymphomas, particularly central nervous system lymphomas, are most frequently associated with EBV and most are high-grade malignancies with short survival times (Ambinder et al. 1999; Fauci and Lane 1999; Freedman 1999; Kieff and Rickinson 2001; Rickinson and

Kieff 2001). In both diseases, the malignant cells resemble LCLs that are cultured in vitro. Thus, EBV expresses the full latency gene expression program, which comprises latent infection membrane proteins LMP1, LMP2A, or LMP2B; EBV nuclear antigens EBNA1, EBNA LP (leader protein), EBNA2, EBNA3A, EBNA3B, or EBNA3C; complementary strand transcripts (CSTs) known also as BamHI A rightward transcript (BARTs); and EBV-encoded RNAs EBER1 or EBER2 (Allday et al. 1989; Alfieri et al. 1991; Kieff and Rickinson 2001; Rickinson and Kieff 2001).

While the endemic form of Burkitt's lymphoma (BL) seen in Africa is strongly associated with EBV, the childhood or sporadic form, which has a worldwide distribution, is less strongly but still significantly associated with EBV (Lenoir et al. 1984; IARC 1997; Habeshaw et al. 1999). Typically, these cells have an aberrant Ig gene rearrangement that places an Ig enhancer near the c-myc gene resulting in deregulated expression of this transcriptional activator (reviewed in Magrath 1990). In EBV-positive BL, gene expression is muted with continued expression of EBNA1, CSTs, and EBERs (Rowe et al. 1987). While the role of EBV in the pathogenesis of BL is not fully understood, one possibility is that latent EBV gene expression provides a survival advantage either to B-cells, to enable accumulation of the Ig/c-myc translocation, or to translocation positive B-cells, to adapt a malignant phenotype by further accumulation of mutations. Deregulated c-myc expression can substitute for EBV-transforming proteins (Polack et al. 1996), thus allowing EBV to downregulate protein expression (Kieff and Rickinson 2001; Rickinson and Kieff 2001).

The association of Hodgkin's lymphoma with EBV is circumstantial and based upon frequent though not uniform detection of EBV in the malignant cells (Weiss et al. 1987; Wu et al. 1990; Herbst et al. 1991; Pallesen et al. 1991). Hodgkin's lymphoma consists of a mixture of infiltrating leukocytes and malignant mononuclear Hodgkin and multi-nuclear Reed-Sternberg cells (Harris et al. 1994). The malignant cells appear to be of B-cell origin and typically have aberrant, non-productive Ig gene rearrangements, which should result in apoptosis (Kuppers et al. 1994; Braeuninger et al. 1997; Kuppers and Rajewsky 1998; Kanzler et al. 2000). The involvement of EBV in the development of Hodgkin's lymphoma is that LMP2A or LMP1 may provide stimuli similar to Ig receptors and CD40 co-receptors that enable cell survival. As with BL, inappropriate cell survival may allow the accumulation of profound mutations (Rickinson and Kieff 2001).

While the non-keratinizing, poorly differentiated form of nasopharyngeal carcinoma (NPC) is highly associated with EBV, the keratinizing, squamous cell differentiated form is less so. This malignant disease of epithelial cells is worldwide in distribution; however, some ethnic groups such as southern Chinese show higher incidence. The association with EBV was detected based on elevated antibody titers to EBV (Gunven et al. 1970; zur Hausen et al. 1970; Henle et al. 1973). In EBV-positive cells, gene expression similar to Hodgkin's lymphoma comprises LMP1, LMP2s, EBNA1, EBERs, and CSTs (Brooks et al. 1992; Busson et al. 1992; Karran et al. 1992; Smith and Griffin 1992). Progress in understanding the molecular contribution of EBV to the development of NPC has been slowed by the difficulty in propagating NPC tumor tissue in vitro and developing a tissue culture model similar to infecting primary B-cells in vitro (Rickinson and Kieff 2001).

3
Latent EBV Gene Expression in Growth-Transformed B-Cells

Investigations of EBV-mediated transformation of primary B-cells are typically performed in vitro by infecting B-cells from healthy human donors. EBV efficiently transforms B-cells into continuously proliferating LCLs, which are then cultivated and expanded. These aspects combined with advances in genetically manipulating EBV DNA have led to progress in delineating the mechanism by which EBV proteins transform cell growth and provide a substantial model for investigating lymphoproliferative disease.

In latent EBV growth-transformed LCLs, EBV expresses EBNA LP, EBNA1, EBNA2, EBNA3A, EBNA3B, EBNA3C, CSTs, EBER1, EBER2, LMP1, LMP2A, and LMP2B. Recombinant virus genetic analyses have revealed the critical importance of LMP1, EBNA2, EBNA3A, EBNA3C, or EBNA LP to transformation (reviewed in Kieff and Rickinson 2001). EBNA1 and its cis-acting sequence, the origin of plasmid replication (OriP), are required to maintain the circular EBV DNA in the nucleus of the dividing cells (Yates et al. 1984; Lee et al. 1999). EBNA2 is required for transformation by regulating transcription of viral and cell genes (Cohen et al. 1989; Hammerschmidt and Sugden 1989). The BamHI C promoter, which enables expression of all of the EBNAs, and the shared LMP1/LMP2B promoter are EBNA2-regulated genes (Abbot et al. 1990; Fahraeus et al. 1990; Ghosh and Kieff 1990; Wang et al. 1990b; Sung et al. 1991; Zimber-Strobl et al. 1991). EBNA2 regulates expression of cell genes that likely have positive effects on cell growth including c-myc (Kaiser et al. 1999) or c-fgr (Knutson 1990) and surface receptors CD21 or CD23 (Aman et al. 1990; Wang et al. 1990a; Cordier-Bussat et al. 1993). EBNA LP has co-activating effects on EBNA2-mediated transactivation of the LMP1/LMP2B promoter and the EBV BamHI C promoter (Harada and Kieff 1997; Nitsche et al. 1997). This provides a partial explanation as to its critical role in transformation (Mannick et al. 1991). EBNA 3C has transcriptional effects that are essential for EBV-mediated B-cell transformation whereas EBNA 3A effects are critical and EBNA 3B effects are not required (Tomkinson et al. 1993). EBNA 3C is different from EBNA 3A or 3B in that it can co-activate with EBNA2 or activate the LMP1 promoter in G1 growth-arrested cells (Allday and Farrell 1994; Marshall and Sample 1995; Zhao and Sample 2000).

While EBNA effects are mainly enforced at the nuclear level, the latent infection membrane proteins LMP1 and LMP2s function as constitutively active membrane receptors. The LMP2s are not essential for transformation; however, they may provide survival or regulatory signals that are important in vivo (Longnecker 2000). The LMP2A N-terminus contains an immuno-receptor tyrosine activation motif (ITAM) that activates Src-family kinases including Fyn, Lyn, or Syk (Burkhardt et al. 1992; Murray et al. 1996; Merchant et al. 2000). By appropriating the Ig receptor signaling pathway, LMP2A may activate anti-apoptotic Akt/PKB through the PI3 kinase pathway (Scholle et al. 2000; Swart et al. 2000), desensitize cells to signals that activate EBV lytic infection to maintain latency (Miller et al. 1993, 1994a,b, 1995a), or provide survival signals in B-cells that have not productively rearranged their Ig gene loci (Caldwell et al. 1998, 2000).

4 Early Analyses of LMP1 Structure

LMP1 was first analyzed by cloning and sequencing a viral transcript from LCLs. Thus, LMP1 is a 386-residue, integral membrane protein with several distinctive domains (Fig. 1). The 24-residue N-terminal domain is rich in arginines or prolines but lacks the residues indicative of a signal peptide. The next segment consists of six groups of hydrophobic residues separated by five groups of hydrophilic residues that form six membrane-spanning domains separated by short reverse turns. While reminiscent of a voltage-gated ion conductance channel, none of the transmembrane domains contains the characteristic residues. The 200-residue carboxyl terminus, which consists mainly of hydrophilic or acidic residues, also contains four imperfect direct repeats that make up the central third of the C-terminus (Fennewald et al. 1984).

Biochemical analyses reveal most LMP1 is in the plasma membrane and oriented with the N-terminal and C-terminal domains in the cytoplasm leaving only three of the five reverse turns exposed to the extracellular environment. A conspicuous aspect of LMP1 is its ability to form tight aggregates in the plasma membrane that coalesce into a single cap-like structure (Hennessy et al. 1984; Mann et al. 1985; Liebowitz et al. 1986). Aggregation is spontaneous and is mediated by the six membrane-spanning domains. The half-life of LMP1 in the plasma membrane is under 2 h. LMP1 then associates with cell cytoskeleton proteins (Baichwal and Sugden 1987; Liebowitz et al. 1987; Mann et al. 1987; Moorthy and Thorley-Lawson 1990) and is post-translationally modified by phosphorylation of the C-terminus principally on serines and at lower ratio threonines, but not on tyrosines (Moorthy and Thorley-Lawson 1993b). LMP1 is then proteolytically cleaved releasing a nearly complete C-terminus as a cytoplasmic protein of 25 kDa (Moorthy and Thorley-Lawson 1990).

These analyses, which evoke models of LMP1 as a plasma membrane receptor, were further impelled by epidemiological data that support the hypothesis that LMP1 is a critical effector of cell transformation. LMP1 is expressed in most malignancies associated with EBV, including Burkitt's lymphoma, Hodgkin's lymphoma, nasopharyngeal carcinoma, and lymphoproliferative disease (Fauci and Lane 1999; Kieff and Rickinson 2001; Rickinson and Kieff

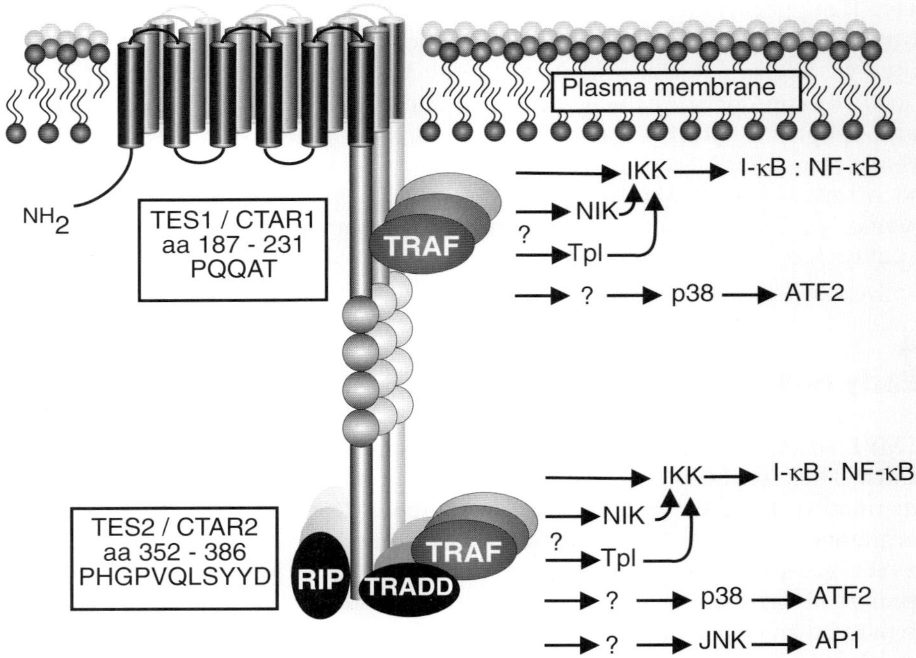

Fig. 1. Diagram of the Epstein-Barr virus transforming protein latent infection membrane protein 1 (LMP1). LMP1 consists of an amino (NH_2) terminal cytoplasmic domain, six membrane-spanning domains separated by short reverse turns, and a carboxyl-terminal cytoplasmic domain. Transformation effector site 1 or C-terminal NF-κB activation region 1 (*TES1/CTAR1*) transduces signals via the indicated pathways through tumor necrosis factor (TNF) receptor associated factors TRAF1, -2, -3, or -5. TES2/CTAR2 transduces signals via TNF receptor associated death domain protein (TRADD) or receptor interacting protein (RIP) to the indicated pathways

2001). However, other than the general similarity to proteins with multiple membrane-spanning domains, LMP1 has no extensive homology with other proteins with which a specific a role in transformation could be inferred.

5
Gene Transfer Experiments Reveal Critical LMP1 Functions

In rodent fibroblast transformation assays, LMP1 transgenes have oncogene-like activities by altering cell morphology, reducing serum dependence, and causing the loss of contact inhibition. Further, LMP1 induces anchorage-independent cell growth and increases cell tumorigenicity in nude mice. These functions are dependent on the membrane-spanning domains, which spontaneously aggregate LMP1 in the plasma membrane (Baichwal and Sugden 1988;

Hammerschmidt et al. 1989; Moorthy and Thorley-Lawson 1993a; Wang et al. 1985). Further attempts to delineate the critical transforming domains, however, were impeded by conflicting or uncertain effects using these assays.

LMP1-transforming effects on rodent fibroblasts are likely related to activation of NF-κB. LMP1 has two C-terminal activating regions (CTAR), and both require aggregation mediated by the membrane-spanning domains. In brief, C-terminal residues 187 to 231 constitute CTAR1 whereas the residues 352 to 386 constitute CTAR2 (Hammarskjold and Simurda 1992; Huen et al. 1995; Mitchell and Sugden 1995). Further relevant to NF-κB, LMP1 transgenes introduced into EBV-negative B-cell lines recapitulate many of the changes induced by activating Ig receptors or infecting cells with EBV, including increased DNA synthesis and release from G0/G1 cell cycle arrest (Peng and Lundgren 1992). Other changes include increased villous projections and cell clumping due in part to increased expression of adhesion molecules LFA-1, LFA-3, or ICAM-1 and cell cytoskeleton protein vimentin (Wang et al. 1988, 1990a; Birkenbach et al. 1989; Peng and Lundgren 1992). LMP1 upregulates expression of cell surface receptors including CD21, CD23, CD39, CD40, CD44, or class II MHC. LMP1 also induces expression of IL-6 or IL-10 (Wang et al. 1990a; Peng and Lundgren 1992; Nakagomi et al. 1994; Eliopoulos et al. 1997). Increased expression of growth factors or receptors is likely to have positive effects on cell growth, which are enhanced by adhesion to other cells that can cross-feed through autocrine or paracrine growth factors expressed on the cell surface. LMP1 expression reduces susceptibility to apoptosis primarily due to upregulated expression of anti-apoptotic proteins. These include the zinc-finger containing protein A20, the Bcl-2 family of proteins including Bcl-2, A1/Bfl-1, or Mcl-1, and cellular inhibitors of apoptosis (cIAPs; Henderson et al. 1991; Laherty et al. 1992; Rowe et al. 1994; Wang et al. 1996; D'Souza et al. 2000; Hong et al. 2000). Mice with LMP1 transgenes can develop B-cell tumors (Kulwichit et al. 1998) or have altered keratin gene expression and epidermal skin hyperplasia (Wilson et al. 1990; Curran et al. 2001). Further relevant to epithelial cells, LMP1 alters cell morphology, modulates cytokeratin expression, and inhibits cell differentiation (Dawson et al. 1990). LMP1 upregulates expression of epidermal growth factor receptor (EGFR), IL-8, or matrix metalloproteinases, which may be involved in transformed cell invasion (Miller et al. 1995b, 1997; Yoshizaki et al. 1998; Eliopoulos et al. 1999b). Although these alterations in cell gene expression are prominent, mutations that affected LMP1 aggregation had only partial loss of some effects and retention of others and thus confounded analyses intended to characterize the critical effector domains.

6
LMP1 Is a Key Effector of B-Lymphocyte Transformation

An assay that clearly reveals the importance of LMP1 to transformation is to analyze EBV recombinants with mutated LMP1 DNA for their ability to growth transform primary B-cells into LCLs. Thus, recombinants with a stop codon insertion in after codon 9 in the LMP1 N-terminus have a non-transforming phenotype. This result provided the first evidence that LMP1 is essential for EBV-mediated B-cell transformation (Kaye et al. 1993). While these recombinants are precluded from expressing the full-length LMP1, a downstream methionine enables expression of LMP1 that lacks the N-terminus and first transmembrane domain. This truncated LMP1 accumulates to wild-type levels and associates with the plasma membrane but does not form aggregates. These results further indicate LMP1 membrane aggregation is a critical aspect of transformation (Kaye et al. 1993).

The effects of the N-terminal truncation are not due to ablation of a critical effector domain because mutational analysis of the N-terminus reveals no specific sequence is required for transformation. Deletion of arginines and prolines, which typically function to tether transmembrane proteins to the cytoplasm, had the most severe effect on LMP1's ability to form aggregates and the ability of EBV recombinants to transform B-cells. Once established, LCLs transformed by these recombinants grew similar to LCLs transformed by EBV with wild-type LMP1 genes. Thus, it is unlikely there is a specific protein interaction with the N-terminus that is required for transformation. The likely critical function of the N-terminal domain is to tether the first membrane-spanning domain to the cytoplasm to enable aggregation. Aggregation is associated with a transforming phenotype, and further supports the hypothesis that LMP1 is a receptor that transduces signals to transform cell growth (Izumi et al. 1997).

In contrast to these results, insertion of a stop codon after codon 187, which allows for expression of the N-terminus and six membrane-spanning domains but none of the C-terminus, abolishes EBV-transforming abilities. This result indicates the C-terminus is essential for transformation. Surprisingly, recombinants that express LMP1 codons 1 to 231 are able to transform B-cells into LCLs that are dependent on co-cultivation with fibroblast feeder layers for further outgrowth (Kaye et al. 1995). LMP1 (1–231) consists of the N-terminus, six membrane-spanning domains, and the first 45 residues of the C-terminus. These recombinants are similar to wild-type EBV for initiating the proliferation of primary B-cells into nascent LCLs. However, most of these LCLs cannot be amplified and established as long-term cultures of indefinitely proliferating cell lines. Surviving cultures also demonstrated an unusual requirement to be cultured at high density (Kaye et al. 1999). EBV recombinants deleted of codons 185 to 211, which comprises the first 25 residues of the C-terminus, are unable to transform primary B-lymphocytes (Izumi et al. 1997). In brief, the first 45 residues of the LMP1 C-terminus constitute an essential transformation effector site since deletion of residues internal to this site abrogates EBV-transforming abilities (Izumi et al. 1997). When linked to the N-terminus and six transmembrane domains, the first 45 residues are also sufficient for initial transformation of B-cell growth (Kaye et al. 1999).

The delineation of the first LMP1 transformation effector site (TES1) within residues 187–231 of the C-terminus also implicated the remaining 155 residues with a function that enables efficient, long-term LCL proliferation. The carboxyl boundary of this function was delineated by mutation of the most distal LMP1 C-terminal residues $Y_{384}YD_{386}$. Although these residues are at the end of the protein, mutation from YYD to ID had a severe adverse effect on transforming abilities. By comparison, recombinants mutated at the YYD sequence to FFD were indistinguishable from wild-type EBV for the ability to transform B-cell growth, for efficiency of nascent LCL amplification into established cell lines, or lytic infection. These results and those described below defined a second transformation effector site (TES2) within residues 352–386. While this site is accessory to TES1, TES2 is critical for EBV to mediate efficient transformation of B-cell growth into LCLs capable of indefinite proliferation (Izumi and Kieff 1997).

The domain intermediate of TES1 and TES2 contains several sites that are highly conserved in most EBV isolates. This includes four direct, imperfect copies of an 11-residue repeat (Fennewald et al. 1984), a site for protease-specific cleavage (Moorthy and Thorley-Lawson 1990), sites proposed to activate Janus kinase 3 (JAK3) and signal transduction and activation of transcription (STAT) proteins (Gires et al. 1999), or sites for specific serine/threonine phosphorylation (Moorthy and Thorley-Lawson 1993b). Also contained in this domain are sequences reported to be important in rodent fibroblast transformation assays (Moorthy and Thorley-Lawson 1992, 1993a; Li et al. 1996; Mehl et al. 1998). Despite the expected importance, EBV recombinants deleted of codons 232–351 are indistinguishable from wild-type EBV in transforming primary B-cell growth, in the efficiency of nascent LCL outgrowth into long-term LCLs, in lytic virus infection, or in JAK/STAT activation (Izumi et al. 1999a; Higuchi et al. 2002). These results indicate that residues 187–231 and residues 352–386 are the only C-terminal residues that are critical for B-cell transformation, and affirm their designation as transformation effector sites 1 and 2 (TES1 and TES2).

7
LMP1-Transforming Signals Are Mediated Through the TNF Receptor Signaling Pathway

LMP1 has two C-terminal activating regions (CTAR1 and CTAR2) that also mediate the activation of mitogen-activated protein kinases (MAPK; Hammarskjold and Simurda 1992; Huen et al. 1995; Mitchell and Sugden 1995; Eliopoulos et al. 1999a,b). TES1 and CTAR1 are interrelated since deletion of residues 185–211 abolishes EBV-transforming abilities and NF-κB-activating signals from this site (Kaye et al. 1995, 1999; Izumi et al. 1997). Further, TES2 is interrelated with CTAR2 since mutation of the terminal YYD to ID adversely affects EBV-transforming abilities and abrogates NF-κB-activating signals from this site (Izumi and Kieff 1997). Because receptors that transduce signals that activate NF-κB or MAPK are critical regulators of cell growth, differentiation, or survival, biochemical investigations were prompted to delineate the underlying mechanisms.

Using TES1/CTAR1 residues 187–231 as probe in a yeast two-hybrid screen, proteins known as TNF receptor associated factors (TRAFs) were retrieved (Mosialos et al. 1995). TRAFs were previously identified as the signaling adapters through which type 2 tumor necrosis factor receptors (TNFR2) mediate NF-κB activation (Rothe et al. 1994). Six TRAFs have been identified, and TRAF1, -2, -3, -5, or -6 are implicated as the signaling adapters of TNF receptor family members including lymphotoxin β receptor (LTβR), TNFR1, TNFR2, CD40, CD30, or CD27. TRAFs consist of an N-terminal zinc finger and in some cases a RING finger domain, which are important for signaling. The C-terminus of TRAFs contains a 200-residue TRAF homology domain, which is further divided into a well-conserved C-terminal domain of about 150 residues (TRAF-C) and a less well-conserved N-terminal domain (TRAF-N) that contains a coiled-coil motif (reviewed in Bradley and Pober 2001; Locksley et al. 2001).

LMP TES1/CTAR1 signaling through TRAFs induces effects in B-cells that are similar to CD40, an Ig receptor co-stimulatory molecule. The sequence that enables LMP1 and CD40 to associate with TRAFs and share signaling effects is a Pro-X-Gln-X-Thr motif. The function of this motif is to stably associate TRAF trimers with TES1/CTAR1 or the CD40 receptor tail. Whereas CD40 and other TNFR family members require ligand binding to aggregate their cytoplasmic tails and enable signaling, the six membrane-spanning domains spontaneously aggregate LMP1 to enable constitutive signaling. TES1/CTAR1 associates with TRAF1 or TRAF3 with higher affinity than with TRAF2 or TRAF5, and residues flanking the core motif influence this behavior. Further, LMP1 association with TRAF1, TRAF2, and possibly TRAF5, mediates signals that activate NF-κB whereas TRAF3 may modulate these signals (Mosialos et al. 1995; Devergne et al. 1996, 1998; Brodeur et al. 1997; Sandberg et al. 1997; Floettmann et al. 1998; Hatzivassiliou et al. 1998; Henriquez et al. 1999). Since deletion of this motif blocks TES1/CTAR1 interaction with TRAFs, abolishes NF-κB or MAPK activating signals from this site, and adversely affects EBV-mediated transformation, TES1/CTAR1 signaling via TRAFs is likely the critical effector pathway responsible for transforming effects on B-cell growth.

LMP1 TES2/CTAR2 also mediates its effects through the TNF receptor-signaling pathway involving TRAFs. Using residues 355–386 as probe in a yeast two-hybrid screen, the TNFR1-associated death domain containing protein TRADD was retrieved. Further analyses revealed TRADD binding is abolished when the probe is mutated at the wild-type YYD sequence to the transformation defective ID sequence, underscoring the likely importance of this result (Izumi and Kieff 1997; Izumi et al. 1999b).

TRADD was previously identified as the proximal signaling adapter of TNFR1. TRADD consists of an N-terminal TRAF binding domain and C-terminal death domain that enables aggregation with other death domain proteins (Hsu et al. 1995). In brief, TNF binding to TNFR1 induces receptor aggregation and TRADD recruitment. Aggregated TRADD then recruits receptor interacting protein (RIP). RIP contains a C-terminal death domain and two N-terminal domains that bind TRAFs to mediate signaling. Aggregated TNFR1 or TRADD also associates with the pro-apoptotic signal transducing adapter Fas-associated death domain protein (FADD), which mediates pro-apoptotic signals through activation of caspases (reviewed in Bradley and Pober 2001; Chen and Goeddel 2002; Locksley et al. 2001).

While TES2/CTAR2 associates with TRADD and the six membrane-spanning domains enable constitutive signaling, there are significant differences with TNFR1. TES2/CTAR2 is not homologous with known death domain proteins including TNFR1, and the TRADD death domain residues that are critical for binding are not identical. The terminal 11 residues of TES2/CTAR2 (PHGPVQLSYYD) are sufficient to signal through TRADD whereas TNFR1 requires a more complex 70-residue death domain to engage TRADD. TES2/CTAR2 does not require RIP to transduce NF-κB-activating signals through TRADD whereas TNFR1 requires RIP. While TES2/CTAR2 transduces NF-κB-activating signals likely through TRADD direct recruitment of TRAFs, it also interacts with RIP to a lesser extent than TRADD and transduces moderate NF-κB-activating signals. Most significantly, overexpression of LMP1 TES2/CTAR2 does not activate apoptosis even under conditions where NF-κB activation is blocked. Since mutation of the terminal TES2/CTAR2 YYD sequence to ID compromises interaction with TRADD or RIP, abolishes NF-κB or MAPK activating signals from this site, and adversely affects EBV mediated transformation, TES2/CTAR2 signal transduction via TRADD or RIP is likely the critical effector pathway that enables efficient amplification of nascent LCLs into cell lines capable of indefinite proliferation (Izumi and Kieff 1997; Eliopoulos et al. 1999a; Izumi et al. 1999b; Kieser et al. 1999).

LMP1 TES1/CTAR1 and TES2/CTAR2 signaling converge on TRAFs as dominant negative TRAF mutants block NF-κB activation from both sites. Analyses performed using dominant negative mutants of the I-κB kinase (IKK) α or β subunits, NF-κB-inducing kinase (NIK), or Cot/Tpl2 have revealed their involvement in LMP1 signaling (Sylla et al. 1998; Eliopoulos et al. 2002). A working model is that aggregated TRAFs mediate the activation of the IKK complex. The connection between TNFRs, TRAFs, and IKK appears to have many avenues since several protein kinases activate IKK including protein kinase C isoforms, protein kinase R, and MAP3K proteins NIK, Cot/Tpl-2, MEKK1, or TAK1. MEKK1 or NIK also interacts with TRAF2 and thus may provide a direct link between receptor aggregated TRAFs to IKK (Malinin et al. 1997; Baud et al. 1999). The IKK complex, which consists of two enzymatic subunits IKKα or IKKβ and essential regulatory subunit IKKγ, phosphorylates the inhibitors of NF-κB (I-κBs) targeting the protein for ubiquitination and proteasome-mediated degradation. Structural alterations in I-κBs allow their associated NF-κB dimers to enter the nucleus and function as sequence-specific transcription factors (reviewed in Bradley and Pober 2001; Locksley et al. 2001; Chen and Goeddel 2002; Ghosh and Karin 2002).

Also downstream of TRAFs is the MAPK pathway. TES2/CTAR2 mediates JNK activation of the AP-1 transcription factor. TRADD, TRAF2, and the kinase SEK upstream of JNK are involved in this signaling (Kieser et al. 1997; Eliopoulos et al. 1999a). Analyses of TNFR signaling indicate TRAFs may connect to SEK via MEKK1, geminal center kinase GCK, or ASK1 (Shi and Kehrl 1997; Nishitoh et al. 1998; Yuasa et al. 1998; Chadee et al. 2002). LMP1 TES1/CTAR1 or TES2/CTAR2 mediates p38 MAPK activation of the ATF-2 transcription factor (Eliopoulos et al. 1999b). While this signaling pathway involves TRAF2, the LMP1 pathway to p38 requires further investigation.

While LMP1 transformation signaling converges on TRAFs, many aspects remain to be resolved (Kaye et al. 1996). TES1/CTAR1 and TES2/CTAR2 signal through TRAFs to activate the IKK complex, yet the signals mediate differential transforming effects on B-cell growth. As examples of the subtle complexity involved, TES1/CTAR1 preferentially activates EGF receptors in epithelial cells and TRAF1 in B-cells (Miller et al. 1995b, 1997, 1998; Devergne et al. 1998). While NF-κB activation is clearly involved with transcriptional effects and critical to LCL survival, inhibition with non-phosphorylatable I-κBα results in apoptosis despite high-level Bcl-2 and Bcl-XL expression (Cahir McFarland et al. 1999). MAPK p38 and JNK are clearly activated; however, their significance in transformation is not understood. Finally, a detailed understanding of the underlying mechanisms through which TRAF signals mediate transcriptional effects requires further investigation.

8 Summary

Latent EBV growth transformation of resting B-cells into indefinitely proliferating cell lines is a successful viral strategy for survival in its host and the basis of several human malignancies. EBV transforms cell growth through viral proteins that modify cell gene expression at the level of transcription or by appropriating signaling pathways. Analyses of the EBV-transforming protein LMP1 have begun to reveal that this receptor transduces critical signals by appropriating the TNF receptor signal transduction pathway to activate NF-κB and MAPK. While this has brought an important aspect into clearer focus, future progress in delineating the underlying mechanism of transformation, which will be essential to devising effective therapies to treat EBV-associated malignancies, will depend on resolving the intricacies of TRAF signal transduction. Since expression of cytokines, receptors, and anti-apoptotic proteins are regulated by TRAF signaling, another critical issue is delineating the genes that are specifically targeted by LMP1 in order to transform B-lymphocyte growth.

Acknowledgements. The author acknowledges the generous support of grants from the Children's Cancer Research Institute, the Howard Hughes Medical Institute, the San Antonio Area Foundation, and the University of Texas Health Science Center at San Antonio.

References

Abbot SD, Rowe M, Cadwallader K, Ricksten A, Gordon J, Wang F, Rymo L, Rickinson AB (1990) Epstein-Barr virus nuclear antigen 2 induces expression of the virus-encoded latent membrane protein. J Virol 64:2126–2134

Alfieri C, Birkenbach M, Kieff E (1991) Early events in Epstein-Barr virus infection of human B lymphocytes [published erratum appears in Virology 1991 Dec;185(2):946]. Virology 181:595–608

Allday MJ, Crawford DH, Griffin BE (1989) Epstein-Barr virus latent gene expression during the initiation of B cell immortalization. J Gen Virol 70:1755–1764

Allday MJ, Farrell PJ (1994) Epstein-Barr virus nuclear antigen EBNA3C/6 expression maintains the level of latent membrane protein 1 in G1-arrested cells. J Virol 68:3491–3498

Aman P, Rowe M, Kai C, Finke J, Rymo L, Klein E, Klein G (1990) Effect of the EBNA-2 gene on the surface antigen phenotype of transfected EBV-negative B-lymphoma lines. Int J Cancer 45:77–82

Ambinder RF, Lemas MV, Moore S, Yang J, Fabian D, Krone C (1999) Epstein-Barr virus and lymphoma. Cancer Treat Res 99:27–45

Babcock GJ, Decker LL, Volk M, Thorley-Lawson DA (1998) EBV persistence in memory B cells in vivo. Immunity 9:395–404

Babcock GJ, Decker LL, Freeman RB, Thorley-Lawson DA (1999) Epstein-barr virus-infected resting memory B cells, not proliferating lymphoblasts, accumulate in the peripheral blood of immunosuppressed patients. J Exp Med 190:567–576

Baichwal VR, Sugden B (1987) Posttranslational processing of an Epstein-Barr virus-encoded membrane protein expressed in cells transformed by Epstein-Barr virus. J Virol 61:866–875

Baichwal VR, Sugden B (1988) Transformation of Balb 3T3 cells by the BNLF-1 gene of Epstein-Barr virus. Oncogene 2:461–467

Baud V, Liu ZG, Bennett B, Suzuki N, Xia Y, Karin M (1999) Signaling by proinflammatory cytokines: oligomerization of TRAF2 and TRAF6 is sufficient for JNK and IKK activation and target gene induction via an amino-terminal effector domain. Genes Dev 13:1297–1308

Birkenbach M, Liebowitz D, Wang F, Sample J, Kieff E (1989) Epstein-Barr virus latent infection membrane protein increases vimentin expression in human B-cell lines. J Virol 63:4079–4084

Bradley JR, Pober JS (2001) Tumor necrosis factor receptor-associated factors (TRAFs). Oncogene 20:6482–6491

Braeuninger A, Kuppers R, Strickler JG, Wacker HH, Rajewsky K, Hansmann ML (1997) Hodgkin and Reed-Sternberg cells in lymphocyte predominant Hodgkin disease represent clonal populations of germinal center-derived tumor B cells. Proc Natl Acad Sci USA 94:9337–9342

Brodeur SR, Cheng G, Baltimore D, Thorley-Lawson DA (1997) Localization of the major NF-kappaB-activating site and the sole TRAF3 binding site of LMP-1 defines two distinct signaling motifs. J Biol Chem 272:19777–19784

Brooks L, Yao QY, Rickinson AB, Young LS (1992) Epstein-Barr virus latent gene transcription in nasopharyngeal carcinoma cells: coexpression of EBNA1, LMP1, and LMP2 transcripts. J Virol 66:2689–2697

Burkhardt AL, Bolen JB, Kieff E, Longnecker R (1992) An Epstein-Barr virus transformation-associated membrane protein interacts with src family tyrosine kinases. J Virol 66:5161–5167

Busson P, McCoy R, Sadler R, Gilligan K, Tursz T, Raab-Traub N (1992) Consistent transcription of the Epstein-Barr virus LMP2 gene in nasopharyngeal carcinoma. J Virol 66:3257–3262

Cahir McFarland ED, Izumi KM, Mosialos G (1999) Epstein-Barr virus transformation: involvement of latent membrane protein 1-mediated activation of NF-kappaB. Oncogene 18:6959–6964

Caldwell RG, Wilson JB, Anderson SJ, Longnecker R (1998) Epstein-Barr virus LMP2A drives B cell development and survival in the absence of normal B cell receptor signals. Immunity 9:405–411

Caldwell RG, Brown RC, Longnecker R (2000) Epstein-Barr virus LMP2A-induced B-cell survival in two unique classes of EmuLMP2A transgenic mice. J Virol 74:1101–1013

Chadee DN, Yuasa T, Kyriakis JM (2002) Direct activation of mitogen-activated protein kinase kinase kinase MEKK1 by the Ste20p homologue GCK and the adapter protein TRAF2. Mol Cell Biol 22:737–749

Chen G, Goeddel DV (2002) TNF-R1 signaling: a beautiful pathway. Science 296:1634–1635

Cohen JI (1999) Epstein-Barr virus infections, including infectious mononucleosis. In: Braunwald E, Fauci AS, Isselbacher KJ, Kasper DL, Hauser SL, Longo DL, Jameson JL (eds) Harrison's principles of internal medicine. McGraw-Hill, Boston, part 7, sect 12, chap 186

Cohen JI, Wang F, Mannick J, Kieff E (1989) Epstein-Barr virus nuclear protein 2 is a key determinant of lymphocyte transformation. Proc Natl Acad Sci USA 86:9558–9562

Cordier-Bussat M, Billaud M, Calender A, Lenoir GM (1993) Epstein-Barr virus (EBV) nuclear-antigen-2-induced up-regulation of CD21 and CD23 molecules is dependent on a permissive cellular context. Int J Cancer 53:153–160

Curran JA, Laverty FS, Campbell D, Macdiarmid J, Wilson JB (2001) Epstein-Barr virus encoded latent membrane protein-1 induces epithelial cell proliferation and sensitizes transgenic mice to chemical carcinogenesis. Cancer Res 61:6730–6738

Dantuma NP, Heessen S, Lindsten K, Jellne M, Masucci MG (2000) Inhibition of proteasomal degradation by the gly-Ala repeat of Epstein-Barr virus is influenced by the length of the repeat and the strength of the degradation signal. Proc Natl Acad Sci USA 97: 8381–8385

Dawson CW, Rickinson AB, Young LS (1990) Epstein-Barr virus latent membrane protein inhibits human epithelial cell differentiation. Nature 344:777–780

Devergne O, Hatzivassiliou E, Izumi KM, Kaye KM, Kleijnen MF, Kieff E, Mosialos G (1996) Association of TRAF1, TRAF2, and TRAF3 with an Epstein-Barr virus LMP1 domain important for B-lymphocyte transformation: role in NF-kappaB activation. Mol Cell Biol 16:7098–7108

Devergne O, McFarland EC, Mosialos G, Izumi KM, Ware CF, Kieff E (1998) Role of the TRAF binding site and NF-kappaB activation in Epstein-Barr virus latent membrane protein 1-induced cell gene expression. J Virol 72:7900–7908

D'Souza B, Rowe M, Walls D (2000) The bfl-1 gene is transcriptionally upregulated by the Epstein-Barr virus LMP1, and its expression promotes the survival of a Burkitt's lymphoma cell line. J Virol 74:6652–6658

Eliopoulos AG, Stack M, Dawson CW, Kaye KM, Hodgkin L, Sihota S, Rowe M, Young LS (1997) Epstein-Barr virus-encoded LMP1 and CD40 mediate IL-6 production in epithelial cells via an NF-kappaB pathway involving TNF receptor-associated factors. Oncogene 14:2899–2916

Eliopoulos AG, Blake SM, Floettmann JE, Rowe M, Young LS (1999a) Epstein-Barr virus-encoded latent membrane protein 1 activates the JNK pathway through its extreme C terminus via a mechanism involving TRADD and TRAF2. J Virol 73:1023–1035

Eliopoulos AG, Gallagher NJ, Blake SM, Dawson CW, Young LS (1999b) Activation of the p38 mitogen-activated protein kinase pathway by Epstein-Barr virus-encoded latent membrane protein 1 coregulates interleukin-6 and interleukin-8 production. J Biol Chem 274:16085–16096

Eliopoulos AG, Davies C, Blake SS, Murray P, Najafipour S, Tsichlis PN, Young LS (2002) The oncogenic protein kinase Tpl-2/Cot contributes to Epstein-Barr virus-encoded latent infection membrane protein 1-induced NF-kappaB signaling downstream of TRAF2. J Virol 76:4567–4579

Fahraeus R, Jansson A, Ricksten A, Sjoblom A, Rymo L (1990) Epstein-Barr virus-encoded nuclear antigen 2 activates the viral latent membrane protein promoter by modulating the activity of a negative regulatory element. Proc Natl Acad Sci USA 87:7390–7394

Fauci AS, Lane HC (1999) Human immunodeficiency virus (HIV) disease: AIDS and related disorders. In: Braunwald E, Fauci AS, Isselbacher KJ, Kasper DL, Hauser SL, Longo DL, Jameson JL (eds) Harrison's principles of internal medicine. McGraw-Hill, Boston, sect 1, chap 308, p 12

Fennewald, S., van Santen, V. and Kieff, E. (1984) Nucleotide sequence of an mRNA transcribed in latent growth- transforming virus infection indicates that it may encode a membrane protein. J Virol 51(2), 411–9.

Floettmann JE, Eliopoulos AG, Jones M, Young LS, Rowe M (1998) Epstein-Barr virus latent membrane protein-1 (LMP1) signalling is distinct from CD40 and involves physical cooperation of its two C-terminus functional regions. Oncogene 17:2383–2392

Freedman AS (1999) Malignancies of lymphoid cells. In: Braunwald E, Fauci AS, Isselbacher KJ, Kasper DL, Hauser SL, Longo DL, Jameson JL (eds) Harrison's principles of internal medicine. McGraw-Hill, Boston, part 6, sect 2, chap 113

Ghosh S, Karin M (2002) Missing pieces in the NF-kappaB puzzle. Cell 109:S81–96

Ghosh D, Kieff E (1990) cis-acting regulatory elements near the Epstein-Barr virus latent-infection membrane protein transcriptional start site. J Virol 64:1855–1858

Gires O, Kohlhuber F, Kilger E, Baumann M, Kieser A, Kaiser C, Zeidler R, Scheffer B, Ueffing M, Hammerschmidt W (1999) Latent membrane protein 1 of Epstein-Barr virus interacts with JAK3 and activates STAT proteins. EMBO J 18:3064–3073

Gunven P, Klein G, Henle G, Henle W, Clifford P (1970) Epstein-Barr virus in Burkitt's lymphoma and nasopharyngeal carcinoma. Antibodies to EBV associated membrane and viral capsid antigens in Burkitt lymphoma patients. Nature 228:1053–1056

Habeshaw G, Yao QY, Bell AI, Morton D, Rickinson AB (1999) Epstein-Barr virus nuclear antigen 1 sequences in endemic and sporadic Burkitt's lymphoma reflect virus strains prevalent in different geographic areas. J Virol 73:965–975

Hammarskjold ML, Simurda MC (1992) Epstein-Barr virus latent membrane protein transactivates the human immunodeficiency virus type 1 long terminal repeat through induction of NF-kappa B activity. J Virol 66:6496–6501

Hammerschmidt W, Sugden B (1989) Genetic analysis of immortalizing functions of Epstein-Barr virus in human B lymphocytes. Nature 340:393–397

Hammerschmidt W, Sugden B, Baichwal VR (1989) The transforming domain alone of the latent membrane protein of Epstein-Barr virus is toxic to cells when expressed at high levels. J Virol 63:2469–2475

Harada S, Kieff E (1997) Epstein-Barr virus nuclear protein LP stimulates EBNA-2 acidic domain-mediated transcriptional activation. J Virol 71:6611–6618

Harris NL, Jaffe ES, Stein H, Banks PM, Chan JK, Cleary ML, Delsol G, De Wolf-Peeters C, Falini B, Gatter KC (1994) A revised European-American classification of lymphoid neoplasms: a proposal from the International Lymphoma Study Group. Blood 84:1361–1392

Hatzivassiliou E, Miller WE, Raab-Traub N, Kieff E, Mosialos G (1998) A fusion of the EBV latent membrane protein-1 (LMP1) transmembrane domains to the CD40 cytoplasmic domain is similar to LMP1 in constitutive activation of epidermal growth factor receptor expression, nuclear factor-kappa B, and stress-activated protein kinase. J Immunol 160:1116–1121

Henderson S, Rowe M, Gregory C, Croom-Carter D, Wang F, Longnecker R, Kieff E, Rickinson A (1991) Induction of bcl-2 expression by Epstein-Barr virus latent membrane protein 1 protects infected B cells from programmed cell death. Cell 65:1107–1115

Henle W, Ho HC, Henle G, Kwan HC (1973) Antibodies to Epstein-Barr virus-related antigens in nasopharyngeal carcinoma. Comparison of active cases with long-term survivors. J Natl Cancer Inst 51:361–369

Hennessy K, Fennewald S, Hummel M, Cole T, Kieff E (1984) A membrane protein encoded by Epstein-Barr virus in latent growth-transforming infection. Proc Natl Acad Sci USA 81:7207–7211

Henriquez NV, Floettmann E, Salmon M, Rowe M, Rickinson AB (1999) Differential responses to CD40 ligation among Burkitt lymphoma lines that are uniformly responsive to Epstein-Barr virus latent membrane protein 1. J Immunol 162:3298–3307

Herbst H, Dallenbach F, Hummel M, Niedobitek G, Pileri S, Muller-Lantzsch N, Stein H (1991) Epstein-Barr virus latent membrane protein expression in Hodgkin and Reed-Sternberg cells. Proc Natl Acad Sci USA 88:4766–4770

Higuchi M, Kieff E, Izumi KM (2002) The Epstein-Barr virus latent membrane protein 1 putative Janus kinase 3 (JAK3) binding domain does not mediate JAK3 association or activation in B-lymphoma or lymphoblastoid cell lines. J Virol 76:455–459

Hong SY, Yoon WH, Park JH, Kang SG, Ahn JH, Lee TH (2000) Involvement of two NF-kappa B binding elements in tumor necrosis factor alpha-, CD40-, and Epstein-Barr virus latent membrane protein 1-mediated induction of the cellular inhibitor of apoptosis protein 2 gene. J Biol Chem 275:18022–18028

Hsu H, Xiong J, Goeddel DV (1995) The TNF receptor 1-associated protein TRADD signals cell death and NF-kappa B activation. Cell 81:495–504

Huen DS, Henderson SA, Croom-Carter D, Rowe M (1995) The Epstein-Barr virus latent membrane protein-1 (LMP1) mediates activation of NF-kappa B and cell surface phenotype via two effector regions in its carboxy-terminal cytoplasmic domain. Oncogene 10:549–560

IARC (1997) Proceedings of the IARC Working Group on the Evaluation of Carcinogenic Risks to Humans. Epstein-Barr virus and Kaposi's sarcoma herpesvirus/human herpesvirus 8. Lyon, France, 17–24 June, 1997. IARC Monogr Eval Carcinog Risks Hum 70:1–492

Izumi KM, Kieff ED (1997) The Epstein-Barr virus oncogene product latent membrane protein 1 engages the tumor necrosis factor receptor-associated death domain protein to mediate B lymphocyte growth transformation and activate NF-kappaB. Proc Natl Acad Sci USA 94:12592–12597

Izumi KM, Kaye KM, Kieff ED (1997) The Epstein-Barr virus LMP1 amino acid sequence that engages tumor necrosis factor receptor associated factors is critical for primary B lymphocyte growth transformation. Proc Natl Acad Sci USA 94:1447–1452

Izumi KM, McFarland EC, Riley EA, Rizzo D, Chen Y, Kieff E (1999a) The residues between the two transformation effector sites of Epstein-Barr virus latent membrane protein 1 are not critical for B-lymphocyte growth transformation. J Virol 73:9908–9916

Izumi KM, McFarland EC, Ting AT, Riley EA, Seed B, Kieff ED (1999b) The Epstein-Barr virus oncoprotein latent membrane protein 1 engages the tumor necrosis factor receptor-associated proteins TRADD and receptor-interacting protein (RIP) but does not induce apoptosis or require RIP for NF-kappaB activation. Mol Cell Biol 19:5759–5767

Kaiser C, Laux G, Eick D, Jochner N, Bornkamm GW, Kempkes B (1999) The proto-oncogene c-myc is a direct target gene of Epstein-Barr virus nuclear antigen 2. J Virol 73: 4481–4484

Kanzler H, Kuppers R, Helmes S, Wacker HH, Chott A, Hansmann ML, Rajewsky K (2000) Hodgkin and Reed-Sternberg-like cells in B-cell chronic lymphocytic leukemia represent the outgrowth of single germinal-center B-cell-derived clones: potential precursors of Hodgkin and Reed-Sternberg cells in Hodgkin's disease. Blood 95:1023–1031

Karran L, Gao Y, Smith PR, Griffin BE (1992) Expression of a family of complementary-strand transcripts in Epstein-Barr virus-infected cells. Proc Natl Acad Sci USA 89:8058–8062

Kaye KM, Izumi KM, Kieff E (1993) Epstein-Barr virus latent membrane protein 1 is essential for B-lymphocyte growth transformation. Proc Natl Acad Sci USA 90:9150–9154

Kaye KM, Izumi KM, Mosialos G, Kieff E (1995) The Epstein-Barr virus LMP1 cytoplasmic carboxy terminus is essential for B-lymphocyte transformation; fibroblast cocultivation complements a critical function within the terminal 155 residues. J Virol 69: 675–683

Kaye KM, Devergne O, Harada JN, Izumi KM, Yalamanchili R, Kieff E, Mosialos G (1996) Tumor necrosis factor receptor associated factor 2 is a mediator of NF- kappa B activation by latent infection membrane protein 1, the Epstein-Barr virus transforming protein. Proc Natl Acad Sci USA 93:11085–11090

Kaye KM, Izumi KM, Li H, Johannsen E, Davidson D, Longnecker R, Kieff E (1999) An Epstein-Barr virus that expresses only the first 231 LMP1 amino acids efficiently initiates primary B-lymphocyte growth transformation. J Virol 73:10525–10530
Khan G, Miyashita EM, Yang B, Babcock GJ, Thorley-Lawson DA (1996) Is EBV persistence in vivo a model for B cell homeostasis? Immunity 5:173–179
Kieff E, Rickinson A (2001) Epstein-Barr virus and its replication. In: Knipe DM, Howley PM, Griffen DE, Lamb RA, Martin MA, Roizman B, Straus SE (eds) Virology, 4th edn. Lippincott, Williams, and Wilkins: Philadelphia, pp 2511–2573
Kieser A, Kilger E, Gires O, Ueffing M, Kolch W, Hammerschmidt W (1997) Epstein-Barr virus latent membrane protein-1 triggers AP-1 activity via the c-Jun N-terminal kinase cascade. EMBO J 16:6478–6485
Kieser A, Kaiser C, Hammerschmidt W (1999) LMP1 signal transduction differs substantially from TNF receptor 1 signaling in the molecular functions of TRADD and TRAF2. EMBO J 18:2511–2521
Knutson JC (1990) The level of c-fgr RNA is increased by EBNA-2, an Epstein-Barr virus gene required for B-cell immortalization. J Virol 64:2530–2536
Kulwichit W, Edwards RH, Davenport EM, Baskar JF, Godfrey V, Raab-Traub N (1998) Expression of the Epstein-Barr virus latent membrane protein 1 induces B cell lymphoma in transgenic mice. Proc Natl Acad Sci USA 95:11963–11968
Kuppers R, Rajewsky K (1998) The origin of Hodgkin and Reed/Sternberg cells in Hodgkin's disease. Annu Rev Immunol 16:471–493
Kuppers R, Rajewsky K, Zhao M, Simons G, Laumann R, Fischer R, Hansmann ML (1994) Hodgkin disease: Hodgkin and Reed-Sternberg cells picked from histological sections show clonal immunoglobulin gene rearrangements and appear to be derived from B cells at various stages of development. Proc Natl Acad Sci USA 91:10962–10966
Laherty CD, Hu HM, Opipari AW, Wang F, Dixit VM (1992) The Epstein-Barr virus LMP1 gene product induces A20 zinc finger protein expression by activating nuclear factor kappa B. J Biol Chem 267:24157–24160
Lee MA, Diamond ME, Yates JL (1999) Genetic evidence that EBNA-1 is needed for efficient, stable latent infection by Epstein-Barr virus. J Virol 73:2974–2982
Lenoir G, Philip P, Sonier R (1984) Burkitt's type lymphoma: EBV association and cytogenetic markers in cases from various geographic origins. In: Magrath I, O'Conor G, Ramot B (eds) Environmental influences in the pathogenesis of leukaemias and lymphomas. Raven, New York, pp 283–295
Levitskaya J, Coram M, Levitsky V, Imreh S, Steigerwald-Mullen PM, Klein G, Kurilla MG, Masucci MG (1995) Inhibition of antigen processing by the internal repeat region of the Epstein-Barr virus nuclear antigen-1. Nature 375:685–688
Levitskaya J, Sharipo A, Leonchiks A, Ciechanover A, Masucci MG (1997) Inhibition of ubiquitin/proteasome-dependent protein degradation by the Gly-Ala repeat domain of the Epstein-Barr virus nuclear antigen 1. Proc Natl Acad Sci USA 94:12616–12621
Li SN, Chang YS, Liu ST (1996) Effect of a 10-amino acid deletion on the oncogenic activity of latent membrane protein 1 of Epstein-Barr virus. Oncogene 12:2129–2135
Liebowitz D, Wang D, Kieff E (1986) Orientation and patching of the latent infection membrane protein encoded by Epstein-Barr virus. J Virol 58:233–237
Liebowitz D, Kopan R, Fuchs E, Sample J, Kieff E (1987) An Epstein-Barr virus transforming protein associates with vimentin in lymphocytes. Mol Cell Biol 7:2299–2308
Locksley RM, Killeen N, Lenardo MJ (2001) The TNF and TNF receptor superfamilies: integrating mammalian biology. Cell 104:487–501
Longnecker R (2000) Epstein-Barr virus latency: LMP2, a regulator or means for Epstein-Barr virus persistence? Adv Cancer Res 79:175–200
Magrath I (1990) The pathogenesis of Burkitt's lymphoma. Adv Cancer Res 55:133–270
Malinin NL, Boldin MP, Kovalenko AV, Wallach D (1997) MAP3K-related kinase involved in NF-kappaB induction by TNF, CD95 and IL-1. Nature 385:540–544

Mann KP, Thorley-Lawson D (1987) Posttranslational processing of the Epstein-Barr virus-encoded p63/LMP protein. J Virol 61:2100–2108
Mann KP, Staunton D, Thorley-Lawson DA (1985) Epstein-Barr virus-encoded protein found in plasma membranes of transformed cells. J Virol 55:710–720
Mannick JB, Cohen JI, Birkenbach M, Marchini A, Kieff E (1991) The Epstein-Barr virus nuclear protein encoded by the leader of the EBNA RNAs is important in B-lymphocyte transformation. J Virol 65:6826–6837
Marshall D, Sample C (1995) Epstein-Barr virus nuclear antigen 3C is a transcriptional regulator. J Virol 69:3624–3630
Mehl AM, Fischer N, Rowe M, Hartmann F, Daus H, Trumper L, Pfreundschuh M, Muller-Lantzsch N, Grasser FA (1998) Isolation and analysis of two strongly transforming isoforms of the Epstein-Barr-virus(EBV)-encoded latent membrane protein-1 (LMP1) from a single Hodgkin's lymphoma. Int J Cancer 76:194–200
Merchant M, Caldwell RG, Longnecker R (2000) The LMP2A ITAM is essential for providing B cells with development and survival signals in vivo. J Virol 74:9115–9124
Miller CL, Longnecker R, Kieff E (1993) Epstein-Barr virus latent membrane protein 2A blocks calcium mobilization in B lymphocytes. J Virol 67:3087–3094
Miller CL, Lee JH, Kieff E, Burkhardt AL, Bolen JB, Longnecker R (1994a) Epstein-Barr virus protein LMP2A regulates reactivation from latency by negatively regulating tyrosine kinases involved in sIg-mediated signal transduction. Infect Agents Dis 3:128–136
Miller CL, Lee JH, Kieff E, Longnecker R (1994b) An integral membrane protein (LMP2) blocks reactivation of Epstein-Barr virus from latency following surface immunoglobulin crosslinking. Proc Natl Acad Sci USA 91:772–776
Miller CL, Burkhardt AL, Lee JH, Stealey B, Longnecker R, Bolen JB, Kieff E (1995a) Integral membrane protein 2 of Epstein-Barr virus regulates reactivation from latency through dominant negative effects on protein-tyrosine kinases. Immunity 2:155–166
Miller WE, Earp HS, Raab-Traub N (1995b) The Epstein-Barr virus latent membrane protein 1 induces expression of the epidermal growth factor receptor. J Virol 69:4390–4398
Miller WE, Mosialos G, Kieff E, Raab-Traub N (1997) Epstein-Barr virus LMP1 induction of the epidermal growth factor receptor is mediated through a TRAF signaling pathway distinct from NF-kappaB activation. J Virol 71:586–594
Miller WE, Cheshire JL, Baldwin AS Jr, Raab-Traub N (1998) The NPC derived C15 LMP1 protein confers enhanced activation of NF-kappa B and induction of the EGFR in epithelial cells. Oncogene 16:1869–1877
Mitchell T, Sugden B (1995) Stimulation of NF-kappa B-mediated transcription by mutant derivatives of the latent membrane protein of Epstein-Barr virus. J Virol 69:2968–2976
Miyashita EM, Yang B, Lam KM, Crawford DH, Thorley-Lawson DA (1995) A novel form of Epstein-Barr virus latency in normal B cells in vivo. Cell 80:593–601
Moorthy R, Thorley-Lawson DA (1990) Processing of the Epstein-Barr virus-encoded latent membrane protein p63/LMP. J Virol 64:829–837
Moorthy R, Thorley-Lawson DA (1992) Mutational analysis of the transforming function of the EBV encoded LMP-1. Curr Top Microbiol Immunol 182:359–365
Moorthy RK, Thorley-Lawson DA (1993a) All three domains of the Epstein-Barr virus-encoded latent membrane protein LMP-1 are required for transformation of rat-1 fibroblasts. J Virol 67:1638–1646
Moorthy RK, Thorley-Lawson DA (1993b) Biochemical, genetic, and functional analyses of the phosphorylation sites on the Epstein-Barr virus-encoded oncogenic latent membrane protein LMP-1. J Virol 67:2637–2645
Mosialos G, Birkenbach M, Yalamanchili R, VanArsdale T, Ware C, Kieff E (1995) The Epstein-Barr virus transforming protein LMP1 engages signaling proteins for the tumor necrosis factor receptor family. Cell 80:389–399
Murray PG, Niedobitek G, Kremmer E, Grasser F, Reynolds GM, Cruchley A, Williams DM, Muller-Lantzsch N, Young LS (1996) In situ detection of the Epstein-Barr virus-encoded

nuclear antigen 1 in oral hairy leukoplakia and virus-associated carcinomas. J Pathol 178:44–47

Nakagomi H, Dolcetti R, Bejarano MT, Pisa P, Kiessling R, Masucci MG (1994) The Epstein-Barr virus latent membrane protein-1 (LMP1) induces interleukin-10 production in Burkitt lymphoma lines. Int J Cancer 57:240–244

Nishitoh H, Saitoh M, Mochida Y, Takeda K, Nakano H, Rothe M, Miyazono K, Ichijo H (1998) ASK1 is essential for JNK/SAPK activation by TRAF2. Mol Cell 2:389–395

Nitsche F, Bell A, Rickinson A (1997) Epstein-Barr virus leader protein enhances EBNA-2-mediated transactivation of latent membrane protein 1 expression: a role for the W1W2 repeat domain. J Virol 71:6619–6628

Pallesen G, Hamilton-Dutoit SJ, Rowe M, Young LS (1991) Expression of Epstein-Barr virus latent gene products in tumour cells of Hodgkin's disease. Lancet 337:320–322

Peng M, Lundgren E (1992) Transient expression of the Epstein-Barr virus LMP1 gene in human primary B cells induces cellular activation and DNA synthesis. Oncogene 7:1775–1782

Polack A, Hortnagel K, Pajic A, Christoph B, Baier B, Falk M, Mautner J, Geltinger C, Bornkamm GW, Kempkes B (1996) c-myc activation renders proliferation of Epstein-Barr virus (EBV)-transformed cells independent of EBV nuclear antigen 2 and latent membrane protein 1. Proc Natl Acad Sci USA 93:10411–10416

Rickinson A, Kieff E (2001) Epstein-Barr virus. In: Knipe DM, Howley PM, Griffen DE, Lamb RA, Martin MA, Roizman B, Straus SE (eds) Virology, 4th edn. Lippincott, Williams, and Wilkins, Philadelphia, pp 2573–2627

Rothe M, Wong SC, Henzel WJ, Goeddel DV (1994) A novel family of putative signal transducers associated with the cytoplasmic domain of the 75 kDa tumor necrosis factor receptor. Cell 78:681–692

Rowe M, Rowe DT, Gregory CD, Young LS, Farrell PJ, Rupani H, Rickinson AB (1987) Differences in B cell growth phenotype reflect novel patterns of Epstein-Barr virus latent gene expression in Burkitt's lymphoma cells. EMBO J 6:2743–2751

Rowe M, Peng-Pilon M, Huen DS, Hardy R, Croom-Carter D, Lundgren E, Rickinson AB (1994) Upregulation of bcl-2 by the Epstein-Barr virus latent membrane protein LMP1: a B-cell-specific response that is delayed relative to NF-kappa B activation and to induction of cell surface markers. J Virol 68:5602–5612

Sandberg M, Hammerschmidt W, Sugden B (1997) Characterization of LMP-1's association with TRAF1, TRAF2, and TRAF3. J Virol 71:4649–4656

Scholle F, Bendt KM, Raab-Traub N (2000) Epstein-Barr virus LMP2A transforms epithelial cells, inhibits cell differentiation, and activates Akt. J Virol 74:10681–10689

Sharipo A, Imreh M, Leonchiks A, Branden C, Masucci MG (2001) cis-Inhibition of proteasomal degradation by viral repeats: impact of length and amino acid composition. FEBS Lett 499:137–142

Shi CS, Kehrl JH (1997) Activation of stress-activated protein kinase/c-Jun N-terminal kinase, but not NF-kappaB, by the tumor necrosis factor (TNF) receptor 1 through a TNF receptor-associated factor 2- and germinal center kinase related-dependent pathway. J Biol Chem 272:32102–32107

Smith PR, Griffin BE (1992) Transcription of the Epstein-Barr virus gene EBNA-1 from different promoters in nasopharyngeal carcinoma and B-lymphoblastoid cells. J Virol 66:706–714

Sung NS, Kenney S, Gutsch D, Pagano JS (1991) EBNA-2 transactivates a lymphoid-specific enhancer in the BamHI C promoter of Epstein-Barr virus. J Virol 65:2164–2169

Swart R, Ruf IK, Sample J, Longnecker R (2000) Latent membrane protein 2A-mediated effects on the phosphatidylinositol 3-kinase/Akt pathway. J Virol 74:10838–10845

Sylla BS, Hung SC, Davidson DM, Hatzivassiliou E, Malinin NL, Wallach D, Gilmore TD, Kieff E, Mosialos G (1998) Epstein-Barr virus-transforming protein latent infection membrane protein 1 activates transcription factor NF-kappaB through a pathway that includes the NF-kappaB-inducing kinase and the IkappaB kinases IKKalpha and IKKbeta. Proc Natl Acad Sci USA 95:10106–10111

Thorley-Lawson DA, Babcock GJ (1999) A model for persistent infection with Epstein-Barr virus: the stealth virus of human B cells. Life Sci 65:1433–1453

Tomkinson B, Robertson E, Kieff E (1993) Epstein-Barr virus nuclear proteins EBNA-3A and EBNA-3C are essential for B-lymphocyte growth transformation. J Virol 67:2014–2025

Wang D, Liebowitz D, Kieff E (1985) An EBV membrane protein expressed in immortalized lymphocytes transforms established rodent cells. Cell 43:831–840

Wang D, Liebowitz D, Wang F, Gregory C, Rickinson A, Larson R, Springer T, Kieff E (1988) Epstein-Barr virus latent infection membrane protein alters the human B-lymphocyte phenotype: deletion of the amino terminus abolishes activity. J Virol 62:4173–4184

Wang F, Gregory C, Sample C, Rowe M, Liebowitz D, Murray R, Rickinson A, Kieff E (1990a) Epstein-Barr virus latent membrane protein (LMP1) and nuclear proteins 2 and 3C are effectors of phenotypic changes in B lymphocytes: EBNA-2 and LMP1 cooperatively induce CD23. J Virol 64:2309–2318

Wang F, Tsang SF, Kurilla MG, Cohen JI, Kieff E (1990b) Epstein-Barr virus nuclear antigen 2 transactivates latent membrane protein LMP1. J Virol 64:3407–3416

Wang S, Rowe M, Lundgren E (1996) Expression of the Epstein Barr virus transforming protein LMP1 causes a rapid and transient stimulation of the Bcl-2 homologue Mcl-1 levels in B-cell lines. Cancer Res 56:4610–4613

Weiss LM, Strickler JG, Warnke RA, Purtilo DT, Sklar J (1987) Epstein-Barr viral DNA in tissues of Hodgkin's disease. Am J Pathol 129:86–91

Wilson JB, Weinberg W, Johnson R, Yuspa S, Levine AJ (1990) Expression of the BNLF-1 oncogene of Epstein-Barr virus in the skin of transgenic mice induces hyperplasia and aberrant expression of keratin 6. Cell 61:1315–1327

Wu TC, Mann RB, Charache P, Hayward SD, Staal S, Lambe BC, Ambinder RF (1990) Detection of EBV gene expression in Reed-Sternberg cells of Hodgkin's disease. Int J Cancer 46:801–804

Yates J, Warren N, Reisman D, Sugden B (1984) A cis-acting element from the Epstein-Barr viral genome that permits stable replication of recombinant plasmids in latently infected cells. Proc Natl Acad Sci USA 81:3806–3810

Yoshizaki T, Sato H, Furukawa M, Pagano JS (1998) The expression of matrix metalloproteinase 9 is enhanced by Epstein-Barr virus latent membrane protein 1. Proc Natl Acad Sci USA 95:3621–3626

Yuasa T, Ohno S, Kehrl JH, Kyriakis JM (1998) Tumor necrosis factor signaling to stress-activated protein kinase (SAPK)/Jun NH2-terminal kinase (JNK) and p38. Germinal center kinase couples TRAF2 to mitogen-activated protein kinase/ERK kinase kinase 1 and SAPK while receptor interacting protein associates with a mitogen-activated protein kinase kinase kinase upstream of MKK6 and p38. J Biol Chem 273:22681–22692

Zhao B, Sample CE (2000) Epstein-Barr virus nuclear antigen 3C activates the latent membrane protein 1 promoter in the presence of Epstein-Barr virus nuclear antigen 2 through sequences encompassing an spi-1/Spi-B binding site. J Virol 74:5151–5160

Zimber-Strobl U, Suentzenich KO, Laux G, Eick D, Cordier M, Calender A, Billaud M, Lenoir GM, Bornkamm GW (1991) Epstein-Barr virus nuclear antigen 2 activates transcription of the terminal protein gene. J Virol 65:415–423

zur Hausen H, Schulte-Holthausen H, Klein G, Henle W, Henle G, Clifford P, Santesson L (1970) EBV DNA in biopsies of Burkitt tumours and anaplastic carcinomas of the nasopharynx. Nature 228:1056–1058

SV40 and Notch-I: Multi-functionality Meets Pleiotropy

M. Carbone[1] and M. Bocchetta[1]

We have discovered that Simian virus 40 (SV40) infection of human mesothelial cells induces Notch-1. Upregulation of Notch-1 is achieved at the transcriptional level and requires the activity of both the large (90–100 kDa) and the small (20 kDa) SV40 tumor antigens. Notch-1 upregulation is maintained in SV40-transformed mesothelial cell lines in tissue culture, and Notch-1 is overexpressed only in SV40-positive mesotheliomas. Chemical inactivation of Notch-1 causes cell growth arrest of SV40-transformed human mesothelial cells. Our findings indicate that Notch-1 activation plays an important role in SV40-mediated carcinogenesis. Notch-1 is a pleiotropic gene that influences cell differentiation, proliferation and apoptosis. The effects of Notch-1 activation are species and cell type specific, in that, in different species or cell type within the same species, Notch-1 signaling can cause opposite effects. In this chapter, we review some general aspects of Notch signaling, and how Notch signaling regulates differentiation, the cell cycle and apoptosis. Recent data involving Notch and human cancer will be discussed. Then, we will deal with the current information regarding a specific form of lung cancer (malignant mesothelioma) and the involvement of SV40 in the pathogenesis of this form of human cancer. We will enter into the details of SV40-mediated cell transformation, and the specific interaction between SV40 and primary human mesothelial cells when infected with SV40 in vitro. Finally, we will discuss the recently discovered link between SV40-mediated carcinogenicity and Notch-1 signaling, the interaction between SV40 and Notch-1 in SV40 infected mesothelial cells and in SV40 mesotheliomas. We discuss how Notch-1 induction is required for the growth of SV40-transformed human cells. Finally, we discuss how strategies that interfere with Notch-1 expression may cause apoptosis of SV40-positive transformed cells, and therefore may represent new possible therapeutic approaches for SV40-positive mesothelioma patients.

[1]Cancer Immunology Program, Department of Pathology, Cardinal Bernardin Cancer Center, Loyola University Chicago, Maywood, Illinois 60153, USA, e-mail: mbocche@lumc.edu

Progress in Molecular and Subcellular Biology
C. Alonso (Ed.): Viruses and Apoptosis
© Springer-Verlag Berlin Heidelberg 2004

1
Notch Proteins and Their Ligands

LIN-12/Notch receptors are single-pass, transmembrane proteins conserved from *C. elegans* to humans. In vertebrates there are four known Notch receptors (Notch-1 through -4). Notch receptors are heterodimeric proteins, consisting of a large, modular N-terminal extracellular portion non-covalently bound to a transmembrane domain, followed by an intracellular, C-terminal portion. Most of the extracellular potion of Notch proteins consists of tandem repeats of epidermal growth factor-like (EGFL), cysteine-reach domains. The number and the spacing between EGFL repeats are approximately conserved between Notch-1 to -3, while Notch-4 has lost and rearranged the structure of some of the EGFL repeats (Egan et al. 1998). EGFL repeats 11 and 12 mediate interactions with Notch ligands (Rebay et al. 1991). Downstream from the EGFL repeats there are three copies of evolutionarily conserved Notch/LIN-12 repeats. Deletion of the latter domains produces constitutively active Notch proteins (Greenwald 1994). Below the transmembrane domain, Notch proteins contain six cdc10/ankyrin repeats involved in the interaction of activated Notch with nuclear transcription factors (Fortini and Artavanis-Tsakonas 1994). Ankyrin repeats are followed by a long, glutamine-rich domain containing a nuclear localization signal (Egan et al. 1998). The very C-terminal portion of Notch contains PEST sequences, thought to regulate protein turnover (Rechsteiner 1988). After their synthesis, Notch precursors (the precursor of Notch-1 is of about 300 kDa) are cleaved in the trans-Golgi by a furyn-like convertase (Logeat et al. 1998) giving rise to mature proteins, which are exposed on the plasma membrane, where they can interact with their ligands. Truncated Notch proteins harboring deletions of the extracellular, N-terminal domain, are constitutively active. This and other evidence suggest that the extracellular portion operates as an inhibitory domain of Notch activation. Conformational changes due to binding with Notch ligands hinder this inhibitory function, thus allowing Notch to be proteolitically activated (Egan et al. 1998).

In *Drosophila*, Notch has two ligands, Delta and Serrate. Humans have two Serrate-like proteins (Jagged-1 and -2), and a Delta-like protein family (Delta-1 to Delta-4). Notch ligands are single-pass transmembrane proteins that interact with Notch when presented to Notch by a neighboring cell. This mechanism of receptor–ligand interaction has been well described during the process of lateral specification in *Drosophila* (Heitzler and Simpson 1991). According to this model, a signaling cell presents ligands to a juxtaposed Notch-expressing cell. Upon ligand binding, Notch is cleaved in its cytoplasmic portion by a Presenilin/gamma-secretase, nicastrin complex (Lai 2002) in proximity of the transmembrane domain. This process leads to the release in the cell of the activated form of Notch (intracellular Notch, or N^{IC}). Besides this mechanism for Notch activation requiring contiguous signaling cells, there could be alternative pathways leading to Notch activation. Some studies have pointed out that individual cells can express both Notch and its ligands, hence implicating that receptor–ligand interactions and consequent induction of the Notch signaling pathway may not require the participation of two cells (reviewed in Artavanis-Tsakonas et al. 1999). Furthermore, recent data suggest that Notch can interact with its ligands in the cytoplasm (Sakamoto et al. 2002). Cleaved, soluble extracellular fragments of Delta may also participate in Notch signaling, although the role of these soluble ligands is still poorly understood (Artavanis-Tsakonas et al. 1999).

1.1
The Notch Signaling Network

The precise functions and interactions of intracellular Notch are still not perfectly defined. Nevertheless, there is a large amount of experimental data indicating that in *Drosophila* N^{IC} targets the transcriptional repressor *Suppressor of Hairless* [*Su*(*H*)], and its homologue CBF-1 in mammals. Notch targeting of CBF-1 appears mediated through direct binding of N^{IC} to CBF-1 (Zhou et al. 2000). Such binding leads to the release of transcriptional corepressors from interaction with CBF-1 (Kao et al. 1998; Zhou et al. 2000). Furthermore, N^{IC} interacts with p300, thus complexing the latter to CBF-1, so that p300 can function as a transcriptional coactivator for CBF-1 (Oswald et al. 2001). After interacting with N^{IC}, *Su*(*H*)/CBF-1 promotes transcription of a number of helix-loop-helix transcriptional factor genes collectively named as *Enhancer of Spilt* [*E(spl)*; reviewed in Artavanis-Tsakonas et al. 1999]. Although the interaction between N^{IC} and *Su*(*H*)/CBF-1 has been demonstrated (Fortini and Artavanis-Tsakonas 1994; Kao et al. 1998; Zhou et al. 2000), it is still unclear where this interaction takes place. According to a currently accepted model, after Presenilin-mediated cleavage, N^{IC} translocates to the nucleus where it interacts with *Su*(*H*)/CBF-1, thus activating transcription of *E(spl)* genes. However, there is a significant amount of data that does not support this model in several systems (reviewed in Artavanis-Tsakonas et al. 1999). In essence, several studies in which activation of full-length Notch was analyzed failed to detect nuclear localization of N^{IC}. Rather, mostly cytoplasmic localization of N^{IC} was common. To explain this apparent discrepancy, some have shown that minute amounts of nuclear Notch are sufficient for Notch signaling (Schroeter et al. 1998; De Strooper et al. 1999; Struhl and Greenwald 1999). On the other hand, nuclear localization of N^{IC} is readily detected after transfection of cells with truncated mutants of Notch, which are constitutively active (Jeffries and Capobianco 2000). These mutants do not necessarily follow the precise trafficking of N^{IC} produced by the proteolytic activation of full-length Notch, since they are neither exposed on the plasma membrane, nor do they require proteolytic cleavage for activation. With these qualifications in mind, it is safe to state that the precise details of Notch interaction with *Su*(*H*)/CBF-1 still need to be elucidated.

Activation of *E(spl)* causes repression of differentiation factors, such as the *Achaete-Scute* genes. These events are able to suppress differentiation of muscle cells (Kopan et al. 1994) and of neuronal cells (Nye et al. 1994). Repression of *Achaete-Scute* genes also causes downregulation of Delta. Together with the positive feedback loop of Notch activation on Notch expression, these mechanisms are thought to account for the choice of juxtaposed cells to acquire either the signaling phenotype (expression of Delta) or the responder phenotype (expression of Notch) during lateral specification (Seuget et al. 1997).

Another target of Notch signaling is NFκB. Notch-1 seems to affect NFκB through binding of members of the IκB family of inhibitors (Aster et al. 1994). Additionally, N^{IC} appears to posses IκB functions itself (Guan et al. 1996). N^{IC} appears to bind and inhibit the p50 subunit and not the p65 subunit of NFκB (Wang et al. 2001). Through its interactions with the NFκB network, Notch-1 influences cell fate decision in a number of systems (Cheng et al. 2001; Li et al. 2001; Nickoloff et al. 2002).

1.2 Notch Influences Cell Differentiation, Proliferation and Apoptosis

Notch signaling regulates critical cell fate decision during development. One general aspect of Notch activation is the maintenance of a pre-committed state that allows the cell to properly interpret differentiation or proliferation stimuli. This critical tuning of incoming differentiation clues is regulated by Notch through affecting cell proliferation, apoptosis and, obviously, differentiation programs. During post-natal life, Notch regulation of cell proliferation and apoptosis is highly context dependent. Notch-1 induces growth arrest and promotes differentiation of mouse keratinocytes (Rangarajan et al. 2001a). Notch-1 also inhibits proliferation of myeloid progenitor cells (Schroeder and Just 2000), and causes growth arrest and apoptosis in B cells (Morimura et al. 2000). Both Notch-1 and Notch-2 cause suppression of growth of small cell lung cancer cells, a phenomenon related to G1 arrest mediated through increased levels of the cyclin-dependent kinase inhibitors $p21^{WAF1/Cip1}$ and $p27^{kip1}$ (Sriuranpong et al. 2001). The latter piece of evidence indicates that the oncogenic potential of deregulated Notch expression (see below) is also context dependent. On the other hand, Notch signaling promotes proliferation in hematopoietic progenitor cells (Carlesso et al. 1999; Karanu et al. 2000, 2001) and in primary mammary epithelial cells (Soriano et al. 2000). Notch induction also promotes cell proliferation of rabbit skin fibroblasts, a process mediated by increased cyclin-dependent kinase-2 (CDK-2) due to lowered expression of the CDK inhibitor $p27^{kip1}$ (Cereseto and Tsai 2000). The incongruous effects mediated by Notch activation on $p27^{kip1}$ expression levels between the latter system and small cell lung cancer cells are an exquisite example of cell- and context-dependent Notch-mediated cellular events. Notch induction triggers apoptosis in B cells (Morimura et al. 2000), but has anti-apoptotic effects in T cell lines (Jehn et al. 1999), and in murine erythroleukemia cells (Shelly et al. 1999).

1.3 Notch and Cancer

The Notch protein family is a critical regulator of differentiation programs. Its ability to affect the cell cycle kinetics and the response to apoptotic signals suggest that Notch proteins may be involved in the malignant transformation of some cell systems. In the last decade evidence has accumulated casting light on Notch participation in carcinogenesis and human tumors. Truncated Notch-1 genes harboring deletions in the extracellular, N-terminal portion of Notch-1 (mutations that produce constitutively active Notch-1 proteins) have been implicated in T-cell acute lymphoblastic leukemia (T-ALL; Ellisen et al. 1991). Introduction of these mutants into mice produces T cell leukemias in about half of the experimental animals, thus demonstrating that truncated, constitutively active Notch-1 is an oncoprotein (Pear et al. 1996; Hoemann et al. 2000). Similarly, Notch-4 truncated at its N-terminal domain causes mammary tumors in mice (Robbins et al. 1992). In addition, the transforming activities of constitutively active Notch-1 and Notch-2 proteins have been demonstrated in vitro (Capobianco et al. 1997; Ronchini and Capobianco 2001).

Despite their well-established role in oncogenesis, truncated Notch proteins are not common in human cancer. However, the levels of expression of Notch proteins and their ligands are tightly regulated in normal cells, and deregulated expression of full-length Notch is expected to profoundly affect downstream cell fate decisions (Artavanis-Tsakonas et al. 1999). Accordingly, aberrant expression of Notch receptors, their ligands, and downstream effectors has been described in a number of human tumors. Using a laser microdissection procedure, Leethanakul et al. (2000) demonstrated that several members of the Notch pathway are specifically overexpressed in squamous cell carcinomas of the head and neck. Notch-3 appears specifically overexpressed in renal carcinoma (Rae et al. 2000), while overexpression of Notch-1 (and Jagged-1) has been described in acute myeloid leukemia (Tohda and Nara 2001), and in lymphomas (Jundt et al. 2002). More controversial is the involvement of Notch signaling (particularly Notch-1) in cervical cancer. Early investigations showed overexpression of Notch-1 in cervical carcinoma (Zagouras et al. 1995). These findings have been confirmed by others (Daniel et al. 1997), indicating that there is an increase in Notch-1 expression through progression from cervical intraepithelial neoplastic lesions (CIN) to metastatic carcinoma. Furthermore, Notch-1 was found to synergize with human papilloma virus16 (HPV16) oncoproteins in a p53-null environment (Rangarajan et al. 2001b). Oncogenic papilloma viruses are associated with about 98% of cervical carcinomas worldwide, and are believed to cause cervical cancer. A recent report, however, did not confirm previous findings, indicating that specific loss of Notch-1 is required for progression from CIN to HPV16-positive cervical carcinomas. In this study, Notch-1 antagonized HPV16 oncoprotein expression through a mechanism involving components of the activator protein-1 (AP-1) complex and wild-type p53 (Talora et al. 2002). Although the participation of Notch-1 in cervical cancer still needs to be fully elucidated, there is a consistent amount of data indicating that full-length Notch-1 is overexpressed in a number of human cancers, especially those arising from cells of the mesoderm (Talora et al. 2002). Overexpression of Notch-1 appears to actively contribute to the malignant phenotype, as artificial downregulation of Notch-1, or pharmaceutical inhibition of Notch-1 activation, suppresses the neoplastic phenotype of cells containing the SV40 early genes (see below), active telomerase, and oncogenic *ras* (Weijzen et al. 2002). In conclusion, it appears that aberrant Notch signaling (through expression of both constitutively active and overexpression of wild-type forms of Notch proteins) participates in the maintenance of the malignant phenotype.

2
Malignant Mesothelioma and SV40

Malignant mesothelioma (MM) is a tumor of the lining of the lungs, heart, and abdomen. It is among the most aggressive human tumors (average survival from diagnosis is less than 1 year), and no effective therapeutic strategies have been developed so far for the treatment of MM patients. MM is firmly linked to asbestos exposure according to epidemiological investigations; however, the mechanisms of asbestos-induced carcinogenicity are still rather undefined. MM develops in about 5% of heavily exposed individuals (such as asbestos miners), and about 20% of MMs occur in individuals with no history of exposure to asbestos (Carbone et al. 2002). These data suggest that other factors may cause MM, alone or in cooperation with asbestos. Another puzzling aspect of asbestos-induced carcinogenicity in MM is the apparent lack of a dose-response effect. MM preferentially develops in subjects with above-average asbestos exposure. Highly exposed individuals mainly develop asbestosis and/or lung cancer. Even more obscure is the effect of asbestos when assayed in vitro. Although several studies have pointed out the potential mutagenic effects of asbestos (reviewed in Manning et al. 2002), and have linked asbestos exposure to increased AP-1 and EGF activity (Mossman et al. 1997; Manning et al. 2002), asbestos does not transform primary mesothelial cells in vitro. When exposed to asbestos, human mesothelial cells reproducibly die of apoptosis (Xu et al. 1999; Bocchetta et al. 2000; Liu et al. 2000).

During the last decade, a large number of studies conducted in different countries have linked Simian virus 40 (SV40) to MM. The overall consensus is that about 50% of human MM cases contain SV40 (Klein et al. 2002). The association is highly specific: laser capture microdissection experiments have demonstrated that SV40 is present only in mesothelioma cancer cells and not in the tissues surrounding the tumor (Shivapurkar et al. 1999). Causation is suggested by the fact that hamsters injected with SV40 develop mesothelioma, lymphomas, brain, and bone tumors. This is precisely the panel of human tumors in which SV40 has been detected (Jasani et al. 2001). SV40-related causation of MM is supported by the fact that SV40 is biologically active in MM specimens (Carbone et al. 1997; De Luca et al. 1997), that antisense treatments targeting SV40 causes growth arrest and apoptosis of SV40 positive MM cell lines (Waheed et al. 1999), and by the high susceptibility to cell transformation of primary human mesothelial cells when infected with SV40 in vitro (Bocchetta et al. 2000).

SV40 is a small DNA virus identified as a contaminant in early poliovaccine preparations (Sweet and Hilleman 1960). Soon after its discovery it was recognized that SV40 is able to transform cells of different origin in vitro and to cause tumors in experimental animals. Over the last few decades, SV40 has become the most common agent to transform and/or immortalize cells. SV40 has a small genome (about 5.2 kb), but the information is densely packed. Through alternative spicing and alternate reading frames, a cell productively infected by SV40 synthesizes at least six virally encoded proteins. Early after entry of SV40 into the nuclei of infected cells, transcription of the so-called early genes takes place. The protein products of these genes (the large T tumor antigen, or Tag, and the small t tumor antigen, or tag) are responsible for the deregulation of the cell cycle of the host cell and entry into S-phase (an event eventually leading to cell division in some cell systems). During the S-G2 phase the viral genome is actively replicated. Then transcription at the late promoter is efficiently carried out, and large amounts of the late gene products (VP-1 through 3, components of the viral capsid) are synthesized. Massive amounts of SV40 viral particles accumulate in the nuclei, and the infected host undergoes cell lysis, with subsequent release of infective SV40 viral particles. This pattern of SV40 infection is termed permissive and it is typical of monkey kidney cells in vitro. Human cells are semi-permissive to SV40 infection in vitro, meaning that only a fraction of cells exposed to SV40 are infected, but infected cells actively produce viral particles and undergo SV40-induced cell lysis. Some epidermal human cells escape SV40-mediated cell lysis after infection by secreting large amounts of viral particles (Clayson et al. 1989). Rodent cells are termed non-permissive to SV40 infection. In these cells SV40 DNA replication does not take place because Tag (which, among other functions, initiates the SV40 genome replication) cannot engage in proper interactions with the DNA polymerase I/α primase of the host (Schneider et al. 1994). Cell transformation of infected rodent cells can occur if the viral DNA integrates in the host genome in such a way that expression of the SV40 oncoproteins is maintained.

2.1 SV40-Mediated Oncogenicity

SV40 is a small, selfish machinery. Being a DNA virus, and encoding only for a few proteins, it needs cooperation from the host in order to replicate its genome and to thrive. As mentioned above, the SV40 early gene products disrupt the control of the cell cycle in infected cells. This "transforming program" is executed both in the nucleus and in the cytoplasm of the host. Tag is mainly a nuclear protein, and its known functions take place in the nucleus. The SV40 Tag is an extraordinarily multifunctional protein, and the most potent oncogene ever discovered. Besides its implications in SV40 replication, transcription, DNA and RNA helicase activities, Tag engages in a number of interactions with host proteins. Tag enters into TFIIB complexes (Damania and Alwine 1994), a function that may influence transcriptional activity in a still undefined number of cellular promoters. Among the interactions with cellular proteins (including p300, HSP70 and others), Tag binds and inactivates both p53 and pRb protein family members (Fanning and Knippers 1992; Ali and DeCaprio 2001; Testa and Giordano 2001). Through these interactions Tag simultaneously knocks off the two major tumor suppressor pathways of the cell. The SV40 Tag also transcriptionally upregulates insulin growth factor I (IGF-I) and its receptor (IGF-IR; Porcu et al. 1994). The latter activity provides a mytogenic stimulus in an environment in which the cell cycle checkpoints are compromised. Indeed, SV40 loses its transforming capabilities in mouse cells treated with an antisense oligodeoxynucleotide to the IGF-1 receptor mRNA (Porcu et al. 1992).

The transforming functions of Tag are complemented in the cytoplasm by the SV40 tag. The best-characterized function of tag is the binding and inactivation of protein phosphatase 2A (PP2A; Mateer at al. 1998; Rundell and Parakati 2001). PP2A dephosphorylates a number of substrates, including members of the mitogen-activated protein kinases (MAP-Ks), proteins involved in the signal transduction pathway shared by several growth factor receptors, and also downstream from *ras*. Through inactivation of PP2A, tag indirectly reinforces mitogenic extracellular stimuli, an event eventually leading to immediate early genes (AP-1) induction (Rundell and Parakati 2001). How SV40-infected cells escape apoptosis is still debated. Potentially, Tag should elicit both pro- and anti-apoptotic effects. By binding and inactivating pRb with consequent release of E2F-1, cells may undergo E2F-1-mediated apoptosis, since E2F-1 can either promote cell cycle progression or trigger apoptosis (Rogoff et al. 2002). On the other hand, Tag inhibits p53 functions, thus bypassing p53-dependent apoptosis. However, recent data indicate that Tag-expressing cells are still able to execute p53-dependent apoptosis (Cole and Tevethia 2002). Some studies have pointed out that Tag contains a Bcl 2 domain, which could interfere with the intrinsic apoptotic program (Tsai et al. 2000). However, several reports have shown that both Tag and tag have pro-apoptotic effects, both p53-dependent and p53-independent (Gjoerup et al.

2001; Cole and Tevethia 2002). Overall, the problem concerning how SV40 does not trigger apoptosis in the majority of the infected cells still awaits elucidation.

2.2
SV40 and Primary Human Mesothelial Cells

The study of SV40-mediated oncogenicity in human cells has been routinely performed introducing in various cell systems different kinds of mutated SV40 genomes lacking the viral origin of replication, and often also lacking the late genes. This is normally required in order to overcome the virally induced cell lysis that would otherwise be the preponderant response of human cells hosting replicating SV40. Since mesotheliomas contain SV40 virus, and not artificially engineered plasmids containing the SV40 early genes, we and others initiated a number of experiments aimed at the understanding of the reaction of primary human mesothelial cells to SV40 infection. We found that human mesothelial cells are particularly susceptible to SV40 infection. Infected mesothelial cells survive SV40 infection because they synthesize SV40 in amounts compatible with cell survival, a mechanism related to mesothelial cell p53 expression levels (Bocchetta et al. 2000). Cell survival in the presence of oncogenic SV40 tumor antigens poses a threat for cell transformation. Indeed, human mesothelial cells undergo SV40-mediated transformation at a very high rate, since about 1 in 5000 infected cells will develop into a tridimensional focus of fully transformed cells (Bocchetta et al. 2000). This frequency of cell transformation is quite unprecedented; no other cell systems undergo SV40-mediated transformation at such high rate (the rate of SV40-mediated transformation of human cells is, on average 10^{-7}; Bryan and Reddel 1994). The SV40-induced mesothelial foci can be cultured in 100% of cases, and display a complete transformed phenotype in tissue culture. About 90% of these SV40-induced mesothelial foci are immortal from the beginning of the culturing process (Bocchetta et al. 2000). All this evidence suggests that SV40 must elicit specific effects on human mesothelial cells in order to produce such impressive outcomes. Immortalization is mediated by the specific induction of telomerase activity early after SV40 infection of mesothelial cells (Foddis et al. 2002). It still needs to be elucidated, however, whether SV40 activates telomerase in the entire population of infected cells, or whether telomerase activation is the determinant factor for focus formation. In fact, telomerase activity increases from early infection of primary mesothelial cells through culturing of SV40-transformed mesothelial foci (Foddis et al. 2002).

As stated above, SV40 induces cell transformation in about 1 in 5000 infected cells. This result has been widely reproduced in our laboratory. Foci become visible in the infected cell population between 2 and 3 weeks after infection of different primary cells cultures. Such a high transformation rate and the speed at which foci develop indicate that SV40 must affect a broad range of cellular pathways all converging in the orchestrated cell proliferation of the infected mesothelial cell. Furthermore, the frequency and the speed of focus formation suggest that SV40 may be sufficient for mesothelial complete cell transformation and immortalization, and additional mutations may play only a limited role. SV40-specific methylation of the tumor suppressor gene RASSF1A in SV40-transformed mesothelial clones has been recently described (Toyooka et al. 2002). This evidence supports the notion that SV40 produces specific effects in mesothelial cells other than the well-known interactions with the major regulators of cell proliferation p53 and pRb protein family members. Moreover, a recent study demonstrated that SV40 induces phosphorylation of the *met* oncogene and production of hepatocyte growth factor (HGF) in human mesothelial cells (Cacciotti et al. 2001). The latter SV40-induced activity not only contributes to the whole picture of SV40-mediated transformation of human mesothelial cells, but also opens new perspectives on SV40-mediated carcinogenicity. SV40 may work in a tumor not only as a multi-functional oncogene, but may also have landscaping effects on neighboring cells. We now know that human cells are turned into transplantable tumors in Scid mice by the simultaneous action of the SV40 tumor antigens, active telomerase, and oncogenic *ras* (Hahn et al. 1999). In mesothelial cells SV40 alone may indeed be sufficient for cell transformation, since it activates telomerase after infection (Foddis et al. 2002), and promotes phosphorylation and activation of the proto-oncogene *met* (Cacciotti et al. 2001). Met is upstream from *ras*, and activates both the MAP-K cascade and phosphatidylinositol 3-kinase (PI3-K) and Akt activation. So, it appears that a mesothelial cell containing SV40 virus possesses all the genetic determinants required to achieve complete malignant transformation of human cells (Hahn et al. 1999): SV40 tumor antigens, telomerase activity, and activated *ras* signaling (through activation of *met*). So, why does focus formation affect only 1 in 5000 infected cells? Mesothelial cells cultured from pleural fluids obtained from non-cancerous donors grow in tissue culture for few passages (5–8). This may implicate that a large percentage of the cell population may be terminally impaired in initiating cell division. A more likely explanation of the fact that not all SV40-infected mesothelial cells become fully transformed may lie in that activated *met* and IGF pathways do not functionally complement oncogenic *ras*. Thus, an exacerbated local level of mitogenetic signaling through the met and IGF receptor (IGF-R) pathways may determine the event of focus formation. Conversely, oncogenic *ras* may be complemented by activation of *met* and IGF-R in all infected cells, and focus formation may be determined by the activation of telomerase in only a fraction of infected cells. This hypothesis is supported by recent findings indicating that telomerase does participate in cell transformation, other than in immortalization of the cell (Stewart et al. 2002; Wong et al. 2002). To further clarify this issue, it will be essential to elucidate whether SV40 induces telomerase in all or just a portion of the infected cell population.

2.3
SV40 Induces Notch-1 Expression

All the activities described above explain well why mesothelial cells are induced to proliferate when they harbor SV40. However, the process of SV40-mediated cell transformation can potentially directly affect also the differentiation program of infected cells. Human mesothelial cells are considered among the least differentiated cells in the body. They represent the remnant of the embryonic mesoderm (Carbone et al. 2002). Notch-1 critically regulates cell fate decision in precommited, or not terminally differentiated cells. For this reason, we investigated whether SV40 infection of primary human mesothelial cells could have interfered with Notch-1 signaling. We found that SV40 infection readily upregulates Notch-1 protein expression, and that this induction is achieved at the transcriptional level (Bocchetta et al. 2003). Both SV40 tumor antigens are required for SV40-mediated Notch-1 induction. The SV40 tag exerts its function by interfering with the MAP-K signal transduction pathway. Interestingly, Notch-1 induction appears dependent on either the SV40 gene dosage or levels of expression of the SV40 tumor antigens, since non-replicating plasmids in which transcription is regulated from the SV40 promoter fail to induce Notch-1 (Bocchetta et al. 2003). The latter piece of evidence emphasizes the importance of (and the difference between) infection and transfection experiments when assessing the relevance of a virus in the pathogenesis of certain tumors. Notch-1 activation appears necessary for growth of SV40-transformed mesothelial foci. Chemical inhibition of Notch-1 cleavage by gamma-secretases causes complete growth arrest of foci, and the cells appear to be blocked in G2/M (Bocchetta et al. 2003). All the above observations gain further importance from the finding that Notch-1 is specifically upregulated in SV40-positive mesotheliomas. Therefore, it appears that the link between Notch-1 signaling and SV40 is highly specific. We propose that interaction with Notch signaling may be a general property of DNA viruses. Besides the aforementioned interaction of HPV with Notch-1, Kaposi sarcoma associated herpes virus (KSHV) requires the activation of Notch signaling for productive infection. KSHV does not upregulate Notch receptors directly, but rather synthesizes a virally encoded protein (RTA protein) that binds CBF-1, thus mimicking N^{IC} functions (Liang et al. 2002). A similar mechanism is adopted by adenovirus through the adenoviral oncoprotein 13S E1A (Ansieau et al. 2001), and by the Epstein-Barr virus (EBV) nuclear antigen 2 (EBVNA2; Fuchs et al. 2001). At the same time, N^{IC} can complement EBVNA2-dependent expression of EBV genes and EBVNA2-induced transformation of lymphocytes (Hofelmayr et al. 2001). In conclusion, it appears that there is a broader connection between Notch signaling and the life cycle of DNA tumor viruses.

3 Future Directions

The connection between Notch signaling and SV40-mediated mesothelial cell transformation opens new and unknown scenarios for the understanding of SV40 infection and carcinogenesis. Since mesothelial foci overexpress the Notch ligands Jagged-1 and Jagged-2 in conjunction with Notch-1 (Bocchetta et al. 2003), it will be important to study whether cell contact plays any role for Notch-1 activation. It will also be important to analyze genes downstream of Notch-1, and how they respond to SV40 infection, if and how they affect the differentiation program of early SV40-infected mesothelial cells. The study of these phenomena will not only provide a better understanding of SV40-mediated malignant transformation of mesothelial cells, but could also contribute to the identification of therapeutic targets for the treatment of SV40-positive mesothelioma patients. Because Notch-1 is required for the growth of SV40-transformed mesothelial cells, and chemical inhibition of Notch-1 activation causes growth arrest in these cells, Notch-1 appears a useful target to develop new therapeutic approaches. More specifically, by causing G2/M growth arrest, chemical inhibition of Notch-1 may sensitize SV40 positive mesothelioma cells to the proapoptotic activity of the SV40 Tag (Cole and Tevethia 2002).

Acknowledgements. Supported by NIH CA92657-01A1 to M.C., and by NIH CA91122 to M.B.

References

Ali SH, DeCaprio JA (2001) Cellular transformation by SV40 large T antigen: interaction with host proteins. Semin Cancer Biol 11:15–23

Ansieau S, Strobl LJ, Leutz A (2001) Activation of the Notch-regulated transcription factor CBF1/RBP-Jkappa through the 13SE1A oncoprotein. Genes Dev 15:380–385

Artavanis-Tsakonas S, Rand MD, Lake RJ (1999) Notch signaling: cell fate control and signal integration in development. Science 284:770–776

AsterJ, Pear W, Hasserjian R, Erba H, Davi F, Luo B, Scott M, Baltimore D, Sklar J (1994) Functional analysis of the TAN-1 gene, a human homolog of *Drosophila* notch. Cold Spring Harb Symp Quant Biol LIX:125–136

Bocchetta M, Di Resta I, Powers A, Fresco R, Tosolini A, Testa JR, Pass HI, Rizzo P, Carbone M (2000) Human mesothelial cells are unusually susceptible to simian virus 40-mediated transformation and asbestos cocarcinogenicity. Proc Natl Acad Sci USA 97:10214–10219

Bocchetta M, Miele L, Pass HI, Carbone M (2003) Notch-1 induction, a novel activity of SV40 required for growth of SV40-transformed human mesothelial cells. Oncogene 22:81–89

Bryan TM, Reddel RR (1994) SV40-induced immortalization of human cells. Crit Rev Oncog 5:331–357

Cacciotti P, Libener R, Betta P, Martini F, Porta C, Procopio A, Strizzi L, Penengo L, Tognon M, Mutti L, Gaudino G (2001) SV40 replication in human mesothelial cells induces HGF/Met receptor activation: a model for viral-related carcinogenesis of human malignant mesothelioma. Proc Natl Acad Sci USA 98:12032–12037

Capobianco AJ, Zagouras P, Blaumueller CM, Artavanis-Tsakonas S, Bishop JM (1997) Neoplastic transformation by truncated alleles of human NOTCH1/TAN1 and NOTCH2. Mol Cell Biol 17:6265–6273
Carbone M, Rizzo P, Grimley PM, Procopio A, Mew DJ, Shridhar V, de Bartolomeis A, Esposito V, Giuliano MT, Steinberg SM, Levine AS, Giordano A, Pass HI (1997) Simian virus-40 large-T antigen binds p53 in human mesotheliomas. Nat Med 3:908–912
Carbone M, Kratzke RA, Testa JR (2002) The pathogenesis of mesothelioma. Semin Oncol 29:2–17
Carlesso N, Aster JC, Sklar J, Scadden DT (1999) Notch1-induced delay of human hematopoietic progenitor cell differentiation is associated with altered cell cycle kinetics. Blood 93:838–848
Cereseto A, Tsai S (2000) Jagged2 induces cell cycling in confluent fibroblasts susceptible to density-dependent inhibition of cell division. J Cell Physiol 185:425–431
Cheng P, Zlobin A, Volgina V, Gottipati S, Osborne B, Simel EJ, Miele L, Gabrilovich DI (2001) Notch-1 regulates NF-kappaB activity in hemopoietic progenitor cells. J Immunol 167:4458–4467
Clayson ET, Brando LV, Compans RW (1989) Release of simian virus 40 virions from epithelial cells is polarized and occurs without cell lysis. J Virol 63:2278–2288
Cole SL, Tevethia MJ (2002) Simian virus 40 large T antigen and two independent T-antigen segments sensitize cells to apoptosis following genotoxic damage. J Virol 76:8420–8432
Damania B, Alwine JC (1994) TAF-like function of SV40 large T antigen. Genes Dev 10:1369–1381
Daniel B, Rangarajan A, Mukherjee G, Vallikad E, Krishna S (1997) The link between integration and expression of human papillomavirus type 16 genomes and cellular changes in the evolution of cervical intraepithelial neoplastic lesions. J Gen Virol 78:1095–1101
De Luca A, Baldi A, Esposito V, Howard CM, Bagella L, Rizzo P, Caputi M, Pass HI, Giordano GG, Baldi F, Carbone M, Giordano A (1997) The retinoblastoma gene family pRb/p105, p107, pRb2/p130 and simian virus-40 large T-antigen in human mesotheliomas. Nat Med 3:913–916
De Strooper B, Annaert W, Cupers P, Saftig P, Craessaerts K, Mumm JS, Schroeter EH, Schrijvers V, Wolfe MS, Ray WJ, Goate A, Kopan R (1999) A presenilin-1-dependent gamma-secretase-like protease mediates release of Notch intracellular domain. Nature 398:518–522
Egan SE, St-Pierre B, Leow CC (1998) Notch receptors, partners and regulators: from conserved domains to powerful functions. Curr Top Microbiol Immunol 228:273–324
Ellisen LW, Bird J, West DC, Soreng AL, Reynolds TC, Smith SD, Sklar J (1991) TAN-1, the human homolog of the *Drosophila* notch gene, is broken by chromosomal translocations in T lymphoblastic neoplasms. Cell 66:649–661
Fanning E, Knippers R (1992) Structure and function of simian virus 40 large tumor antigen. Annu Rev Biochem 61:55–85
Foddis R, De Rienzo A, Broccoli D, Bocchetta M, Stekala E, Rizzo P, Tosolini A, Grobelny JV, Jhanwar SC, Pass HI, Testa JR, Carbone M (2002) SV40 infection induces telomerase activity in human mesothelial cells. Oncogene 21:1434–1442
Fortini ME, Artavanis-Tsakonas S (1994) The suppressor of hairless protein participates in notch receptor signaling. Cell 79:273–282
Fuchs KP, Bommer G, Dumont E, Christoph B, Vidal M, Kremmer E, Kempkes B (2001) Mutational analysis of the J recombination signal sequence binding protein (RBP-J)/Epstein-Barr virus nuclear antigen 2 (EBNA2) and RBP-J/Notch interaction. Eur J Biochem 268:4639–4646
Gjoerup O, Zaveri D, Roberts TM (2001) Induction of p53-independent apoptosis by simian virus 40 small t antigen. J Virol 75:9142–9155
Guan E, Wang J, Laborda J, Norcross M, Baeuerle PA, Hoffman T (1996) T cell leukemia-associated human Notch/translocation-associated Notch homologue has IB-like activity and physically interacts with nuclear factor-B proteins in T cells. J Exp Med 183:2025–2032
Greenwald I (1994) Structure/function studies of lin-12/Notch proteins. Curr Opin Genet Dev 4:556–562
Hahn WC, Counter CM, Lundberg AS, Beijersbergen RL, Brooks MW, Weinberg RA (1999) Creation of human tumor cells with defined genetic elements. Nature 400:464–468
Heitzler P, Simpson P (1991) The choice of cell fate in the epidermis of *Drosophila*. Cell 64:1083–1092

Hoemann CD, Beaulieu N, Girard L, Rebai N, Jolicoeur P (2000) Two distinct Notch1 mutant alleles are involved in the induction of T-cell leukemia in c-myc transgenic mice. Mol Cell Biol 20:3831–3842

Hofelmayr H, Strobl LJ, Marschall G, Bornkamm GW, Zimber-Strobl U (2001) Activated Notch1 can transiently substitute for EBNA2 in the maintenance of proliferation of LMP1-expressing immortalized B cells. J Virol 75:2033–2040

Jasani B, Cristaudo A, Emri SA, Gazdar AF, Gibbs A, Krynska B, Miller C, Mutti L, Radu C, Tognon M, Procopio A (2001) Association of SV40 with human tumours. Semin Cancer Biol 11:49–61

Jeffries S, Capobianco AJ (2000) Neoplastic transformation by Notch requires nuclear localization. Mol Cell Biol 20:3928–3941

Jehn BM, Bielke W, Pear WS, Osborne BA (1999) Cutting edge: protective effects of notch-1 on TCR-induced apoptosis. J Immunol 162:635–638

Jundt F, Anagnostopoulos I, Forster R, Mathas S, Stein H, Dorken B (2002) Activated Notch1 signaling promotes tumor cell proliferation and survival in Hodgkin and anaplastic large cell lymphoma. Blood 99:3398–3403

Kao HY, Ordentlich P, Koyano-Nakagawa N, Tang Z, Downes M, Kintner CR, Evans RM, Kadesch T (1998) A histone deacetylase corepressor complex regulates the Notch signal transduction pathway. Genes Dev 12:2269–2277

Karanu FN, Murdoch B, Gallacher L, Wu DM, Koremoto M, Sakano S, Bhatia M (2000) The notch ligand jagged-1 represents a novel growth factor of human hematopoietic stem cells. J Exp Med 192:1365–1372

Karanu FN, Murdoch B, Miyabayashi T, Ohno M, Koremoto M, Gallacher L, Wu D, Itoh A, Sakano S, Bhatia M (2001) Human homologues of Delta-1 and Delta-4 function as mitogenic regulators of primitive human hematopoietic cells. Blood 97:1960–1967

Klein G, Powers A, Croce C (2002) Association of SV40 with human tumors. Oncogene 21:1141–1149

Kopan R, Nye JS, Weintraub H (1994) The intracellular domain of mouse Notch: a constitutively activated repressor of myogenesis directed at the basic helix-loop-helix region of MyoD. Development 120:2385–2396

Lai EC (2002) Notch cleavage: Nicastrin helps Presenilin make the final cut. Curr Biol 12:R200–R202

Leethanakul C, Patel V, Gillespie J, Pallente M, Ensley JF, Koontongkaew S, Liotta LA, Emmert-Buck M, Gutkind JS (2000) Distinct pattern of expression of differentiation and growth-related genes in squamous cell carcinomas of the head and neck revealed by the use of laser capture microdissection and cDNA arrays. Oncogene 19:3220–3224

Li J, Peet GW, Balzarano D, Li X, Massa P, Barton RW, Marcu KB (2001) Novel NEMO/IkappaB kinase and NF-kappa B target genes at the pre-B to immature B cell transition. J Biol Chem 276:18579–18590

Liang Y, Chang J, Lynch SJ, Lukac DM, Ganem D (2002) The lytic switch protein of KSHV activates gene expression via functional interaction with RBP-Jkappa (CSL), the target of the Notch signaling pathway. Genes Dev 16:1977–1989

Liu W, Ernst JD, Courtney Broaddus V (2000) Phagocytosis of crocidolite asbestos induces oxidative stress, DNA damage, and apoptosis in mesothelial cells. Am J Respir Cell Mol Biol 23:371–378

Logeat F, Bessia C, Brou C, LeBail O, Jarriault S, Seidah NG, Israel A (1998) The Notch1 receptor is cleaved constitutively by a furin-like convertase. Proc Natl Acad Sci USA 95:8108–8112

Manning CB, Cummins AB, Jung MW, Berlanger I, Timblin CR, Palmer C, Taatjes DJ, Hemenway D, Vacek P, Mossman BT (2002) A mutant epidermal growth factor receptor targeted to lung epithelium inhibits asbestos-induced proliferation and proto-oncogene expression. Cancer Res 62:4169–4175

Manning CB, Vallyathan V, Mossman BT (2002) Diseases caused by asbestos: mechanisms of injury and disease development. Int Immunopharmacol 2:191–200

Mateer SC, Fedorov SA, Mumby MC (1998) Identification of structural elements involved in the interaction of simian virus 40 small tumor antigen with protein phosphatase 2A. J Biol Chem 273:35339–35346

Morimura T, Goitsuka R, Zhang Y, Saito I, Reth M, Kitamura D (2000) Cell cycle arrest and apoptosis induced by Notch1 in B cells. J Biol Chem 275:36523–36531

Mossman BT, Faux S, Janssen Y, Jimenez LA, Timblin C, Zanella C, Goldberg J, Walsh E, Barchowsky A, Driscoll K (1997) Cell signaling pathways elicited by asbestos. Environ Health Perspect 105(Suppl 5):1121–1125

Nickoloff BJ, Qin JZ, Chaturvedi V, Denning MF, Bonish B, Miele L (2002) Jagged-1 mediated activation of notch signaling induces complete maturation of human keratinocytes through NF-kappaB and PPARgamma. Cell Death Differ 9:842–855

Nye JS, Kopan R, Axel R (1994) An activated Notch suppresses neurogenesis and myogenesis but not gliogenesis in mammalian cells. Development 120:2421–2430

Oswald F, Tauber B, Dobner T, Bourteele S, Kostezka U, Adler G, Liptay S, Schmid RM (2001) p300 acts as a transcriptional coactivator for mammalian Notch-1. Mol Cell Biol 21:7761–7774

Pear WS, Aster JC, Scott ML, Hasserjian RP, Soffer B, Sklar J, Baltimore D (1996) Exclusive development of T cell neoplasms in mice transplanted with bone marrow expressing activated Notch alleles. J Exp Med 183:2283–2291

Porcu P, Ferber A, Pietrzkowski Z, Roberts CT, Adamo M, LeRoith D, Baserga R (1992) The growth-stimulatory effect of simian virus 40 T antigen requires the interaction of insulinlike growth factor 1 with its receptor. Mol Cell Biol 12:5069–5077

Porcu P, Grana X, Li S, Swantek J, De Luca A, Giordano A, Baserga R (1994) An E2F binding sequence negatively regulates the response of the insulin-like growth factor 1 (IGF-I) promoter to simian virus 40T antigen and to serum. Oncogene 9:2125–2134

Rae FK, Stephenson SA, Nicol DL, Clements JA (2000) Novel association of a diverse range of genes with renal cell carcinoma as identified by differential display. Int J Cancer 88:726–732

Rangarajan A, Syal R, Selvarajah S, Chakrabarti O, Sarin A, Krishna S (2001b) Activated Notch1 signaling cooperates with papillomavirus oncogenes in transformation and generates resistance to apoptosis on matrix withdrawal through PKB/Akt. Virology 286:23–30

Rebay I, Fleming RJ, Fehon RG, Cherbas L, Cherbas P, Artavanis-Tsakonas S (1991) Specific EGF repeats of Notch mediate interactions with Delta and Serrate: implications for Notch as a multifunctional receptor. Cell 67:687–699

Rechsteiner M (1988) Regulation of enzyme levels by proteolysis: the role of pest regions. Adv Enzyme Regul 27:135–151

Robbins J, Blondel BJ, Gallahan D, Callahan R (1992) Mouse mammary tumor gene int-3: a member of the notch gene family transforms mammary epithelial cells. J Virol 66:2594–2599

Rogoff HA, Pickering MT, Debatis ME, Jones S, Kowalik TF (2002) E2F1 induces phosphorylation of p53 that is coincident with p53 accumulation and apoptosis. Mol Cell Biol 22:5308–5318

Ronchini C, Capobianco AJ (2001) Induction of cyclin D1 transcription and CDK2 activity by Notch(ic): implication for cell cycle disruption in transformation by Notch(ic). Mol Cell Biol 21:5925–5934

Rundell K, Parakati R (2001) The role of the SV40 ST antigen in cell growth promotion and transformation. Semin Cancer Biol 11:5–13

Sakamoto K, Ohara O, Takagi M, Takeda S, Katsube K (2002) Intracellular cell-autonomous association of Notch and its ligands: a novel mechanism of Notch signal modification. Dev Biol 241:313–326

Schroeter EH, Kisslinger JA, Kopan R (1998) Notch-1 signalling requires ligand-induced proteolytic release of intracellular domain. Nature 393:382–386

Schneider C, Weisshart K, Guarino LA, Dornreiter I, Fanning E (1994) Species-specific functional interactions of DNA polymerase alpha-primase with simian virus 40 (SV40) T antigen require SV40 origin DNA. Mol Cell Biol 14:3176–3185

Schroeder T, Just U (2000) mNotch1 signaling reduces proliferation of myeloid progenitor cells by altering cell-cycle kinetics. Exp Hematol 28:1206–1213

Seugnet L, Simpson P, Haenlin M (1997) Transcriptional regulation of Notch and Delta: requirement for neuroblast segregation in *Drosophila*. Development 124:2015–2025

Shelly LL, Fuchs C, Miele L (1999) Notch-1 inhibits apoptosis in murine erythroleukemia cells and is necessary for differentiation induced by hybrid polar compounds. J Cell Biochem 73:164–175

Shivapurkar N, Wiethege T, Wistuba II, Salomon E, Milchgrub S, Muller KM, Churg A, Pass HI, Gazdar AF (1999) Presence of simian virus 40 sequences in malignant mesotheliomas and mesothelial cell proliferations. J Cell Biochem 76:181–188

Soriano JV, Uyttendaele H, Kitajewski J, Montesano R (2000) Expression of an activated Notch4(int-3) oncoprotein disrupts morphogenesis and induces an invasive phenotype in mammary epithelial cells in vitro. Int J Cancer 86:652–659

Sriuranpong V, Borges MW, Ravi RK, Arnold DR, Nelkin BD, Baylin SB, Ball DW (2001) Notch signaling induces cell cycle arrest in small cell lung cancer cells. Cancer Res 61: 3200–3205

Stewart SA, Hahn WC, O'Connor BF, Banner EN, Lundberg AS, Modha P, Mizuno H, Brooks MW, Fleming M, Zimonjic DB, Popescu NC, Weinberg RA (2002) Telomerase contributes to tumorigenesis by a telomere length-independent mechanism. Proc Natl Acad Sci USA 99:12606–12611

Struhl G, Greenwald I (1999) Presenilin is required for activity and nuclear access of Notch in *Drosophila*. Nature 398:522–525

Sweet BH, Hilleman MR (1960) The vacuolating virus, S.V.40. Proc Soc Exp Biol Med 105:420–427

Talora C, Sgroi DC, Crum CP, Dotto GP (2002) Specific down-modulation of Notch1 signaling in cervical cancer cells is required for sustained HPV-E6/E7 expression and late steps of malignant transformation. Genes Dev 16:2252–2263

Testa JR, Giordano A (2001) SV40 and cell cycle perturbations in malignant mesothelioma. Semin Cancer Biol 11:31–38

Tohda S, Nara N (2001) Expression of Notch1 and Jagged1 proteins in acute myeloid leukemia cells. Leuk Lymphoma 42:467–472

Toyooka S, Carbone M, Toyooka KO, Bocchetta M, Shivapurkar N, Minna JD, Gazdar AF (2002) Progressive aberrant methylation of the RASSF1A gene in simian virus 40 infected human mesothelial cells. Oncogene 21:4340–4344

Tsai SC, Pasumarthi KB, Pajak L, Franklin M, Patton B, Wang H, Henzel WJ, Stults JT, Field LJ (2000) Simian virus 40 large T antigen binds a novel Bcl-2 homology domain 3-containing proapoptosis protein in the cytoplasm. J Biol Chem 275:3239–3246

Waheed I, Guo ZS, Chen GA, Weiser TS, Nguyen DM, Schrump DS (1999) Antisense to SV40 early gene region induces growth arrest and apoptosis in T-antigen-positive human pleural mesothelioma cells. Cancer Res 59:6068–6073

Wang J, Shelly L, Miele L, Boykins R, Norcross MA, Guan E (2001) Human Notch-1 inhibits NF-kappa B activity in the nucleus through a direct interaction involving a novel domain. J Immunol 167:289–295

Wong JM, Kusdra L, Collins K (2002) Subnuclear shuttling of human telomerase induced by transformation and DNA damage. Nat Cell Biol 4:731–736

Weijzen S, Rizzo P, Braid M, Vaishnav R, Jonkheer SM, Zlobin A, Osborne BA, Gottipati S, Aster JC, Hahn WC, Rudolf M, Siziopikou K, Kast WM, Miele L (2002) Activation of Notch-1 signaling maintains the neoplastic phenotype in human Ras-transformed cells. Nat Med 8:979–986

Xu L, Flynn BJ, Ungar S, Pass HI, Linnainmaa K, Mattson K, Gerwin BI (1999) Asbestos induction of extended lifespan in normal human mesothelial cells: interindividual susceptibility and SV40 T antigen. Carcinogenesis 20:773–783

Zagouras P, Stifani S, Blaumueller CM, Carcangiu ML, Artavanis-Tsakonas S (1995) Alterations in Notch signaling in neoplastic lesions of the human cervix. Proc Natl Acad Sci USA 92:6414–6418

Zhou S, Fujimuro M, Hsieh JJ, Chen L, Miyamoto A, Weinmaster G, Hayward SD (2000) SKIP, a CBF1-associated protein, interacts with the ankyrin repeat domain of NotchIC to facilitate NotchIC function. Mol Cell Biol 20:2400–2410

Signal Transduction and Apoptosis Pathways as Therapeutic Targets

P.F. Valerón[1], S. Aznar-Benitah[1] and J.C. Lacal[1]

1 Introduction

The knowledge of the cellular events that take place during the development of a disease is crucial for establishing effective therapies. During the last three decades, the design of antitumoral agents has been primarily based on the fact that tumor cells exhibit a higher proliferation rate than "normal" cells. As a consequence, drugs that are cytotoxic either by affecting DNA replication, cellular morphological integrity, or cellular metabolism, such as alkylating, antimetabolites and microtubule-destabilizing agents, have been approved as valid antitumoral therapies. Although these compounds are still used nowadays as the principal weapons against tumor development and progression, many types of tumors of epithelial origin (the main kind of solid tumors in humans) are resistant to these conventional drugs. In addition, most of these therapies have considerable side effects as a consequence of non-specific cytotoxic effects on normal non-tumorigenic cells that usually exhibit a high proliferation rate, such as the intestinal epithelium or the bone marrow. A second serious clinical issue is the acquired resistance shown by many advanced tumors towards conventional chemotherapeutics that ultimately renders the therapeutic strategy useless.

The increasing effort in elucidating the molecular basis for the development of cancer has provided an extensive body of information that has led to the development of target and type of tumor-based therapies. In a healthy tissue, the net balance of cell death and cell replication (unless a specific stimulus is received) is maintained to assure that no increase in cell number takes place. However, any genetic alteration of these processes in a subpopulation of the cells can lead to a rupture of this delicate balance contributing to neoplastic growth. In this sense, genes that control cell proliferation and cell death either positively or negatively are the primary candidates to suffer a genetic alteration that ultimately leads to the development of a tumor. Other "secondary" genes would constitute those that encode for detoxifying proteins, DNA repair machinery, cell adhesion and cell motility that per se do not cause cellular

[1]Instituto de Investigaciones Biomédicas, CSIC, Arturo Duperier 4, , 28029 Madrid, Spain, e-mail: jclacal@iib.uam.es

Progress in Molecular and Subcellular Biology
C. Alonso (Ed.): Viruses and Apoptosis
© Springer-Verlag Berlin Heidelberg 2004

transformation. Instead, they might render the cells prone to an aberrant growth upon mutation of a second gene, or allow progression to a more malignant phenotype (Johnstone et al. 2002).

The vast majority of proto-oncogenes codify for proteins that participate in signal transduction pathways involved in mitogenesis, cell migration and cell survival. These proteins include certain growth factors and cytokines as well as their receptors, intracellular signaling proteins and transcription factors that upon activation lead to cell growth. Among these are receptor tyrosine kinases (EGFR, PDGFR), G-coupled receptors, serine/threonine and tyrosine intracellullar kinases (Abl, Src, GSK2, Akt, Raf, JNK), small G proteins (Ras, Rho) and transcription factors (Stats, Smads, AP1, etc.). Accumulation of some of these genes carrying mutations that constitutively activate their protein products ultimately leads to aberrant cell growth (Arteaga et al. 2002). Determining the exact role and incidence of these proteins in cell proliferation in specific human tumors and their contribution to the different steps that occur during tumor progression has led to the identification of targets for anticancer therapies (see below).

On the other hand, the apoptotic program relies on a family of cysteine-proteases that hydrolyze specific substrates on aspartic acid residues, named caspases. The activation of these proteases takes place via an extrinsic pathway due to an external stimulus such as tumor necrosis factors, and a second intrinsic pathway in which the mitochondria participate. The extrinsic pathway initiates with the ligation of the transmembrane death receptors (CD95, TNFalpha, and TRAIL) that activate caspase 8. This activation can lead to two responses depending on the cell type (Krammer 2000). Whereas type I cells generate enough levels of caspase 8 to directly initiate the effector caspase pathway, type II cells that contain low levels of caspase 8 hydrolyze Bid, a member of the Bcl-2 family of proteins, to yield the truncated form tBid that interacts with Bax and Bak to induce mitochondrial damage and activation of effector caspases (Tilly 2001).

The intrinsic pathway requires mitochondrial membrane depolarization and subsequent liberation of proteins contained within, such as Smac2/DIABLO, Htr2 and cytochrome C. Cytochrome C and Apaf1 induce caspase 9 activation, whereas Smac2/DIABLO and Htr2 interact with and inhibit IAPs (Susuki et al. 2001; Wang 2001). Mitochondrial membrane depolarization takes place via members of the Bcl-2 family of anti- and pro-apoptotic proteins. Bax and Bak can be directly activated by interaction with Bid or by inactivation of the anti-apoptotic members Bcl-2, Bcl-X_L with other BH3-only proteins such as Noxa, Puma, Bad or Bim (Adams and Cory 1998; Huang and Strasser 2000). Bax/Bak activation then leads to mitochondrial membrane depolarization and cytoplasmic localization of the mitochondrial proteins mentioned above.

Although both intrinsic and extrinsic pathways coexist depending on the cell type, there is a considerable amount of cross-talk among them. In this sense, caspase 8 can proteolytically activate Bid, facilitating cytochrome C liberation, whereas activators of the intrinsic pathway can sensitize cells to extrinsic death ligands (Green 2000).

Finally, many cellular stresses such as oncogenic activation, DNA damage, hypoxia, and lack of survival factors can lead to cell cycle arrest and ultimately to apoptosis. The tumor suppressor p53 is a general sensor of cellular stresses that controls cell cycle and apoptosis via upregulation of p21 and CDK/Cyclin inhibition (Lowe and Lin 2000). The p53 gene is mutated in a high percentage of human tumors and is the cause of increased survival, DNA-instability tolerance and resistance to chemotherapeutics, and therefore constitutes a plausible target for cancer therapy.

In this chapter, an attempt to describe some of the new generation of target-based designed drugs, validation and proof of concept of targets for anticancer therapy, their mode of action and clinical relevance will be made. Due to space limitations, and because other chapters in this book thoroughly address cell proliferation or apoptotic signaling pathways, only a brief introduction to some aspects necessary for the proper understanding of the action of the compounds will be made. We are also aware that many new drugs targeted at signaling proteins with very promising preclinical results are emerging, but due to space limitation we will only consider those that have entered clinical trials.

2 Strategies for Modulation of Apoptosis

2.1 Tumor Suppressor p53

p53 is a nuclear phosphoprotein that functions as a transcription factor inducible in response to DNA damage (Levine 1997; Agarwal et al. 1998). In the absence of genetic damage its transcriptional activity is kept null via rapid ubiquitination and proteosome-dependent degradation by interacting with mdm2 (Yuan et al. 1996). Upon DNA damage a signaling cascade is activated that leads to p53 phosphorylation, protein stabilization and transcriptional activation (Caspari 2000). Depending on the intensity and length of the insult to the cell, p53 can subsequently induce the expression of several genes involved in cell cycle arrest, DNA repair, differentiation or apoptosis. The G1-S phase is a tightly regulated mitotic checkpoint susceptible to low DNA damage, heat shock, hyperoxia, hypoxia and oncogenic transformation. Among the different genes involved in G1-S cycle arrest are the inhibitor of cyclin-dependent kinases p21WAF-1, 14-3-3 and reprimo (Kastan et al. 1991; Kuerbitz et al. 1992; McKay et al. 1999). Under extreme conditions, p53 triggers the transcription of proapoptotic genes such as Bax, DR5, Fas/APO1, PTEN and antiapoptotic genes that include Bcl-2, Bcl-X_L and IAPs, among others that ultimately lead to efficient cell death (Miyashita and Reed 1995; Owen-Schaub et al. 1995; Jayaraman et al. 1997; Polyak et al. 1997; Ryan et al. 2001). In this sense, p53 inactivation in a large number of human cancers with varying incidence leads to a reduced response to cytotoxic treatments due to impaired ability of tumor cells to efficiently turn on the apoptotic machinery (Yu et al. 2002).

P53 mutation is the most common genetic aberration in human cancers with more than 50% of tumors carrying some mutation in its coding gene (Attardi and Jacks 1999; Ryan et al. 2001). Approximately 1700 different mutations in p53 have been described in the past decade, some being somatic others in the germ-line (Lane and Lain 2002). In addition, other p53-related signaling components, such as mdm2, ATM, Chk2 and p19ARF, or p53-regulated genes including PTEN, Bax, Bak and Apaf-1, are altered in human tumors to varying degrees with the same final outcome. This implies that p53 might exist in its wild-type form in the tumor, yet still having an aberrant apoptotic signaling capability (Schmitt et al. 1999).

This knowledge has led to the development of different approaches to restore wild-type 53 function in tumor cells. As mentioned above, many different mutations have been detected in the ORF of p53, primarily within the DNA-binding domain; however, over 13% of mutations are found outside this region. These mutations affect DNA-binding activity, protein stability, and altered folding, among others (Lane and Lain 2002). Thus, the precise effect of these mutations over p53 activity needs to be clarified if one intends to selectively target the mutated protein for cancer therapy.

One general approach has been the use of gene therapy which has been extensively studied both in preclinical and clinical models. Retroviral-mediated p53 restoration in cell lines derived from hepatocellular carcinoma (HCC) leads to inhibition of cell proliferation and efficient activation of the apoptotic machinery. These results have led to the proposal of intra-tumoral injection of retroviral wt-p53 as a promising strategy for treatment of patients with HCC (Havlik et al. 2002). Other models have been tested with promising results (i.e. low toxicity and efficient tumor cell apoptosis). However, clinical trials have shown that these strategies must be improved.

A second, more sophisticated, gene therapy approach consists of placing a "killer gene" under a promoter of a repressor protein. In turn, this repressor protein is only expressed in the presence of wild-type p53 (normal cells). Therefore, tumor cells that lack normal p53 function do not synthesize the repressor thereby expressing the toxic genes. In non-transformed cells, p53 is able to efficiently trigger the repressor preventing the expression of the cytotoxic gene (Zhu 2000). However, the absence of p53 functionality in normal cells with no inflicted damage would argue against the suitability of this approach, since the cytotoxic gene product will be then produced also in normal cells, resulting in unspecific toxicity.

A third new approach is based on the development of small interacting molecules. These molecules are generally selected on their ability to affect a downstream pathway regulated by p53. Compounds that can stabilize the DNA-binding domain of p53 in the active conformation, therefore promoting p53 transcriptional activity, have been identified and have been shown to negatively affect tumor growth in mice (Foster et al. 1999). With respect to downstream signaling, small molecules that affect p21, Cyclin-dependent kinases (CDK), and the proapoptotic protein Bax have been identified (for a thorough review, see Lane and Lain 2002). In addition, peptides that restore p53 activity by mediating an Mdm2-dependent mechanism have been developed with promising results with breast tumor cell lines.

Although many different therapeutic approaches have been developed to affect p53-signaling in human tumors, there still is a long way to travel until some of these are incorporated as valid antitumoral treatments for specific cancers. The next decade promises to provide further information on p53 activity and the development of new drugs and enhancement of existing ones that will surely have a beneficial outcome in the clinic.

2.2 Bcl-2

As mentioned above, Bcl2 is involved in the protection of apoptosis upon several stimuli. The Bcl-2 gene was originally discovered as a proto-oncogene in low-grade B-cell non-Hodgkin's lymphoma located at the breakpoints of t(14;18) translocations. Later works have shown that Bcl2 is overexpressed in chronic lymphocytic lymphomas, in some diffuse large-cell lymphomas and in most follicular lymphomas (Adams and Cory 1998). Furthermore, besides being involved in the development of cancer, Bcl2 has also been associated with resistance to cancer treatment.

An antisense oligodeoxynucleotide (ODN) approach has been developed to target Bcl2 expression in human malignancies. Genta (San Diego, CA, USA) has developed an antisense ODN against Bcl2 mRNA, termed G3139. In mouse models of human follicular lymphoma, G3139 treatment resulted in complete remission of the disease. Phase I and II trials are being carried out against several type of tumors. These include non-Hodgkin's lymphoma, melanoma, breast carcinoma, hormone-refractory prostate cancer, chemorefractory small-cell lung cancer (SCLC), and acute myelogenous leukemia (Jansen et al. 2000; Waters et al. 2000; Marcucci et al. 2002). Of special interest are the trials of G3139 against human melanoma, a type of tumor where Bcl2 is expressed in up to 90% of all cases. G3139 was administered intravenously in combination with dacarbazine to 14 patients with advanced malignant melanoma in a phase I trial (Jansen et al. 2000). Except hepatic abnormalities in four patients that discontinued within 1 week, no major toxicities were observed. Bcl2 and apoptosis levels were measured during the treatment with reduced expression of Bcl2 and increased apoptosis in 40% of samples. Furthermore, one complete, two partial and three minor responses were observed in 6 out of 14 patients. Based on these promising results, a multicenter phase III trial has been approved and is currently being carried out for patients with malignant melanoma.

Recently, results from a 20 patient phase I trial of G3139 in combination with conventional chemotherapy, with fludarabine, cytarabine (ARA-C), and G-CSF [FLAG] (salvage chemotherapy), in refractory or relapsed acute leukemia have been reported (Marcucci et al. 2002). Thus, this line of research seems to be rather promising on both single treatments and in combination.

3 Signal Transduction Targets

3.1 The Family of EGFR

Four members of the family of EGF receptors have been cloned to date with very similar structure and function: HER1 (epidermal growth factor receptor, EGFR), HER2 (c-erbB-2, neu), HER3 (c-erbB-3), and HER4 (c-erbB-4). They all are transmembrane glycoproteins of approximately 170 kDa (Carpenter and Cohern 1990). The most-studied isoform is the EGFR receptor.

The EGFR receptor responds to several ligands that include epidermal growth factor (EGF), transforming growth factor alpha (TGFα), heparin-binding EGF, and betacellulin, although it's response is most prominent upon EGF and TGFα coupling. Upon ligand binding, EGFR follows the common scheme of RTK signaling, with receptor homodimerization, tyrosine transautophosphorylation, coupling to intracellular adapter proteins and tyrosine kinases that ultimately propagate the signal towards the inside of the cell (Salomon et al. 1995; Yarden and Ullrich 1988). EGFR and c-erbB2 (HER2) share an 82% sequence homology in their tyrosine kinase domain, and can heterodimerize in certain scenarios (Pringent and Lemoine 1992). EGFR is normally expressed in the epithelial fraction of the skin, the gastrointestinal tract and the liver. However, in many carcinomas, EGFR expression is enhanced leading to constitutive signaling independent of ligand (Thompson and Gill 1985; Schlessinger 1988; Gullick 1991).

EGFR overexpression leads to increased cell cycle progression and proliferation, reduced apoptosis, angiogenesis, and cell motility, all of which permit tumor formation and progression in several human tumors that include colorectal carcinomas, head and neck squamous cell carcinomas, pancreatic, non-small cell lung, breast, kidney, ovaries, bladder carcinomas, as well as gliomas. Many studies have shown that there is a good correlation between EGFR expression and reduced mean survival with poor prognosis, increased relapse, and higher risk of metastasis (Salomon et al. 1995; Grandis et al. 1996; Rubin Grandis et al. 1997). In breast cancer, the HER2-neu somatic gene amplification takes place in 25–30% of studied cases and is responsible for both tumor development and progression to a more malignant phenotype (Slamon et al. 1987, 1989; Van de Vijer et al. 1988; Schroeder et al. 1997).

Several different strategies are being developed that aim to target the activity of the EGFR family of receptors in several types of tumors. These include small molecules that compete with ATP binding, monoclonal antibodies, conjugated ligands, immunoconjugates and antisense oligonucleotides. For instance, ZD1839 (Iressa) and OSI-774 (Erlotinib, Tarceva), two potent small molecules that inhibit EGFR kinase activity, have promising antitumoral activities over patients with refractory non-small cell lung cancer and patients with advanced head and neck cancer with disease stabilization after 3 months of oral intake, respectively (Ciardiello et al. 2000; Hidalgo et al. 2001). ZD1839 is a synthetic anilinoquinazoline that selectively inhibits EGFR via competitive binding at the ATP-binding pocket. This compound has been tested either alone or in combination with conventional chemotherapeutics (cisplatin, carboplatin, paclitaxel, docetaxel, etoposide, doxorubicin, among others) with additive antitumoral activity that results in complete tumor regression in several xenograft tumors (Sausville et al. 2002). Phase I and II clinical trials with oral administration of Iressa have been carried out either alone or in combination with carboplatin and paclitaxel for prostate cancer and NSCLC, with partial response and stable disease (over 4 months) and relatively mild side effects. OSI-774 is a quinazoline derivative inhibitor that also competes for ATP binding, thus inhibiting tyrosine kinase activity. A phase I study has shown that OSI-774 showed partial responses in patients with colorectal carcinomas and renal cell carcinoma with disease stabilization (over 5 months) in cervical, NSCLC, colon, prostate, and head and neck carcinomas. Phase II studies have shown that with cisplatin-refractory NSCLC, 11% patients had partial response, and 34% tumor stabilization (Goel et al. 2002).

A second strategy has involved the use of specific monoclonal antibodies targeted to the extracellular region of the receptor. A wide variety of antibodies have been tested in vitro and in vivo in animal models which have displayed an effective antitumoral activity. Consequently, some of these entered clinical trials, including IMC-C225 (chimeric mouse-human monoclonal antibody also known as cetuximab), EMD 55900 (monoclonal antibody 425), a murine anti-EGFR monoclonal antibody, ICR 62, a rat monoclonal antibody, and trastuzumab (Herceptin), a fully human anti-EGFR antibody (Hoffman et al. 1997; Yang et al. 2001). However, the major limitations to some of these antibodies and other mouse-related antibodies is their relatively high immunogenicity when administered in humans, and a low affinity towards the EGF receptor.

With this problem in mind, a humanized antibody has been developed termed IMC-C225 (Cetuximab). This antibody has entered phase I and II clinical trials for colorectal carcinomas (Wu et al. 1995), pancreatic carcinoma (Bruns et al. 2000), breast carcinoma (Moyer et al. 1997), prostatic carcinoma (Prewett et al. 1996), renal cell carcinoma (Prewett et al. 1998), and squamous cell carcinoma of the head and neck (Baselga et al. 1993). IMC-C225 in combination with conventional chemotherapy against colon cancer (together with CPI-11), head and neck cancer (combined with cisplatin) and in pancreatic cancers (combined with gemcitabine) has resulted in additive antitumoral activity (Herbst and Langer 2002).

A totally humanized anti-EGFR antibody has also been developed, termed Trastuzumab (Herceptin). Recent clinical results have been published that show that Herceptin leads to clinical benefit of first line chemotherapy in advanced breast cancer in terms of overall survival without unacceptable side effects (Slamon et al. 2001). This antibody has been approved for clinical use against breast cancer.

3.2 VEGF Receptor

Angiogenesis is a complex process that is tightly regulated by the action of pro- and antiangiogenic factors. De novo synthesis of blood vessels is a physiological process that takes place during tissue growth and repair, female reproductive cycle, and fetal development. However, angiogenesis is intimately ligated to several pathological processes such as cancer and coronary heart disease.

Among the best characterized is the vascular endothelial growth factor (VEGF), a heterodimeric glycoprotein highly conserved among species with multiple isoforms. VEGF constitutes one of the most potent endothelial growth factors that participates in the development and differentiation of the vascular system. The most abundant isoform is VEGF165 that binds to two endothelial receptors with intrinsic tyrosine kinase activity, VEGFR-1 (Flt1) and VEGFR-2 (KDR/Flk1; Pegram and Reese 2002). VEGF is elevated in several human tumors and its expression responds to hypoxia, oncogenes, and is associated with poor prognosis, thus constituting an important target for targeted antitumoral therapy (Rosen 2002).

Several compounds are known to exhibit potent anti-VEGF activity. Thalidomide is a glutamic acid derivative that some 40 years ago was used in the clinic as a sedative hypnotic whose activity has been associated with congenital abnormalities when administered to gestating women (Kumar et al. 2002). At the present time, due to its antiangiogenic activity, it has been reintroduced as an anticancer therapy with several ongoing clinical trials for advanced multiple myeloma and renal-cell carcinoma (Escudier et al. 2002; Yakoub-Agha et al. 2002).

A second interesting compound is AE-941 (Neovastat), a natural product obtained from shark cartilage with antiangiogenic activity that inhibits matrix metalloproteinases (MMP-2, MMP-9 and MMP-12), and VEGF coupling to endothelial cells. Neovastat is being evaluated in various clinical trials for the NSCLC and renal-cell carcinoma (Falardeau et al. 2001; Gingras et al. 2001).

However, from a more directed therapy point of view, Bevacizumab (anti-VEGFrhuMab) needs to be mentioned. This monoclonal antibody is thought to be a valid therapy for myelodysplasia (List 2002). Combinations of Bevacizumab together with other antiangiogenic compounds (TNP-470, Thalidomide, CC5013, carboxyamidotriazole, endostatin, SU5416, SU6668, and 2-methoxyestradiol) are being tested for lung cancer or very aggressive tumors such as metastatic prostate cancer refractory to other therapies (Shepherd 2001; Figg et al. 2002).

Small molecules that target VEGFR activity have also been developed. This is the case for SU5416, which is a competitive inhibitor of ATP binding for the VEGF receptor 1. This drug has been used in a phase I clinical trial against colorectal, lung, renal carcinomas and Kaposi's sarcoma, with stable disease in some cases (over 6 months), but with no observable benefit. However, some severe side effects such as increased liver enzymes, projectile vomiting, and vascular complications such as thrombotic events argue against the suitability of this drug.

3.3
BCR-ABL and CML

Perhaps this might be one of the most direct targets for cancer therapy. This chimeric protein results from the reciprocal translocation between chromosome 9 and 22 in chronic myelogenous leukemia (CML), with a constitutive tyrosine kinase activity and signaling. The interesting aspect of BCR-ABL in CML is the fact that this is the main genomic alteration in this type of human cancer.

In a screen for compounds that inhibit PKC, a derivative of phenylaminopiridine was found, termed ST1571 (Imatinib mesylate, Gleevec), capable of inhibiting PDGFR autophosphorylation. It was then optimized for BCR-ABL and c-Kit. These three kinases are the only substrates to this drug known to date. Given that CML development relies on BCR-ABL activity, this compound was tested as a possible therapeutic drug against this type of cancer. In several clinical trials Gleevec has exhibited a potent antiproliferative activity, which has led to its approval by the FDA for the treatment of CML, only 3 years after the first clinical trial had started (Kantarjian et al. 2002; Sawyers et al. 2002; Talpaz et al. 2002). In this sense, 98% of interferon-refractory CML patients responded to Gleevec treatment with complete cytogenetic response seen in 13% and very significant cytogenetic responses in 31%. However, 70% of patients with lymphoid phenotype, ALL (Philadelphia chromosome positive) or lymphoid blast crisis, had a minor response, and most of the patients relapsed. In addition, a good response in patients with progressive gastrointestinal stromal tumors (GIST), soft-tissue sarcomas, and other tumors has been observed in phase II clinical trials suggesting that Gleevec might be useful for other human tumors besides CML (Sausville et al. 2002).

3.4 CDK Inhibitors

A wide variety of non-receptor tyrosine kinases, serine/threonine kinases and lipid kinases are involved in mitogenic signaling. The specific effects of many of these kinases have been extensively studied in many different tumor models. Accordingly, selective drugs have been developed that target these signaling proteins, some of which have entered clinical trials with promising results.

Cyclin-dependent kinases (CDKs) are highly conserved general regulators of the cell cycle. The kinase activity of these proteins is turned-on by promitogenic stimuli that stimulate their binding to cyclins. Several regulatory mechanisms of CDK activity exist that include phosphorylation by CDK-activating kinases (CAK), dephosphorylation of threonine residues by CDC25 phosphatase, inhibition of INK4 and CIP/KIP families, and cyclin binding. Active CDK subsequently phosphorylates a number of substrates that lead to cell cycle progression.

A compound known to affect CDK activity is flavopiridol, a natural alkaloid derivative with potent antiproliferative activity over a wide panel of tumor-derived cell lines (Senderowicz and Sausville 2000). Flavopiridol inhibits the activity of almost all CDKs. According to this, flavopiridol causes cell cycle arrest at G1/S and G2/M transition, as a consequence of inhibition of CDK4/2/6 and CDK1, respectively (Kaur et al. 1992). In addition, flavopiridol may affect transcription of certain genes, such as cyclin D1. This would take place by inhibition of CDK9, a member of the CDK family known to function as a transcription factor (Chao and Price 2001).

Cytotoxicity and apoptosis in mouse tumor xenograft models of various origins in preclinical studies have led to two phase I clinical trials, although few responses have been observed in these trials and some phase II trials, with the exception of mantle cell lymphoma. However, combination of flavopiridol with gemtabicine, taxol or irinotecan may prove to potentiate the antineoplastic activity of both combined drugs (Sausville et al. 2002).

Besides flavopiridol, CDK inhibition has been achieved by targeting the ATP-binding pocket via small molecules. Some of these compounds include butyrolactone I, olomucine, roscovitine, purvanolol and paullone derivatives that selectively inhibit CDK1 and CDK2 with no effect over CDK4/6 (Carnero 2002). One such compound that has entered clinical trials is UCN-01 with activity against melanoma, lung cancer, and non-Hodgkin's lymphoma. Accordingly, UCN-01 has entered phase II trials.

Although other approaches, such as peptide- and protein-based inhibitors, are being developed to target specific CDKs and CDK signaling, all are still being evaluated in preclinical studies and therefore have not entered the clinic.

3.5 Ras and Rho Small GTPases

The Ras and Rho family of GTPases are proteins that cycle between an active GTP-bound state and an inactive GDP-bound state. Both Ras and Rho GTPases mediate cellular processes essential for proper development and differentiation, as well as tissue homeostasis. These include cell–cell and cell–extracellular matrix (ECM) adhesion, actin cytoskeleton reorganization, cell growth and apoptosis. Although these GTPases mediate housekeeping functions in development and tissue homeostasis, dysregulatory processes such as point mutations or overexpression lead to malignant phenotypes. It is estimated that 25–30% of human cancers contain a mutated version of a Ras protein, most frequently K-Ras followed by N-Ras and, at a much lower rate, H-Ras. Pancreatic cancers, cholangiocarcinomas, adenocarcinoma of the lung, squamous head and neck tumors, and acute leukemia among others have been described to contain to varying degrees mutations in Ras. With respect to the subfamily of Rho GTPases, no point mutations in human tumors have been found; however, overexpression of either the GTPase itself or some upstream or downstream element of Rho signaling occurs in human tumors leading to an increase in downstream signaling. Pancreatic cancers, breast cancer (both inflammatory and non-inflammatory), melanomas, colorectal carcinomas, testicular germ cancer, head and neck squamous cell carcinomas, leukemias, osteosarcomas, gastric cancer, thyroid papillary carcinomas, hepatocellular carcinomas, ovarian cancers, neuroblastomas, prostate cancer, bladder cancer and renal carcinomas have been reported to contain some aberration in Rho signaling (Aznar and Lacal 2001; Aznar et al., 2003). Accordingly, several members of this family of GTPases have been considered potential targets for treatment of several human tumors.

3.6 Farnesyltransferase Inhibitors

The function of Ras and Rho requires membrane localization achieved by a complex mechanism that involves C-terminal prenylation (farnesylation of geranylgeranylation), protease cleavage, carboxymethylation and palmitoylation. Given the involvement of these proteins in tumor development, a strategy was set up to target Ras prenylation preventing membrane association and activation. Initially, inhibition of the enzymes farnesyltransferase (FT) and geranylgeranyl transferase (GGT) responsible for processing of the Harvey Ras protein was the aim. Several types of FT and GGT inhibitors (FTIs, GGTIs) have been developed in recent years, but the best-characterized ones belong to the family of peptidomimetic drugs. Treatment of tumor-derived cells and xenograft models with FTIs exhibited non-toxic antiproliferative and antitumor activities via induction of apoptotsis (Gibbs et al. 1994; Johnston 2001).

However, the fact that K-Ras follows alternative prenylation methods to H-Ras and that cell lines transformed by oncogenes dependent or independent of Ras signaling are sensitive to FTIs argues against the specificity of these drugs. Recent reports have suggested that perhaps a member of the Rho family, RhoB, might account for some but not all of FTI-induced cytotoxic events (Aznar and Lacal 2001; Aznar et al. 2003; Prendergast 2001). In any case, some of these drugs have entered clinical trials with different outcomes.

Some of these include BMS-214662, L-778, SCH66336, R115777, all of which have shown acceptable toxicities with different administration protocols. However, the tumor response has been at most disease stabilization in a small percentage of patients with some individual cases of tumor regression with a reduction in FT activity in the tumor. This indicates that, although the responses have been mild, the drugs efficiently target farnesyltransferase activity in vivo, indicating that the strategy is valid and probably awaits newly developed compounds with higher activity in vivo (Sausville et al. 2002).

3.7 Raf Kinase

Raf is the first member of a multistep kinase cascade implicated in a large number of cellular processes, and constitutes one of the most studied proteins downstream of Ras. Sequential activation of downstream kinases that ultimately results in the activation of p42/p44 ERK/MAPK by Raf promotes nuclear migration of these and phosphorylates and activates several transcription factors which regulate transcription of early-immediate genes involved in proliferation.

Extensive evidence has implicated the Raf cascade in Ras-mediated transformation, although new evidence using primary human cell lines has indicated that Raf is not necessary for Ras-mediated transformation (Hamad et al. 2002). In addition, the B-Raf gene has been found mutated in 66% of malignant melanomas and at lower frequency in a wide range of human cancers (Davies et al. 2002).

One approach to inhibit Raf activity was based on an antisense oligodeoxynucleotide, and is already in phase I trials with promising results (Monia et al. 1996; Geiger et al. 1997; Gokhale et al. 1999; O'Dwyer et al. 1999; Stevensonm et al. 1999; Cunningham et al. 2000). This antisense oligodeoxynucleotide, termed ISIS 5132 or CGP 69846A, targets the 3′-untranslated region of human Raf-1 mRNA and suppresses tumor progression of a variety of human tumor cell lines (breast, small cell lung, large cell lung, colon, and squamous lung carcinoma) in mice. The drug is well tolerated at doses up to 6.0 mg/kg with fatigue and fever as side effects, when administered during 3 weeks. Accordingly to its effect, ISIS 5132 correlates in some patients with clinical benefits. Also, continuous administration of ISIS 5132 during 21 days at a 4.0 mg/kg dose has been carried out with mild side effects and with disease stabilization in 2 out of 36 patients.

Acknowledgements. Pilar F. Valerón and Salvador Aznar-Benitah have contributed equally to this work.

References

Adams JM, Cory S (1998) The Bcl-2 protein family: arbiters of cell survival. Science 281:1322–1326

Agarwal ML, Taylor WR, Chernov MV, Chernova OB, Stark GR (1998) The p53 network. J Biol Chem 273:1–4

Arteaga CL, Khuri F, Krystal G, Sebti S (2002) Overview of rationale and clinical trials with signal transduction inhibitors in lung cancer. Semin Oncol 29:15–26

Attardi LJ, Jacks T (1999) The role of p53 in tumor suppression: lessons from mouse models. Cell Mol Life Sci 55:48–63

Aznar S, Lacal JC (2001) Searching new targets for anticancer drug design: the families of Ras and RhoGTPases and their effectors. Prog Nucleic Acid Res Mol Biol 67:193–234

Aznar Benitah S, Espina C, Valerón PF, Lacal JC (2003) Rho GTPases in human carcinogenesis: a tale of excess. Rev Oncol 5:70–78

Baselga J, Norton L, Masui H, et al (1993) Antitumor effects of doxorubicin in combination with anti-epidermal growth factor receptor monoclonal antibodies. J Natl Cancer Inst 85:1327–1333

Bruns CJ, Harbison MT, Davis SW, et al (2000) Epidermal growth factor receptor blockade with C225 plus gemcitabine results in regression of human pancreatic carcinoma growing orthotopically in nude mice by antiangiogenic mechanisms. Clin Cancer Res 6:1936–1948

Capdeville R, Buchdunger E, Zimmermann J, Matter A (2002) Glivec (STI571, imatinib), a rationally developed, targeted anticancer drug. Nat Rev Drug Discov 1:493–502

Carnero A (2002) Targeting the cell cycle for cancer therapy. Br J Cancer 87:129–133

Carpenter G, Cohen S (1990) Epidermal growth factor. J Biol Chem 265:7709–7712

Caspari T (2000) How to activate p53. Curr Biol 10:R315–317

Chao SH, Price DH (2001) Flavopiridol inactivates P-TEFb and blocks most RNA polymerase II transcription in vitro. J Biol Chem 276:31795–31799

Ciardiello F, Caputo R, Bianco R, Damiano V, Fontanini G, Cuccato S, De Placido S, Bianco AR, Tortora G. (2000) Antitumor effect and potentiation of cytotoxic drug activity in human cancer cells by ZD-1839 (Iressa),an epidermal growth factor receptor-selective tyrosine kinase inhibitor. Clin Cancer Res 6:2053–2063

Cunningham CC, Holmlund JT, Schiller JH, Geary RS, Kwoh TJ, Dorr A, Nemunaitis J (2000) A phase I trial of c-Raf kinase antisense oligonucleotide ISIS 5132 administered as a continuous intravenous infusion in patients with advanced cancer. Clin Cancer Res 6:1626–1631

Davies H, Bignell GR, Cox C, Stephens P, Edkins S, Clegg S, Teague J, Woffendin H, Garnett MJ, Bottomley W, Davis N, Dicks E, Ewing R, Floyd Y, Gray K, Hall S, Hawes R, Hughes J, Kosmidou V, Menzies A, Mould C, Parker A, Stevens C, Watt S, Hooper S, Wilson R, Jayatilake H, Gusterson BA, Cooper C, Shipley J, Hargrave D, Pritchard-Jones K, Maitland N, Chenevix-Trench G, Riggins GJ, Bigner DD, Palmieri G, Cossu A, Flanagan A, Nicholson A, Ho JW, Leung SY, Yuen ST, Weber BL, Seigler HF, Darrow TL, Paterson H, Marais R, Marshall CJ, Wooster R, Stratton MR, Futreal PA (2002) Mutations of the BRAF gene in human cancer. Nature 417:949–954

Escudier B, Lassau N, Couanet D, Angevin E, Mesrati F, Leborgne S, Garofano A, Leboulaire C, Dupouy N, Laplanche A (2002) Phase II trial of thalidomide in renal-cell carcinoma. Ann Oncol 13:1029–1035

Falardeau P, Champagne P, Poyet P, Hariton C, Dupont E (2001) Neovastat, a naturally occurring multifunctional antiangiogenic drug, in phase III clinical trials. Semin Oncol 28:620–625

Figg WD, Kruger EA, Price DK, Kim S, Dahut WD (2002) Inhibition of angiogenesis: treatment options for patients with metastatic prostate cancer. Invest New Drugs 20:183–194

Foster BA, Coffey HA, Morin MJ, Rastinejad F (1999) Pharmacological rescue of mutant p53 conformation and function. Science 286:2507–2510

Fox SB, Gatter KC, Harris AL (1996) Tumour angiogenesis. J Pathol 179:232–237

Geiger T, Muller M, Monia BP, Fabbro D (1997) Antitumor activity of a c-raf antisense oligonucleotide in combination with standard chemotherapy agents against various human tumors transplanted subcutaneously into nude mice. Clin Cancer Res 3:1179–1185

Gibbs J, Oliff A, Kohl NE (1994) Farnesyltransferase inhibitors: Ras research yields a potential cancer therapeutic. Cell 77:175–178

Gingras D, Batist G, Beliveau R (2001) AE-941 (Neovastat): a novel multifunctional antiangiogenic compound. Expert Rev Anticancer Ther 1:341–347

Goel S, Mani S, Perez-Soler R (2002) Tyrosine kinase inhibitors: a clinical perspective. Curr Oncol Rep 4:9–19

Gokhale PC, McRae D, Monia BP, Bagg A, Rahman A, Dritschilo A, Kasid U (1999) Antisense raf oligodeoxyribonucleotide is a radiosensitizer in vivo. Antisense Nucleic Acid Drug Dev 9:191–201

Grandis JR, Melhem MF, Barnes EL, Tweardy DJ (1996) Quantitative immunohistochemical analysis of transforming growth factor and epidermal growth factor receptor in patients with squamous cell carcinoma of the head and neck. Cancer 78:1284–1292

Green DR (2000) Apoptotic pathways: paper wraps stone blunts scissor. Cell 102:1–4

Gullick WJ (1991) Prevalence of aberrant expression of the epidermal growth factor receptor in human cancers. Br Med Bull 47:87–98

Hamad NM, Elconin JH, Karnoub AE, Bai W, Rich JN, Abraham RT, Der CJ, Counter CM (2002) Distinct requirements for Ras oncogenesis in human versus mouse cells. Genes Dev 16:2045–2057

Havlik R, Jiao LR, Nicholls J, Jensen SL, Habib NA (2002) Gene therapy for liver metastasis. Semin Oncol 29:202–208

Herbst RS, Langer CL (2002). Epidermal growth factor receptors as a target for cancer treatment: the emerging role of IMC-C225 in the treatment of lung and head and neck cancers. Semin Oncol 29:27–36

Hidalgo M, Siu LL, Nemunaitis J, Rizzo J, Hammond LA, Takimoto C, Eckhardt SG, Tolcher A, Britten CD, Denis L, Ferrante K, Von Hoff DD, Silberman S, Rowinsky EK (2001) Phase I and pharmacologic study of OSI-774, an epidermal growth factor receptor tyrosine kinase inhibitor, in patients with advanced solid malignancies. J Clin Oncol 19:3267–3279

Hoffmann T, Hafner D, Ballo H, Hass I, Bier H (1997) Antitumor activity of anti-epidermal growth factor receptor monoclonal antibodies and cisplatin in ten human head and neck squamous cell carcinoma lines. Anticancer Res 17:4419–4426

Huang DC, Strasser A (2000) BH3-only proteins – essential initiators of apoptotic cell death. Cell 103:839–842

Jansen B, Wacheck V, Heere-Reess E, et al (2000) Chemosensitization of malignant melanoma by BCL2 antisense therapy. Lancet 356:1728–1733

Jayaraman L, Murthy KGK, Curran T, Xanthoudakis S, Prives C (1997) Identification of redox/repair protein Ref-1 as an activator of p53. Genes Dev 11:558–570

Johnston SR (2001) Farnesyl transferase inhibitors: a novel targeted therapy for cancer. Lancet Oncol 2:18–26

Johnstone RW, Ruefli AA, Lowe SW (2002) Apoptosis: a link between cancer genetics and chemotherapy. Cell 108:153–164

Kantarjian H, Sawyers C, Hochhaus A, Guilhot F, Schiffer C, Gambacorti-Passerini C, Niederwieser D, Resta D, Capdeville R, Zoellner U, Talpaz M, Druker B, Goldman J, O'Brien SG, Russell N, Fischer T, Ottmann O, Cony-Makhoul P, Facon T, Stone R, Miller C, Tallman M, Brown R, Schuster M, Loughran T, Gratwohl A, Mandelli F, Saglio G, Lazzarino M, Russo D,

Baccarani M, Morra E (2002) Hematologic and cytogenetic responses to imatinib mesylate in chronic myelogenous leukemia. N Engl J Med 346:645–652
Kastan MB, Onyekwere O, Sidransky D, Vogelstein B, Craig RW (1991) Participation of p53 protein in the cellular response to DNA damage. Cancer Res 51:6304–6311
Kaur G, Stetler-Stevenson M, Sebers S, Worland P, Sedlacek H, Myers C, Czech J, Naik R, Sausville E (1992) Growth inhibition with reversible cell cycle arrest of carcinoma cells by flavone L86-8275. J Natl Cancer Inst 84:1736–1740
Krammer PH (2000) CD95's deadly mission in the immune system. Nature 407:789–795
Kuerbitz SJ, Plunkett BS, Walsh WV, Kastan MB (1992) Wild-type p53 is a cell cycle checkpoint determinant following irradiation. Proc Natl Acad Sci USA 89:7491–7495
Kumar S, Witzig TE, Rajkumar SV (2002) Thalidomide as an anti-cancer agent. J Cell Mol Med 6:160–174
Kurakowa H, Lengferink AEG, Simpson JF, Pisacane PI, Sliwkowski MX, Forbes JT, et al (2000) Inhibition of HER2/neu (erbB-2)and mitogen-activated protein kinases enhances tamoxifen action against HER2/neu overexpressing, tamoxifen resistant breast cancer cells. Cancer Res 60:5887–5894
Lane DP, Lain S (2002) Therapeutic exploitation of the p53 pathway. Trends Mol Med 8:S38–42
Levine AJ (1997) p53: the cellular gatekeeper for growth and division. Cell 88:323–331
List AF (2002) New approaches to the treatment of myelodysplasia. Oncologist 7:39–49
Lowe SW, Lin AW (2000) Apoptosis in cancer. Carcinogenesis 21:485–495
Lynch DH, Yang XD (2002) Therapeutic potential of ABX-EGF: a fully human anti-epidermal growth factor receptor monoclonal antibody for cancer treatment. Semin Oncol 29:47–50
Marcucci G, Byrd JC, Dai G, Klisovic MI, Young DC, Cataland SR, Fisher DB, Lucas D, Chan KK, Porcu P, Lin ZP, Farag SF, Frankel SR, Zweibel JA, Kraut EH, Balcerzak SP, Bloomfield CD, Grever MR, Caligiuri MA (2002) Phase i and pharmacodynamic studies of g3139, a bcl-2 antisense oligonucleotide, in combination with chemotherapy in refractory or relapsed acute leukemia. Blood 2002 Aug 22; [epub ahead of print]
McKay BC, Ljungman M, Tainbow AJ (1999) Potential roles for p53 in nucleotide excision repair. Carcinogenesis 20:1389–1396
Miyashita T, Reed JC (1995) Tumor suppressor p53 is a direct transcriptional activator of the human bax gene. Cell 80:293–299
Modjtahedi H, Hickish T, Nicolson M, Moore J, Styles J, Eccles S, Jackson E, Solter J, Sloane J, Spencer L, Priest K, Smith I, Dean C, Gore M (1996) Phase I trial and tumour localization of the anti-EGFR monoclonal antibody ICR62 in head and neck or lung cancer. Br J Cancer 73:228–235
Monia BP, Johnston JF, Geiger T, Muller M, Fabbro D (1996) Antitumor activity of a phosphorothioate oligodeoxynucleotide targeted against C-raf kinase. Nat Med 2:668–675
Moyer JD, Barbacci EG, Iwata KK, et al. (1997) Induction of apoptosis and cell cycle arrest by CP-358,774, an inhibitor of epidermal growth factor receptor tyrosine kinase. Cancer Res 57:4838–4848
O'Dwyer PJ, Stevenson JP, Gallagher M, Cassella A, Vasilevska I, Monia BP, Holmlund J, Dorr FA, Yao KS (1999) C-raf-1 depletion and tumor responses in patients treated with the c-raf-1 antisense oligodeoxynucleotide ISIS 5132 (CGP 69846A). Clin Cancer Res 5:3977–3982
Owen-Schaub LB, Zhang W, Cusack JC, Angelo LS, Santee SM, Fuwara T, Roth JA, Deisseroth AB, Zhang WW, Kruzel E, Radinsky R (1995) Wild-type human p53 and a temperature-sensitive mutant induce Fas/APO-1 expression. Mol Cell Biol 15:3032–3040
Pegram MD, Reese DM (2002) Combined biological therapy of breast cancer using monoclonal antibodies directed against HER2/neu protein and vascular endothelial growth factor. Semin Oncol 29:29–37
Pietras RJ, Poen JC, Gallardo D, Wongvipat N, Lee HJ, Slamon DJ (1999) Monoclonal antibody to HER-2/neuroreceptor modulates repair of radiation-induced DNA damage and enhances radiosensitivity of human breast cancer cells overexpressing this oncogene. Cancer Res 59:1347–1355

Polyak K, Xia Y, Zweier JL, Kinzler KW, Vogelstein B (1997) A model for p53-induced apoptosis. Nature 389:300–305

Prendergast GC (2001) Actin' up: RhoB in cancer and apoptosis. Nat Rev Cancer 1:162–168

Prewett M, Rockwell P, Rockwell RF, et al (1996) The biologic effects of C225, a chimeric monoclonal antibody to the EGFR, on human prostate carcinoma. J Immunother Emphasis Tumor Immunol 19:419–427

Prewett M, Rothman M, Waksal H, et al (1998) Mouse-human chimeric anti-epidermal growth factor receptor antibody C225 inhibits the growth of human renal cell carcinoma xenografts in nude mice. Clin Cancer Res 4:2957–2966

Pringent SA, Lemoine NR (1992) The type I (EGFR-related) family of growth factor receptors and their ligands. Prog Growth Factor Res 4:1–24

Rosen LS (2002) Clinical experience with angiogenesis signaling inhibitors: focus on vascular endothelial growth factor (VEGF) blockers. Cancer Control 9:36–44

Rubin Grandis J, Chakraborty A, Melhem MF, Zeng O, Tweardy DJ (1997) Inhibition of epidermal growth factor receptor gene expression and function decreases proliferation of head and neck squamous carcinoma but not normal mucosal epithelial cells. Oncogene 15:409–416

Ryan KM, Phillips AC, Vousden KH (2001) Regulation and function of p53 tumor suppressor protein. Curr Opin Cell Biol 13:332–337

Salomon DS, Brandt R, Ciardiello F, Normanno N (1995) Epidermal growth factor-related peptides and their receptors in human malignancies. Crit Rev Oncol Hematol 19:183–232

Sausville EA, Elsayed Y, Monga M, Kim G (2003) Signal transduction-directed cancer treatments. Annu Rev Pharmacol Toxicol 43:199–231

Sawyers CL, Hochhaus A, Feldman E, Goldman JM, Miller CB, Ottmann OG, Schiffer CA, Talpaz M, Guilhot F, Deininger MW, Fischer T, O'Brien SG, Stone RM, Gambacorti-Passerini CB, Russell NH, Reiffers JJ, Shea TC, Chapuis B, Coutre S, Tura S, Morra E, Larson RA, Saven A, Peschel C, Gratwohl A, Mandelli F, Ben-Am M, Gathmann I, Capdeville R, Paquette RL, Druker BJ. (2002) Imatinib induces hematologic and cytogenetic responses in patients with chronic myelogenous leukemia in myeloid blast crisis: results of a phase II study. Blood 99:3530–3539

Schlessinger J (1988) The epidermal growth factor receptor as a multi-functional allosteric protein. Biochemistry 27:3119–3123

Schmitt CA, McCurrach ME, de Stanchina E, Wallace-Brodeur RR, Lowe SW (1999) INK4a/ARF mutations accelerate lymphomagenesis and promote chemoresistance by disabling p53. Genes Dev 13:2670–2677

Schroeder W, Biesterfield S, Zillessen S, Rath W (1997) Epidermal growth factor receptor – immunohistochemical detection and clinical significance for the treatment of primary breast cancer. Anticancer Res 17:2799–2802

Senderowicz AM, Sausville EA (2000) Preclinical and clinical development of cyclin-dependent kinase modulators. J Natl Cancer Inst 92:376–387

Shepherd FA (2001) Angiogenesis inhibitors in the treatment of lung cancer. Lung Cancer 34:81–89

Slamon DJ, Clark GM, Wong SG, Levin WJ, Ullrich A, McGuire WL (1987) Human breast cancer: correlation of relapse and survival with amplification of the Her-2/neu oncogene. Science 235:177–182

Slamon DJ, Godolphin W, Jones LA, Holt JA, Wong SG, Keith DE, Levin WJ, Stuart SG, Udove J, Ullrich A (1989) Studies of the Her-2/neu proto-oncogene in human breast and ovarian cancer. Science 244:707–712

Slamon DJ, Leyland-Jones B, Shak S et al. (2001) Use of chemotherapy plus a monoclonal antibody against HER2 for metastatic breast cancer that overexpresses HER2. N Engl J Med 344:783–792

Stevensonm JP, Yao KS, Gallagher M, Friedland D, Mitchell EP, Cassella A, Monia B, Kwoh TJ, Yu R, Holmlund J, Dorr FA, O'Dwyer PJ (1999) Phase I clinical/pharmacokinetic and pharmacodynamic trial of the c-raf-1 antisense oligonucleotide ISIS 5132 (CGP 69846A). Clin Oncol 17:2227–2237

Susuki Y, Imai Y, Nakayama H, Takahashi K, Takio K, Takahashi R (2001) A serine protease, HtrA2, is released from the mitochondria and interacts with xiap, inducing cell death. Mol Cell 8:613–621

Talpaz M, Silver RT, Druker BJ, Goldman JM, Gambacorti-Passerini C, Guilhot F, Schiffer CA, Fischer T, Deininger MW, Lennard AL, Hochhaus A, Ottmann OG, Gratwohl A, Baccarani M, Stone R, Tura S, Mahon FX, Fernandes-Reese S, Gathmann I, Capdeville R, Kantarjian HM, Sawyers CL (2002) Imatinib induces durable hematologic and cytogenetic responses in patients with accelerated phase chronic myeloid leukemia: results of a phase 2 study. Blood 99:1928–1937

Thompson DM, Gill GN (1985) The EGF receptor: structure, regulation and potential role in malignancy. Cancer Surv 4:767–788

Tilly JL (2001) Commuting the death sentence: how oocytes strive to survive. Nat Rev 2:838–848

Van de Vijer MJ, Peterse JL, Mooi MJ, Wisman P, Lomans J, Dalesio O, Nusse R (1988) Neu-protein overexpression in breast cancer. Association with comedo-type ductal carcinoma in situ and limited prognostic value in stage II breast cancer. N Engl J Med 1988:1239–1245

Wang X (2001) The expanding role of mitochondria in apoptosis. Genes Dev 15:2922–2933

Waters JS, Webb A, Cunningham D et al. (2000) Phase I clinical and pharmakokinetic study of Bcl-2 antisense oligonucleotide therapy in patients with non-hodgkin's lyphoma. J Clin Oncol 18:1812–1823

Wu X, Fan Z, Masui H et al. (1995) Apoptosis induced by an anti-epidermal growth factor receptor monoclonal antibody in a human colorectal carcinoma cell line and its delay by insulin. J Clin Invest 95:1897–1905

Yakoub-Agha I, Attal M, Dumontet C, Delannoy V, Moreau P, Berthou C, Lamy T, Grosbois B, Dauriac C, Dorvaux V, Bay JO, Monconduit M, Harousseau JL, Duguet C, Duhamel A, Facon T (2002) Thalidomide in patients with advanced multiple myeloma: a study of 83 patients – report of the intergroupe francophone du myelome (IFM). Hematol J 3:185–192

Yang XD, Jia XC, Corvalan JR, Wang P, Davis CG (2001) Development of ABX-EGF, a fully human anti-EGF receptor monoclonal antibody, for cancer therapy. Crit Rev Oncol Hematol 38:17–23

Yarden Y, Ullrich A (1988) Growth factor receptor tyrosine kinases. Annu Rev Biochem 57:443–478

Yu JL, Rak JW, Coomber BL, Hicklin DJ, Kerbel RS (2002) Effect of p53 status on tumor response to antiangiogenic therapy. Science 295:1526–1528

Yuan ZM, Huang Y, Whang Y, Sawyers C, Weichselbaum R, Kharbanda S, Kufe D (1996) Role for c-Abl tyrosine kinase in growth arrest response to DNA damage. Nature 382:272–274

Zhu J (2000) Targetting gene expression to tumor cells with loss of wild type p53 function. Cancer Gene Ther 7:4–12

Subject Index

Printing: Saladruck, Berlin
Binding: Stein+Lehmann, Berlin